Universitext

Universitext

Universitext is a series of textbooks that presents material from a wide variety of mathematical disciplines at master's level and beyond. The books, often well class-tested by their author, may have an informal, personal even experimental approach to their subject matter. Some of the most successful and established books in the series have evolved through several editions, always following the evolution of teaching curricula, to very polished texts.

Thus as research topics trickle down into graduate-level teaching, first textbooks written for new, cutting-edge courses may make their way into *Universitext.*

For further volumes:
http://www.springer.com/series/223

Antonio Galbis • Manuel Maestre

Vector Analysis Versus Vector Calculus

Antonio Galbis
Depto. Análisis Matemático
Universidad de Valencia
Burjasot (Valencia)
Spain

Manuel Maestre
Depto. Análisis Matemático
Universidad de Valencia
Burjasot (Valencia)
Spain

ISSN 0172-5939 e-ISSN 2191-6675
ISBN 978-1-4614-2199-3 e-ISBN 978-1-4614-2200-6
DOI 10.1007/978-1-4614-2200-6
Springer New York Dordrecht Heidelberg London

Library of Congress Control Number: 2012932088

Mathematics Subject Classification (2010): 58-XX, 53-XX, 79-XX

Printed on acid-free paper

Springer is part of Springer Science+Business Media (www.springer.com)

To Bea and María

Preface

This book aims to be useful. This might appear to be a trivial statement (after all, what would the alternative be?), but let us explain how this simple motivation provides us with a rather ambitious goal. The central theme of the book revolves around Stokes's theorem, and it deals with the following associated paradox. There are clear intuitive notions coming from the physical world and our own visual geometric insight that tell us what a closed surface is, what the interior and exterior of that surface are, what is meant by a flux across it, what a normal vector to it is, and whether it points in or out—in other words, how to orient that surface. The student of vector calculus is usually provided with a clear and useful set of rules as to how to orient a surface in applying the divergence theorem, and how to orient the boundary of a surface in the classical Stokes's theorem. However, when this student undertakes a formal study of orientation through mathematical analysis and/or differential geometry, she or he then realizes that orientation is defined in terms of the tangent space at each point of the surface, and the connection with the practical rules of vector calculus is far from clear. To make things worse, the usual closed surfaces used in $\mathbb{R}^3$, that is, those that are required for practical purposes, have vertices and edges, the most natural example being the cube, and they are not regular surfaces. Hence, a student of mathematics, formally at least, cannot apply Stokes's theorem to most natural situations in which it is required.

There is another element that deeply concerned the authors when they were introduced to the subject, and that is the various notational conventions. Actually, the problem is not just with notation. This is a subject that can be approached in many different ways, all of which are equally valid, and each of which has its own particular merits. There is undoubtedly an advantage in seeing a topic treated in different ways, but unfortunately in this instance, it is all too common for a student to become trapped with the particular notation and/or point of view used by one author on the subject. For example, a recommended textbook may choose to use a vectorial point of view, and employ integrals of differential forms, while another may opt for scalar integrals and the use, or not, of tensors. There are several possibilities for the definition of regular surface, from the very abstract notion of differential manifold to the more familiar concept of differential submanifold of $\mathbb{R}^n$, and so on. A student will follow one approach, which uses one of the more or less equivalent definitions available, but when the student tries to clarify an obscure point by studying another good exposition, very frequently the notation is alien, or worse, inconsistent with what the student already knows, so that the only option is to begin from scratch with the alternative approach. Most of the time, the student becomes frustrated and simply gives up.

This book is intended as a text for undergraduate students who have completed a standard introduction to differential and integral calculus of functions of several variables. We have written the book principally having in mind students of mathematics who need a precise and rigorous exposition of Stokes's theorem. This has led us to choose a differential-geometric point of view. However, we have taken great care to bridge the gap between a formal rigorous approach and a concrete presentation of applications in two and three variables. We show how to use the tools from vector calculus and modern methods that help to check, for example, whether a particular set in $\mathbb{R}^3$ is an orientable surface with boundary. In a less formal way, we show how to apply the obtained results on integration over regular surfaces to less amenable (but more practical) situations like the cube. We have looked at most of the definitions of regular surface and shown the equivalence of them. We discuss how one definition may be more convenient for solving exercises while another, equivalent, definition may be more suitable in proving a theorem. We have tried to include in each chapter as many examples and solved exercises as possible.

We have chosen the point of view of k-forms, but in each possible instance we switch to employing vector fields and the classical notation coming from physics. In general, we have made an effort to explain the connection between the usual practical rules from vector calculus and the rigorous theory that is at the core of vector analysis. This means that the book is also addressed to engineering and physics students, who know quite well how to handle the familiar theorems of Green, Stokes's, and Gauss, but who would like to know why they are true and how to recover these familiar useful tools in $\mathbb{R}^2$ or $\mathbb{R}^3$ from the mighty formal Stokes's theorem in $\mathbb{R}^n$. In summary, we have tried hard to show that vector analysis and vector calculus are not always at odds with one another. Perhaps we should have appended a question mark to the title.

The book contains some appendices that are not necessary for the rest of the book, but will offer the student the opportunity to get more deeply involved in the subject at hand.

While we are in great debt to many authors, including Do Carmo, Edwards, Fleming, Rudin, and Spivak, we do believe that our approach is quite original. However, we do not pretend that originality is our principal motivation. Only to be useful.

Burjasot, Spain

Antonio Galbis
Manuel Maestre

Acknowledgments

The authors are very grateful to Michael Mackey (University College Dublin) for his great assistance in revising the text. We also want to thank Michael for his critical reading of the mathematics, giving us many suggestions that have lead to several improvements of this book.

Contents

Chapter 1
Vectors and Vector Fields

The purpose of this book is to explain in a rigorous way Stokes's theorem and to facilitate the student's use of this theorem in applications. Neither of these aims can be achieved without first agreeing on the notation and necessary background concepts of vector calculus, and therein lies the motivation for our introductory chapter.

In the first section we study three operations involving vectors: the dot product of two vectors of $\mathbb{R}^n$, the cross product of two vectors of $\mathbb{R}^3$, and the triple scalar product of three vectors of $\mathbb{R}^3$. These operations have interesting physical and geometric interpretations. For instance, the dot product will be essential in the definition of the line integral (Definition 2.2.1) or work done by a force field in moving a particle along a path. The length of the cross product of two vectors represents the area of the parallelogram spanned by the two vectors, and the triple scalar product of three vectors allows us to evaluate the volume of the parallelepiped that they span, and it plays an important role in calculating the flux of a vector field across a given surface, as we shall see in Chap. 4.

A. Galbis and M. Maestre, *Vector Analysis Versus Vector Calculus*, Universitext,
DOI 10.1007/978-1-4614-2200-6_1,

1.1 Vectors

Definition 1.1.1. The *dot* product (or *scalar* product or *inner product*) of two vectors

$$\boldsymbol{a} = (a_1, a_2, \ldots, a_n), \quad \boldsymbol{b} = (b_1, b_2, \ldots, b_n) \in \mathbb{R}^n$$

is defined as the scalar

$$\boldsymbol{a} \cdot \boldsymbol{b} = \langle \boldsymbol{a}, \boldsymbol{b} \rangle = \sum_{j=1}^{n} a_j b_j.$$

According to the Pythagorean theorem, the length of a vector $\boldsymbol{a} = (a_1, a_2, a_3) \in \mathbb{R}^3$ is $\sqrt{a_1^2 + a_2^2 + a_3^2}$. The next definition is a generalization of the notion of length to vectors of $\mathbb{R}^n$.

Definition 1.1.2. The Euclidean norm of a vector

$$\boldsymbol{a} = (a_1, a_2, \ldots, a_n) \in \mathbb{R}^n$$

is defined as

$$\|\boldsymbol{a}\| = \sqrt{\langle \boldsymbol{a}, \boldsymbol{a} \rangle} = \sqrt{\sum_{j=1}^{n} a_j^2}.$$

Theorem 1.1.1 (Cauchy–Schwarz inequality).

$$|\langle \boldsymbol{a}, \boldsymbol{b} \rangle| \leq \|\boldsymbol{a}\| \cdot \|\boldsymbol{b}\|.$$

Proof. The inequality is trivial if either $\boldsymbol{a}$ or $\boldsymbol{b}$ is zero, so we assume that neither is. If we let $\boldsymbol{x} = \frac{\boldsymbol{a}}{\|\boldsymbol{a}\|}$ and $\boldsymbol{y} = \frac{\boldsymbol{b}}{\|\boldsymbol{b}\|}$, then $\|\boldsymbol{x}\| = \|\boldsymbol{y}\| = 1$. Hence

$$\begin{aligned} 0 \leq \|\boldsymbol{x} - \boldsymbol{y}\|^2 &= \langle \boldsymbol{x} - \boldsymbol{y}, \boldsymbol{x} - \boldsymbol{y} \rangle \\ &= \|\boldsymbol{x}\|^2 - 2\langle \boldsymbol{x}, \boldsymbol{y} \rangle + \|\boldsymbol{y}\|^2 \\ &= 2 - 2\langle \boldsymbol{x}, \boldsymbol{y} \rangle. \end{aligned}$$

So $\langle \boldsymbol{x}, \boldsymbol{y} \rangle \leq 1$, that is,

$$\langle \boldsymbol{a}, \boldsymbol{b} \rangle \leq \|\boldsymbol{a}\| \cdot \|\boldsymbol{b}\|.$$

Replacing $\boldsymbol{a}$ by $-\boldsymbol{a}$, we obtain

$$-\langle \boldsymbol{a}, \boldsymbol{b} \rangle \leq \|\boldsymbol{a}\| \cdot \|\boldsymbol{b}\|$$

also, and the inequality follows. □

As a very important corollary to the Cauchy–Schwarz inequality we have the following proposition.

Proposition 1.1.1 (Triangle inequalities). *Let* $\boldsymbol{x}, \boldsymbol{y} \in \mathbb{R}^n$. *Then*

1. $\|\boldsymbol{x} \pm \boldsymbol{y}\| \le \|\boldsymbol{x}\| + \|\boldsymbol{y}\|$;

2. $\big|\|\boldsymbol{x}\| - \|\boldsymbol{y}\|\big| \le \|\boldsymbol{x} - \boldsymbol{y}\|$.

Proof. 1. As above,

$$\begin{aligned}\|\boldsymbol{x} \pm \boldsymbol{y}\|^2 &= \langle \boldsymbol{x} \pm \boldsymbol{y}, \boldsymbol{x} \pm \boldsymbol{y}\rangle \\ &= \|\boldsymbol{x}\|^2 \pm 2\langle \boldsymbol{x}, \boldsymbol{y}\rangle + \|\boldsymbol{y}\|^2 \\ &\le \|\boldsymbol{x}\|^2 + 2|\langle \boldsymbol{x}, \boldsymbol{y}\rangle| + \|\boldsymbol{y}\|^2 \\ &\le \|\boldsymbol{x}\|^2 + 2\|\boldsymbol{x}\|\,\|\boldsymbol{y}\| + \|\boldsymbol{y}\|^2 \\ &= (\|\boldsymbol{x}\| + \|\boldsymbol{y}\|)^2.\end{aligned}$$

2. Since

$$\|\boldsymbol{x}\| = \|(\boldsymbol{x} - \boldsymbol{y}) + \boldsymbol{y}\| \le \|\boldsymbol{x} - \boldsymbol{y}\| + \|\boldsymbol{y}\|,$$

we see that

$$\|\boldsymbol{x}\| - \|\boldsymbol{y}\| \le \|\boldsymbol{x} - \boldsymbol{y}\|.$$

Interchanging $\boldsymbol{x}$ and $\boldsymbol{y}$, we also get

$$-(\|\boldsymbol{x}\| - \|\boldsymbol{y}\|) = \|\boldsymbol{y}\| - \|\boldsymbol{x}\| \le \|\boldsymbol{y} - \boldsymbol{x}\| = \|-(\boldsymbol{x} - \boldsymbol{y})\| = \|\boldsymbol{x} - \boldsymbol{y}\|,$$

and the inequality follows. □

Let us assume that $\boldsymbol{a}$ and $\boldsymbol{b}$ are two linearly independent vectors in $\mathbb{R}^3$ and that M is the plane spanned by them. The two vectors generate a triangle in M with sides of length $\|\boldsymbol{a}\|$, $\|\boldsymbol{b}\|$, and $\|\boldsymbol{a} - \boldsymbol{b}\|$. If $\theta \in (0, \pi)$ is the angle between the vectors $\boldsymbol{a}$ and $\boldsymbol{b}$ in M, then the *cosine rule* gives

$$\|\boldsymbol{a} - \boldsymbol{b}\|^2 = \|\boldsymbol{a}\|^2 + \|\boldsymbol{b}\|^2 - 2 \cdot \cos(\theta) \cdot \|\boldsymbol{a}\| \cdot \|\boldsymbol{b}\|.$$

However, we also have

$$\|\boldsymbol{a} - \boldsymbol{b}\|^2 = \langle \boldsymbol{a} - \boldsymbol{b},\, \boldsymbol{a} - \boldsymbol{b}\rangle = \|\boldsymbol{a}\|^2 + \|\boldsymbol{b}\|^2 - 2\langle \boldsymbol{a}, \boldsymbol{b}\rangle,$$

and a comparison of these two expressions gives

$$\langle \boldsymbol{a}, \boldsymbol{b}\rangle = \|\boldsymbol{a}\| \cdot \|\boldsymbol{b}\| \cdot \cos(\theta). \tag{1.1}$$

Indeed, (1.1) could be used to *define* the cosine of the angle between two vectors.

Definition 1.1.3. We say that $\boldsymbol{a}, \boldsymbol{b} \in \mathbb{R}^n$ are *orthogonal* if

$$\langle \boldsymbol{a},\, \boldsymbol{b}\rangle = 0.$$

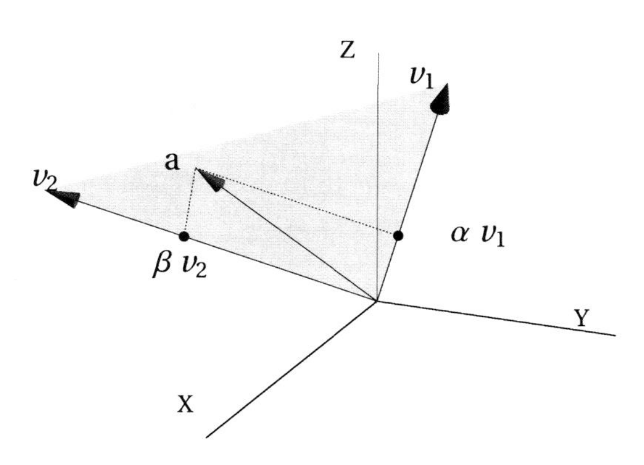

Fig. 1.1 Orthogonal projection

Given two linearly independent vectors $\boldsymbol{a},\boldsymbol{b} \in \mathbb{R}^n$, we want to find the orthogonal projection of $\boldsymbol{a}$ onto the line generated by $\boldsymbol{b}$. To this end, we denote by M the plane generated by $\boldsymbol{a}$ and $\boldsymbol{b}$ and we consider an orthonormal basis $\{\boldsymbol{v}_1,\boldsymbol{v}_2\}$ of M consisting of $\boldsymbol{v}_1 = \frac{\boldsymbol{b}}{\|\boldsymbol{b}\|}$ and a unit vector $\boldsymbol{v}_2 \in M$ orthogonal to $\boldsymbol{v}_1$ (Fig. 1.1). Since $\{\boldsymbol{v}_1,\boldsymbol{v}_2\}$ is a basis of M, there are scalars α and β such that

$$\boldsymbol{a} = \alpha\, \boldsymbol{v}_1 + \beta\, \boldsymbol{v}_2.$$

The projection of $\boldsymbol{a}$ onto the line generated by $\boldsymbol{b}$ is precisely $\alpha \boldsymbol{v}_1$. To determine α, we simply take the inner product with $\boldsymbol{b}$ in the above identity. Since $\langle \boldsymbol{v}_1,\boldsymbol{b}\rangle = \|\boldsymbol{b}\|$ and $\langle \boldsymbol{v}_2,\boldsymbol{b}\rangle = 0$, we obtain

$$\langle \boldsymbol{a},\boldsymbol{b}\rangle = \alpha\|\boldsymbol{b}\|.$$

That is,

$$\alpha = \frac{\langle \boldsymbol{a},\boldsymbol{b}\rangle}{\|\boldsymbol{b}\|}$$

represents the *component of the vector* $\boldsymbol{a}$ *parallel to the vector* $\boldsymbol{b}$. This will be useful in Chap. 2 when we calculate the component of a force in the direction tangent to a given trajectory. Observe that in this way, we can actually construct an orthonormal basis $\{\boldsymbol{v}_1,\boldsymbol{v}_2\}$ of M by taking

$$\boldsymbol{v}_1 := \frac{\boldsymbol{b}}{\|\boldsymbol{b}\|} \quad \text{and} \quad \boldsymbol{v}_2 := \frac{\boldsymbol{a} - \frac{\langle \boldsymbol{a},\, \boldsymbol{b}\rangle}{\|\boldsymbol{b}\|}\frac{\boldsymbol{b}}{\|\boldsymbol{b}\|}}{\left\|\boldsymbol{a} - \frac{\langle \boldsymbol{a},\, \boldsymbol{b}\rangle}{\|\boldsymbol{b}\|}\frac{\boldsymbol{b}}{\|\boldsymbol{b}\|}\right\|}.$$

Definition 1.1.4. The *cross product* of two vectors

$$\boldsymbol{a} = (a_1,a_2,a_3), \quad \boldsymbol{b} = (b_1,b_2,b_3)$$

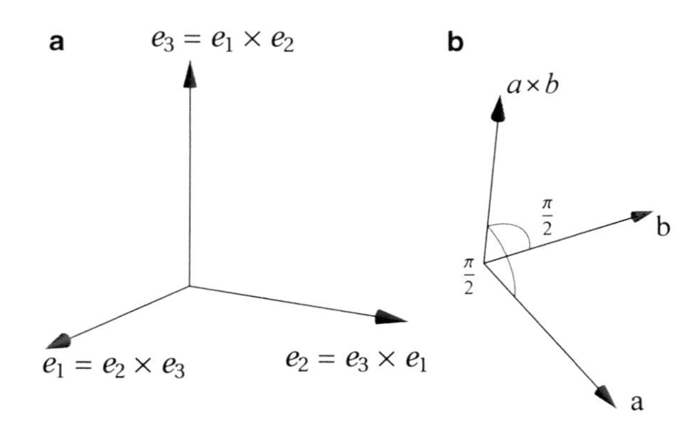

Fig. 1.2 (**a**) Canonical basis; (**b**) cross product

in $\mathbb{R}^3$ is the vector defined by the formal expression

$$\boldsymbol{a} \times \boldsymbol{b} := \begin{vmatrix} \mathbf{e_1} & \mathbf{e_2} & \mathbf{e_3} \\ a_1 & a_2 & a_3 \\ b_1 & b_2 & b_3 \end{vmatrix}.$$

Here,

$$\{\mathbf{e_1}, \mathbf{e_2}, \mathbf{e_3}\}$$

represents the canonical basis (Fig. 1.2) of $\mathbb{R}^3$, namely

$$\mathbf{e_1} = (1,0,0),\ \mathbf{e_2} = (0,1,0),\ \mathbf{e_3} = (0,0,1),$$

and we interpret that the coordinates of the vector $\boldsymbol{a} \times \boldsymbol{b}$ are obtained after expanding the determinant along the first row. That is,

$$\boldsymbol{a} \times \boldsymbol{b} := \left(\begin{vmatrix} a_2 & a_3 \\ b_2 & b_3 \end{vmatrix}, - \begin{vmatrix} a_1 & a_3 \\ b_1 & b_3 \end{vmatrix}, \begin{vmatrix} a_1 & a_2 \\ b_1 & b_2 \end{vmatrix} \right).$$

It is a routine but laborious calculation to check that

$$\begin{aligned} \|\boldsymbol{a} \times \boldsymbol{b}\|^2 &= \|\boldsymbol{a}\|^2 \cdot \|\boldsymbol{b}\|^2 - |\langle \boldsymbol{a}, \boldsymbol{b} \rangle|^2 \\ &= \|\boldsymbol{a}\|^2 \cdot \|\boldsymbol{b}\|^2 \left(1 - \cos^2(\theta)\right) \\ &= \|\boldsymbol{a}\|^2 \cdot \|\boldsymbol{b}\|^2 \sin^2(\theta), \end{aligned}$$

where $\theta \in [0, \pi]$ is the angle between the two vectors $\boldsymbol{a}$ and $\boldsymbol{b}$. Consequently,

$$\|\boldsymbol{a} \times \boldsymbol{b}\| = \|\boldsymbol{a}\| \cdot \|\boldsymbol{b}\| \cdot \sin(\theta).$$

If $\boldsymbol{a}$ and $\boldsymbol{b}$ are linearly independent, this expression gives the area of the parallelogram generated by $\boldsymbol{a}$ and $\boldsymbol{b}$ (see Example 4.1.1).

Definition 1.1.5. The *triple scalar product* of three vectors a, b, and c in $\mathbb{R}^3$ is the scalar defined by

$$\langle \boldsymbol{a},\, \boldsymbol{b}\times \mathbf{c}\rangle .$$

If we write

$$\boldsymbol{a}=(a_1,a_2,a_3),\ \boldsymbol{b}=(b_1,b_2,b_3),\ \mathbf{c}=(c_1,c_2,c_3),$$

then it follows from the definitions that

$$\begin{aligned}\langle \boldsymbol{a},\, \boldsymbol{b}\times \mathbf{c}\rangle &= a_1\begin{vmatrix} b_2 & b_3\\ c_2 & c_3\end{vmatrix} - a_2\begin{vmatrix} b_1 & b_3\\ c_1 & c_3\end{vmatrix} + a_3\begin{vmatrix} b_1 & b_2\\ c_1 & c_2\end{vmatrix}\\ &= \begin{vmatrix} a_1 & a_2 & a_3\\ b_1 & b_2 & b_3\\ c_1 & c_2 & c_3\end{vmatrix}.\end{aligned}$$

From the properties of determinants we have immediately the following properties of cross product of two vectors.

Theorem 1.1.2. *The cross product has the following properties:*
(1) $\boldsymbol{b}\times\boldsymbol{a} = -(\boldsymbol{a}\times\boldsymbol{b})$.
(2) $\boldsymbol{a}\times\boldsymbol{b}$ *is orthogonal to the vectors* $\boldsymbol{a}$ *and* $\boldsymbol{b}$.
(3) $\boldsymbol{a}$ *and* $\boldsymbol{b}$ *are linearly independent if and only if* $\boldsymbol{a}\times\boldsymbol{b} = 0$.

The cross product of vectors in $\mathbb{R}^2$ is not defined. However, as we will see in Sect. 7.2, it is possible to define the cross product of $n-1$ vectors in $\mathbb{R}^n$ whenever $n\geq 3$.

The triple scalar product also has an interesting geometric interpretation. Let $\boldsymbol{a},\boldsymbol{b},\mathbf{c}\in\mathbb{R}^3$ be three linearly independent vectors. These *generate* a parallelepiped, whose base may be taken to be the parallelogram generated by $\boldsymbol{a}$ and $\boldsymbol{b}$ (Fig. 1.3). The vector $\boldsymbol{a}\times\boldsymbol{b}$ is orthogonal to the plane generated by $\boldsymbol{a}$ and $\boldsymbol{b}$, and as a consequence, the height of this parallelepiped (with respect to the aforementioned plane) coincides with the component of $\mathbf{c}$ parallel to the direction $\pm(\boldsymbol{a}\times\boldsymbol{b})$. That is, the height is given by

$$h=\left|\left\langle \mathbf{c},\ \frac{\boldsymbol{a}\times\boldsymbol{b}}{\|\,\boldsymbol{a}\times\boldsymbol{b}\,\|}\right\rangle\right|.$$

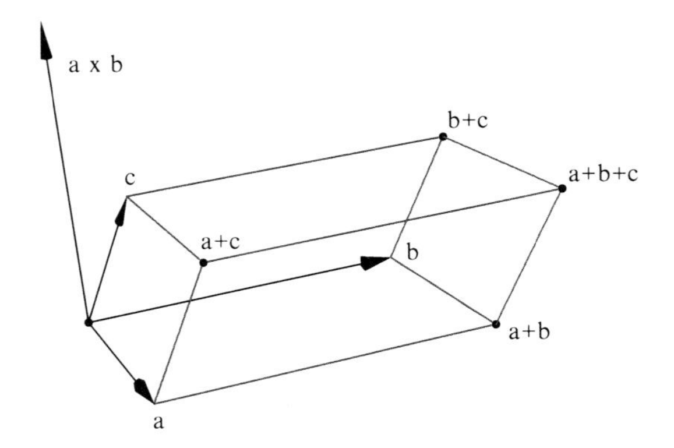

Fig. 1.3 Parallelepiped generated by three vectors

The volume of the parallelepiped can now be calculated by

$$\begin{aligned} \text{volume} &= \text{base} \cdot \text{height} \\ &= \|\boldsymbol{a} \times \boldsymbol{b}\| \cdot h \\ &= |\langle \mathbf{c}, \boldsymbol{a} \times \boldsymbol{b} \rangle| \\ &= |\langle \boldsymbol{a}, \boldsymbol{b} \times \mathbf{c} \rangle| . \end{aligned}$$

So the volume of the parallelepiped is the absolute value of the triple scalar product of the three vectors. We will generalize this result in Chap. 4, Theorem 4.1.1.

1.2 Vector Fields

Throughout, we assume that the reader has a basic knowledge of differential and integral calculus in several variables, but in the interest of convenience and consistency, we will recall the relevant definitions and results when they are first encountered. We recommend to the reader the following excellent references [1–4, 8–10, 12, 13] and [18].

Definition 1.2.1. Given $\boldsymbol{a} \in \mathbb{R}^n$, the *open ball* centered at $\boldsymbol{a}$ and of radius $r > 0$ is the set

$$B(\boldsymbol{a},r) = \{\boldsymbol{x} \in \mathbb{R}^n : \|\boldsymbol{x}-\boldsymbol{a}\| < r\}.$$

The *closed ball* centered at $\boldsymbol{a}$ and of radius $r \geq 0$ is the set

$$D(\boldsymbol{a},r) = \{\boldsymbol{x} \in \mathbb{R}^n : \|\boldsymbol{x}-\boldsymbol{a}\| \leq r\}.$$

Definition 1.2.2. (i) A subset U of $\mathbb{R}^n$ is called *open* if for each $\boldsymbol{x} \in U$ there exists $r > 0$ (which depends on $\boldsymbol{x}$) such that $B(\boldsymbol{x},r) \subset U$.

(ii) A set C in $\mathbb{R}^n$ is *closed* if its complement $\mathbb{R}^n \setminus C$ is an open set.

(iii) Given $\boldsymbol{a} \in \mathbb{R}^n$, a set $G \subset \mathbb{R}^n$ is called a *neighborhood* of $\boldsymbol{a}$ if there exists $r > 0$ such that the ball $B(\boldsymbol{a},r)$ is contained in G. In particular, if the set G is open, then it is an open neighborhood of all of its points.

(iv) If A is a subset of $\mathbb{R}^n$, then the *interior* of A is the set

$$int(A) := \{\boldsymbol{x} \in A : \; A \text{ is a neighborhood of } \boldsymbol{x}\}.$$

(v) The *closure* of A is the set

$$\overline{A} := \{\boldsymbol{x} \in \mathbb{R}^n : \; B(\boldsymbol{x},r) \cap A \neq \varnothing \; \text{ for every } \; r > 0\}.$$

Any open ball $B(\boldsymbol{a},r)$ is an open set. Indeed, if $\boldsymbol{x} \in B(\boldsymbol{a},r)$, then by the triangle inequalities, $B(\boldsymbol{x}, r - \|\boldsymbol{x}-\boldsymbol{a}\|)$ is an open ball centered at $\boldsymbol{x}$ contained in $B(\boldsymbol{a},r)$. Analogously, any closed ball is a closed set. In general, a set A is open if and only if it coincides with its interior, $int(A)$, and it is closed if and only if it coincides with its closure $\overline{A}$.

Now we recall the concepts of continuous and differentiable mappings.

Definition 1.2.3. Let M be a subset of $\mathbb{R}^n$. A mapping $\boldsymbol{f} : M \subset \mathbb{R}^n \to \mathbb{R}^m$ is *continuous* at a point $\boldsymbol{a} \in M$ if given $\varepsilon > 0$ there exists $\delta > 0$ such that for every $\boldsymbol{x} \in M$ with $\|\boldsymbol{x}-\boldsymbol{a}\| < \delta$, we have

$$\|\boldsymbol{f}(\boldsymbol{x}) - \boldsymbol{f}(\boldsymbol{a})\| < \varepsilon.$$

We say that $\boldsymbol{f}$ is continuous on M if it is continuous at every point of M. Usually when the range space is $\mathbb{R}$ we will say that f is a *continuous function*.

Definition 1.2.4. Let M be a subset of $\mathbb{R}^n$ and consider $\boldsymbol{a} \in M$ with the property that $M \cap (B(\boldsymbol{a}, r) \setminus \{\boldsymbol{a}\}) \neq \varnothing$ for every $r > 0$. We say that the mapping $\boldsymbol{f} : M \subset \mathbb{R}^n \to \mathbb{R}^m$ has *limit* $\boldsymbol{b} \in \mathbb{R}^m$ at the point $\boldsymbol{a}$, and write $\lim_{\boldsymbol{x} \to \boldsymbol{a}} \boldsymbol{f}(\boldsymbol{x}) = \boldsymbol{b}$, if given $\varepsilon > 0$ there exists $\delta > 0$ such that for every $\boldsymbol{x} \in M$ with $0 < \|\boldsymbol{x} - \boldsymbol{a}\| < \delta$, we have

$$\|\boldsymbol{f}(\boldsymbol{x}) - \boldsymbol{b}\| < \varepsilon.$$

Definition 1.2.5. Let $f : U \subset \mathbb{R}^n \to \mathbb{R}$ be a function defined on the open set U. We say that f has a *partial derivative* in the ith coordinate at $\boldsymbol{a} \in U$ if the limit

$$\lim_{h \to 0} \frac{f(a_1, \ldots, a_{i-1}, a_i + h, a_{i+1}, \ldots, a_n) - f(a_1, \ldots, a_n)}{h},$$

exists, and when the limit exists, we will denote its value (which is a real number) by $\frac{\partial f}{\partial x_i}(\boldsymbol{a})$.

More generally, for $\boldsymbol{f} : U \subset \mathbb{R}^n \to \mathbb{R}^m$ and $\boldsymbol{v} \in \mathbb{R}^n$, we define the *directional derivative* of f at $\boldsymbol{a} \in U$ in the direction $\boldsymbol{v}$ to be

$$D_{\boldsymbol{v}} \boldsymbol{f}(\boldsymbol{a}) = \lim_{t \to 0} \frac{f(\boldsymbol{a} + t\boldsymbol{v}) - f(\boldsymbol{a})}{t},$$

whenever that limit exists.

Definition 1.2.6. Let $\boldsymbol{f} : U \subset \mathbb{R}^n \to \mathbb{R}^m$ be a mapping defined on the open set U. We say that $\boldsymbol{f}$ is *differentiable* at $\boldsymbol{a} \in U$ if there exists a linear mapping $\boldsymbol{T} : \mathbb{R}^n \to \mathbb{R}^m$ such that

$$\lim_{\boldsymbol{h} \to 0} \frac{\boldsymbol{f}(\boldsymbol{a} + \boldsymbol{h}) - \boldsymbol{f}(\boldsymbol{a}) - \boldsymbol{T}(\boldsymbol{h})}{\|\boldsymbol{h}\|} = \boldsymbol{0}.$$

In that case, we denote the (necessarily unique) linear mapping $\boldsymbol{T}$ by $\mathrm{d}\boldsymbol{f}(\boldsymbol{a})$. We say that $\boldsymbol{f}$ is differentiable on U if it is differentiable at each point of U.

A mapping $\boldsymbol{f} : U \subset \mathbb{R}^n \to \mathbb{R}^m, \boldsymbol{f} = (f_1, \ldots, f_m)$, is differentiable at $\boldsymbol{a} \in U$ if and only if each coordinate function f_j is differentiable at $\boldsymbol{a}$. If we denote by $\boldsymbol{f}'(\boldsymbol{a})$ the matrix (with respect to the canonical basis) of $\mathrm{d}\boldsymbol{f}(\boldsymbol{a})$, the differential of $\boldsymbol{f}$ at $\boldsymbol{a}$, we then have that

$$f'(\boldsymbol{a}) = \begin{pmatrix} \frac{\partial f_1}{\partial x_1}(\boldsymbol{a}) & \frac{\partial f_1}{\partial x_2}(\boldsymbol{a}) & \cdots & \frac{\partial f_1}{\partial x_n}(\boldsymbol{a}) \\ \frac{\partial f_2}{\partial x_1}(\boldsymbol{a}) & \frac{\partial f_2}{\partial x_2}(\boldsymbol{a}) & \cdots & \frac{\partial f_2}{\partial x_n}(\boldsymbol{a}) \\ \vdots & \vdots & \ddots & \vdots \\ \frac{\partial f_m}{\partial x_1}(\boldsymbol{a}) & \frac{\partial f_m}{\partial x_2}(\boldsymbol{a}) & \cdots & \frac{\partial f_m}{\partial x_n}(\boldsymbol{a}) \end{pmatrix}.$$

The matrix $\boldsymbol{f}'(\boldsymbol{a})$ is called the *Jacobian matrix* of $\boldsymbol{f}$ at $\boldsymbol{a}$, and in the case that $m = n$, its determinant is called the *Jacobian* of $\boldsymbol{f}$ at $\boldsymbol{a}$ and denoted by $J\boldsymbol{f}(\boldsymbol{a})$.

In the scalar-valued case where $f : U \subset \mathbb{R}^n \to \mathbb{R}$ is differentiable at $\boldsymbol{a}$, $f'(\boldsymbol{a})$ is called the *gradient* of f at $\boldsymbol{a}$. Usually this is denoted by $\nabla f(\boldsymbol{a})$ and is treated as a (row) vector in $\mathbb{R}^n$, i.e.,

$$\nabla f(\boldsymbol{a}) = \left(\frac{\partial f}{\partial x_1}(\boldsymbol{a}), \frac{\partial f}{\partial x_2}(\boldsymbol{a}), \dots, \frac{\partial f}{\partial x_n}(\boldsymbol{a}) \right).$$

A very useful fact, which can be found in any textbook on analysis in several variables (e.g., [15, p. 217]), is that if a mapping $\boldsymbol{f} : U \subset \mathbb{R}^n \to \mathbb{R}^m$ is differentiable at $\boldsymbol{a} \in U$, then the directional derivative of $\boldsymbol{f}$ at $\boldsymbol{a}$ in the direction $\boldsymbol{v}$ exists for every $\boldsymbol{v} \in \mathbb{R}^n$ and

$$D_{\boldsymbol{v}}\boldsymbol{f}(\boldsymbol{a}) = \mathrm{d}\boldsymbol{f}(\boldsymbol{a})(\boldsymbol{v}).$$

The condition for differentiability given in Definition 1.2.6 is not an easy one to check, but the next theorem, which can be found in any textbook on multivariable calculus, provides a sufficiency condition that is more amenable. We first require another definition.

Definition 1.2.7. Let $\boldsymbol{f} : U \subset \mathbb{R}^n \to \mathbb{R}^m$ be a mapping defined on the open set U. We say that $\boldsymbol{f}$ is *continuously differentiable* at $\boldsymbol{a} \in U$ if there exists $r > 0$ such that the ball $B(\boldsymbol{a}, r)$ is contained in U and all the partial derivatives $\frac{\partial f_i}{\partial x_j}(\boldsymbol{x})$ ($i = 1, \dots, m$, $j = 1, \dots, n$) exist in the ball and are continuous on $\boldsymbol{a}$. Then $\boldsymbol{f}$ is said to be of class C^1 on U if it is continuously differentiable at all points of U.

Theorem 1.2.1. *If $\boldsymbol{f}: U \subset \mathbb{R}^n \to \mathbb{R}^m$ is a mapping that is continuously differentiable at a point $\boldsymbol{a}$ in an open set U, then $\boldsymbol{f}$ is differentiable at $\boldsymbol{a}$. In particular, if $\boldsymbol{f}$ is of class C^1 on the open set U, then $\boldsymbol{f}$ is differentiable on U.*

If $\boldsymbol{f} : U \subset \mathbb{R}^n \to \mathbb{R}^m$ is a mapping of class C^1 on an open set U, then we can consider the continuous functions $\frac{\partial f_i}{\partial x_j} : U \to \mathbb{R}$. We will say that $\boldsymbol{f}$ is of class C^2 on U if each $\frac{\partial f_i}{\partial x_j} : U \to \mathbb{R}$ is of class C^1 on U, i.e., if the functions

$$\frac{\partial}{\partial x_k}\left(\frac{\partial f_i}{\partial x_j}\right)(\boldsymbol{x})$$

(which we call *second-order derivatives* of f_i) all exist and are continuous on U for each $i = 1, \dots, m$ and $j, k = 1, \dots, n$. We denote the above second-order derivative by

$$\frac{\partial^2 f_i}{\partial x_k \partial x_j}(\boldsymbol{x}).$$

One can clearly iterate this process in order to define a function of class C^p on U. If $\boldsymbol{f}$ is of class C^p for every $p \in \mathbb{N}$, then we say that it is of class C^∞ on U.

Definition 1.2.8. A *vector field* is a continuous mapping

$$\mathbf{F} : U \subset \mathbb{R}^n \to \mathbb{R}^n.$$

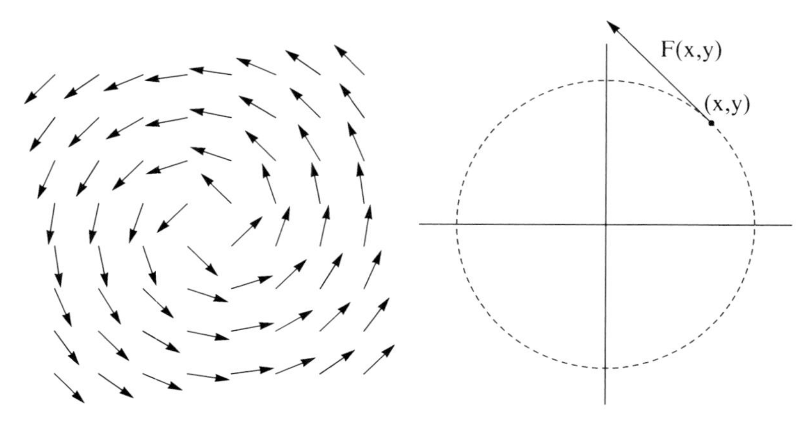

Fig. 1.4 Vector field of Example 1.2.1

The natural interpretation of this definition is that a vector field assigns a vector to a point. Vector fields are useful in representing force fields or velocity fields.

The student of mathematics will find in his/her studies many instances of this situation in which the same abstract concept can be interpreted differently through the simple device of changing the *name* of that concept. Here a mapping from $\mathbb{R}^n$ into itself becomes a physical concept, just by calling it a vector field, and is consequently visualized in a new way, as the following examples show.

Example 1.2.1. Let us consider the vector field (Fig. 1.4)

$$\mathbf{F} : \mathbb{R}^2 \setminus \{0\} \to \mathbb{R}^2$$

defined by

$$\mathbf{F}(x,y) = \left(-\frac{y}{\sqrt{x^2+y^2}}, \frac{x}{\sqrt{x^2+y^2}} \right).$$

Clearly, $\mathbf{F}(x,y)$ is a unit vector, and if we place this vector at the point (x,y), we see that it is a tangent vector at (x,y) to the circle centered at the origin that passes through this point.

Example 1.2.2 (Gravitational fields). Let us consider a particle of mass M located at the origin. The force of attraction exerted on a particle of mass m located at $(x,y,z) \in \mathbb{R}^3 \setminus \{0\}$ is

$$\mathbf{F}(x,y,z) = -\frac{GMm}{x^2+y^2+z^2}\mathbf{u},$$

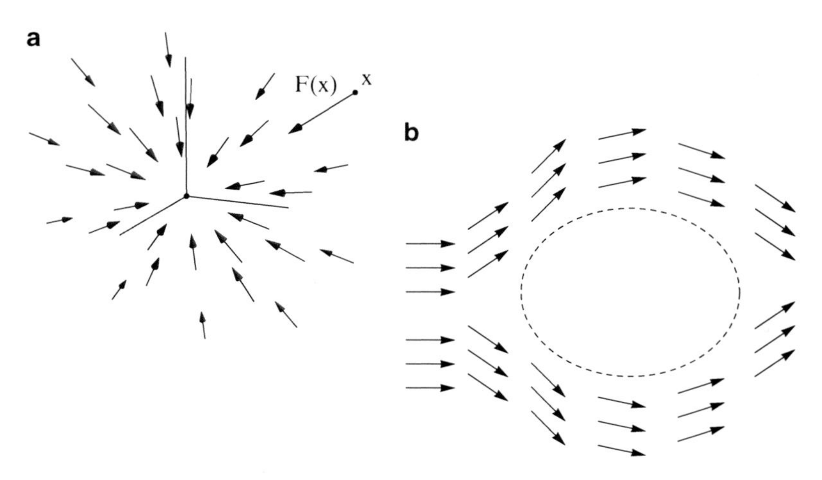

Fig. 1.5 (**a**) Gravitational field; (**b**) particle motion of a fluid

where G is the gravitational constant and $\mathbf{u} = \frac{1}{\sqrt{x^2+y^2+z^2}}(x,y,z)$ is the unit vector in the direction from the origin to (x,y,z). That is, the vector $\mathbf{F}(x,y,z)$ always points from (x,y,z) toward the origin, and its magnitude is inversely proportional to the square of the distance to the origin (Fig. 1.5).

Example 1.2.3 (Velocity field of a fluid). For every point (x,y,z) of an open set $U \subset \mathbb{R}^3$ let $\mathbf{F}(x,y,z)$ represent the velocity of a fluid at the position (x,y,z) at a given fixed time. Then

$$\mathbf{F} : U \subset \mathbb{R}^3 \to \mathbb{R}^3$$

is a vector field.

Example 1.2.4. Let

$$g : U \subset \mathbb{R}^n \to \mathbb{R}$$

be a function of class C^1 on the open set U. Then

$$\mathbf{F} := \nabla g : U \to \mathbb{R}^n,\ \mathbf{F}(\boldsymbol{x}) = \nabla g(\boldsymbol{x}) := \left(\frac{\partial g}{\partial x_1}(\boldsymbol{x}), \ldots, \frac{\partial g}{\partial x_n}(\boldsymbol{x})\right),$$

is a vector field and is called the *gradient field* of g.

Definition 1.2.9. Let

$$\mathbf{F} : U \subset \mathbb{R}^n \to \mathbb{R}^n$$

be a vector field with components

$$\mathbf{F} = (f_1, f_2, \ldots, f_n).$$

We recall that $\mathbf{F}$ is of class C^p (respectively differentiable) on the open set U if each of its components f_j is of class C^p (respectively differentiable).

Next we introduce two basic operations on vector fields; the divergence (which is a scalar function) and the curl, or rotor (which is a vector field). These play a central role in the formulation of two fundamental theorems of vector analysis, namely, the divergence theorem, or Gauss's theorem, and the classical Stokes's theorem. These results will be studied in Chap. 9, and they constitute the main purpose of this book. As we will see later, the divergence and curl of the velocity field of a fluid give pertinent information concerning the behavior of that fluid.

Definition 1.2.10. Let $\mathbf{F} : U \subset \mathbb{R}^n \to \mathbb{R}^n$ be a vector field of class C^1 on the open set U. The *divergence* of $\mathbf{F}$ is the scalar function

$$\text{Div } \mathbf{F} = \sum_{j=1}^{n} \frac{\partial f_j}{\partial x_j}.$$

Definition 1.2.11. Let $\mathbf{F} : U \subset \mathbb{R}^3 \to \mathbb{R}^3$, $\mathbf{F} = (f_1, f_2, f_3)$, be a vector field of class C^1 on the open set U. The *curl* (or *rotor*) of $\mathbf{F}$ is the vector field defined, formally, by the determinant

$$\text{Curl } \mathbf{F} = \begin{vmatrix} \mathbf{e_1} & \mathbf{e_2} & \mathbf{e_3} \\ \frac{\partial}{\partial x} & \frac{\partial}{\partial y} & \frac{\partial}{\partial z} \\ f_1 & f_2 & f_3 \end{vmatrix}.$$

We interpret that the components of $\mathbf{F}$ are obtained after expanding the determinant along the first row. Thus, Curl $\mathbf{F}$ is the vector field

$$\text{Curl } \mathbf{F} = \left(\frac{\partial f_3}{\partial y} - \frac{\partial f_2}{\partial z}, \frac{\partial f_1}{\partial z} - \frac{\partial f_3}{\partial x}, \frac{\partial f_2}{\partial x} - \frac{\partial f_1}{\partial y} \right).$$

Usually we put $\boldsymbol{\nabla}$ to represent the differential operator

$$\boldsymbol{\nabla} = \left(\frac{\partial}{\partial x}, \frac{\partial}{\partial y}, \frac{\partial}{\partial z} \right).$$

This, together with the cross product notation of Definition 1.1.4, suggests the following symbolic representation for the curl of $\mathbf{F}$:

$$\text{Curl } \mathbf{F} = \boldsymbol{\nabla} \times \mathbf{F}.$$

We note that the divergence is defined for vector fields in $\mathbb{R}^n$, while the curl is defined only for vector fields in $\mathbb{R}^3$.

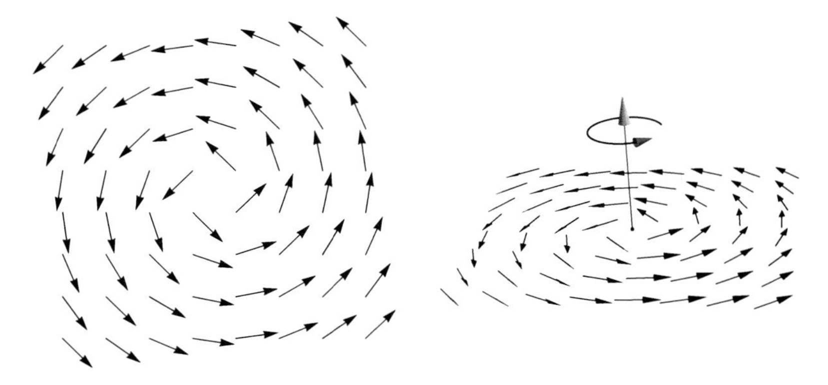

Fig. 1.6 Vector field of Example 1.2.5

Example 1.2.5. The vector field $\mathbf{F}(x,y,z) = (-y,x,0)$ represents a rotation about $\mathbf{N} = (0,0,1)$. We observe that Curl $\mathbf{F} = (0,0,1)$ gives the direction of the rotation axis (Fig. 1.6).We will deduce from Stokes's theorem that this is no coincidence (see Corollary 9.4.1).

We next highlight an important relationship between divergence and curl, but for that, we need the following theorem due to Schwarz about the symmetry of second derivatives of a function of class C^2. It is sometimes also called Clairaut's theorem, or Young's theorem.

Theorem 1.2.2. *Let $f : U \subset \mathbb{R}^n \to \mathbb{R}$ be a function of class C^2 on the open set U. Then for all $\boldsymbol{x} \in U$ and all $i, j = 1, 2, \ldots, n$,*

$$\frac{\partial^2 f}{\partial x_i \partial x_j}(\boldsymbol{x}) = \frac{\partial^2 f}{\partial x_j \partial x_i}(\boldsymbol{x}).$$

Theorem 1.2.3. *Let $\mathbf{F} : U \subset \mathbb{R}^3 \to \mathbb{R}^3$ be a vector field of class C^2 on the open set U. Then*

$$\text{Div}\,(\text{Curl}\,\mathbf{F}) = 0.$$

Proof.[1] Since $\mathbf{F} = (f_1, f_2, f_3)$, Schwarz's theorem concerning the symmetry of second derivatives gives

[1] In Chap. 6 we will present an alternative argument (Corollary 6.3.1) based on properties of the *exterior differential*.

$$\begin{aligned}\text{Div}\,(\text{Curl}\,\mathbf{F}) &= \frac{\partial}{\partial x}\left(\frac{\partial f_3}{\partial y} - \frac{\partial f_2}{\partial z}\right) + \frac{\partial}{\partial y}\left(\frac{\partial f_1}{\partial z} - \frac{\partial f_3}{\partial x}\right) + \frac{\partial}{\partial z}\left(\frac{\partial f_2}{\partial x} - \frac{\partial f_1}{\partial y}\right)\\ &= \left(\frac{\partial^2}{\partial x\,\partial y} - \frac{\partial^2}{\partial y\,\partial x}\right) f_3 + \left(\frac{\partial^2}{\partial z\,\partial x} - \frac{\partial^2}{\partial x\,\partial z}\right) f_2\\ &\quad + \left(\frac{\partial^2}{\partial y\,\partial z} - \frac{\partial^2}{\partial z\,\partial y}\right) f_1\\ &= 0.\end{aligned}$$

□

1.3 Exercises

Exercise 1.3.1. Find the divergence of the vector fields
(1) $\mathbf{F}(x,y) = (\sin(x),\, e^{x-y})$.
(2) $\mathbf{F}(x,y,z) = \big(\sin(y),\, \cos(z),\, z^3\big)$.

Exercise 1.3.2. Find the divergence and the curl at the point $(1,1,0)$ for the vector field

$$\mathbf{F}(x,y,z) = (xyz\ ,\ y\ ,z)\,.$$

Exercise 1.3.3. Check that

$$\text{Div}\ (\text{Curl}\ \mathbf{F}) = 0$$

for the vector field

$$\mathbf{F}(x,y,z) = \big(x^2 z,\ x,\ 2yz\big)\,.$$

Exercise 1.3.4. Let $\mathbf{F}:\mathbb{R}^3 \to \mathbb{R}^3$ be a vector field and let $g:\mathbb{R}^3 \to \mathbb{R}$ be a scalar function, both of class C^1 on $\mathbb{R}^n$. Check that

$$\text{Curl}\ (g\mathbf{F}) = g(\text{Curl}\ \mathbf{F}) + (\nabla g) \times \mathbf{F}.$$

Exercise 1.3.5. If $\mathbf{F},\mathbf{G}:\mathbb{R}^3 \to \mathbb{R}^3$ are vector fields of class C^1, prove that

$$\text{Div}(\mathbf{F} \times \mathbf{G}) = \langle \text{Curl}\ \mathbf{F},\ \mathbf{G}\rangle - \langle \mathbf{F},\ \text{Curl}\ \mathbf{G}\rangle\,.$$

Chapter 2
Line Integrals

In studying the motion of a particle along an *arc* it is convenient to consider the arc as the image of a vector-valued mapping $\gamma : [a,b] \to \mathbb{R}^3$ defined on an interval of the real line and realize $\gamma(t)$ as the position of the particle at time t. This viewpoint is also convenient in analyzing the behavior of a vector field along an arc and is the main motivation for the definitions that follow.

A. Galbis and M. Maestre, *Vector Analysis Versus Vector Calculus*, Universitext,
DOI 10.1007/978-1-4614-2200-6_2,

2.1 Paths

Definition 2.1.1. A *path* is a continuous mapping $\boldsymbol{\gamma} : [a,b] \to \mathbb{R}^n$. We call $\boldsymbol{\gamma}(a)$ the *initial point* and $\boldsymbol{\gamma}(b)$ the *final point*. The image of the path, $\boldsymbol{\gamma}([a,b])$, is called the *arc*[1] of $\boldsymbol{\gamma}$. If $\boldsymbol{\gamma}([a,b]) \subset \Omega$, we say that $\boldsymbol{\gamma}$ is a path in Ω.

Example 2.1.1. The line segment joining two points $\boldsymbol{x},\mathbf{y} \in \mathbb{R}^n$ is the arc $[\boldsymbol{x},\mathbf{y}] := \boldsymbol{\gamma}([0,1])$, where $\boldsymbol{\gamma} : [0,1] \to \mathbb{R}^n$ denotes the path

$$\boldsymbol{\gamma}(t) = \boldsymbol{x} + t(\boldsymbol{y} - \boldsymbol{x}) = t\boldsymbol{y} + (1-t)\boldsymbol{x}.$$

Example 2.1.2. Let $\boldsymbol{\gamma}_j : [0,2\pi] \to \mathbb{R}^2$ be given by $\boldsymbol{\gamma}_j(t) := (\cos(jt), \sin(jt))$. Then for every $j \in \mathbb{Z} \setminus \{0\}$, the arc $\boldsymbol{\gamma}_j([0,2\pi])$ is the unit circle $x^2 + y^2 = 1$ in $\mathbb{R}^2$. As the parameter t increases from 0 to 2π, the point $\boldsymbol{\gamma}_j(t)$ travels around the unit circle $|j|$ times (clockwise when j is negative and counterclockwise when j is positive).

We put

$$\boldsymbol{\gamma}'(t) := \lim_{h \to t \; h \in [a,b]} \frac{\boldsymbol{\gamma}(h) - \boldsymbol{\gamma}(t)}{h - t},$$

if the limit exists. We observe that if $t \in (a,b)$, then $\boldsymbol{\gamma}'(t)$ exists if and only if $\boldsymbol{\gamma}$ is a differentiable mapping at the point t. In this case, $\boldsymbol{\gamma}'(t)$ is the $n \times 1$ column matrix of the differential at t, which is naturally viewed as a vector in $\mathbb{R}^n$. In particular, we may consider $\boldsymbol{\gamma}' : [a,b] \to \mathbb{R}^n$, and where the appropriate limits exist, repeat to find higher-order derivatives of $\boldsymbol{\gamma}$. This prompts the following definitions.

Definition 2.1.2. A path $\boldsymbol{\gamma} : [a,b] \to \mathbb{R}^n$ is said to be a function of class C^q on $[a,b]$ if the qth derivative $\boldsymbol{\gamma}^{(q)}(t)$ exists for every $t \in [a,b]$ and $\boldsymbol{\gamma}^{(q)}$ is continuous on $[a,b]$. The mapping $\boldsymbol{\gamma}$ is said to be a *piecewise* C^q function if there exists a partition $a = t_1 < \cdots < t_k = b$ such that $\boldsymbol{\gamma}_{|[t_i,t_{i+1}]}$ is of class C^q on $[t_i,t_{i+1}]$ for every $1 \leq i \leq k-1$.

Definition 2.1.3. A path $\boldsymbol{\gamma} : [a,b] \to \mathbb{R}^n$ is said to be *smooth* if $\boldsymbol{\gamma}$ is a C^1 function and $\boldsymbol{\gamma}'(t) \neq 0$ for every $t \in [a,b]$.

Unfortunately, the notation is not standard in the literature. We follow Cartan [7, 3.1, p. 48] or Edwards [9, V.1. p. 287]. However Fleming [10, 6.2, pp. 247–249] refers to *curves* as equivalence classes and a path is interpreted as a parametric representation of a curve. Marsden–Tromba [14, 3.1, p. 190] uses the term *trajectory* instead of *path* and imposes no continuity condition, while Do Carmo [5, 1-2. parameterized curves, p. 2] defines *parametric curves* (with values in $\mathbb{R}^3$) and uses the expression *regular curve* instead of smooth path.

The next example, which is a reformulation of Edwards [9, V.1, example 1, p. 287], shows that a path of class C^∞ can have corners. We first need a lemma.

---

[1] Also called the *track* or *trace* of $\boldsymbol{\gamma}$.

Lemma 2.1.1. *Let c be a real number. The function $f:\mathbb{R}\to\mathbb{R}$,*

$$f(t):=\begin{cases} c\,\mathrm{e}^{-\frac{1}{t^2}}, & t<0,\\ 0, & t=0,\\ \mathrm{e}^{-\frac{1}{t^2}}, & t>0,\end{cases}$$

is of class C^∞ on the real line.

Proof. We show that for each $n\in\mathbb{N}\cup\{0\}$ there is a polynomial P_n such that

$$f^{(n)}(t):=\begin{cases} c\,\frac{P_n(t)}{t^{3n}}\mathrm{e}^{-\frac{1}{t^2}}, & t<0,\\ 0, & t=0,\\ \frac{P_n(t)}{t^{3n}}\mathrm{e}^{-\frac{1}{t^2}}, & t>0.\end{cases}$$

In fact, this is obvious for $n=0$. To apply induction, we assume that the result is true for $n=k\in\mathbb{N}\cup\{0\}$. To deduce that the result is also true for $n=k+1$, it is enough to check that $f^{(k+1)}(0)=0$. We are going to use that e^{y^2} diverges to infinity faster than any polynomial in y as y tends to $+\infty$. Indeed, if $P_n(y)=a_ny^n+\cdots+a_1y+a_0$ with $a_n\neq 0$, then

$$\frac{P_n(y)}{\mathrm{e}^{y^2}}=\frac{P_n(y)}{\mathrm{e}^{y}}\cdot\frac{1}{\mathrm{e}^{y^2-y}}.$$

The function e^{-y^2+y} clearly converges to zero. On the other hand, by applying L'Hôpital's rule n times, we obtain

$$\lim_{y\to+\infty}\frac{P_n(y)}{\mathrm{e}^{y}}=\lim_{y\to+\infty}\frac{n!a_n}{\mathrm{e}^{y}}=0.$$

Now by taking $y=\frac{1}{t}$ when $t>0$ or $y=-\frac{1}{t}$ if $t<0$ we obtain that

$$f^{(k+1)}(0)=\lim_{t\to 0}\frac{f^{(k)}(t)-f^{(k)}(0)}{t}=\lim_{t\to 0}\frac{f^{(k)}(t)}{t}=0.$$

□

Example 2.1.3. Let $\boldsymbol{\gamma}:[-1,1]\to\mathbb{R}^2$ be defined by

$$\boldsymbol{\gamma}(t):=\begin{cases}\left(-\mathrm{e}^{-\frac{1}{t^2}+1},\mathrm{e}^{-\frac{1}{t^2}+1}\right), & -1\leq t<0,\\ (0,0), & t=0,\\ \left(\mathrm{e}^{-\frac{1}{t^2}+1},\mathrm{e}^{-\frac{1}{t^2}+1}\right), & 0<t\leq 1.\end{cases}$$

It follows from the above lemma that γ is a path of class C^∞. However, γ is not smooth, since $\gamma'(0) = (0,0)$. We note that $\gamma([-1,1]) = \{(x,|x|) \, : \, x \in [-1,1]\}$ *has a corner* at the point $(0,0)$. Several more examples can be found in Marsden–Tromba [14, 3.2, p. 205].

Our next task is to define and, if possible, to evaluate the length of a path $\gamma : [a,b] \to \mathbb{R}^n$. The basic idea consists in approximating the path by means of line segments whose endpoints are determined by a partition of the interval $[a,b]$. A path with finite length is said to be *rectifiable* or of *bounded variation.* We will show that every piecewise C^1 path is rectifiable and will obtain a formula to evaluate its length.

Definition 2.1.4. Let $\gamma : [a,b] \to \mathbb{R}^n$ be a path and let $P := \{a = t_1 < \cdots < t_k = b\}$ be a partition of $[a,b]$. The *polygonal arc* associated with P is the union of the line segments $[\gamma(t_i), \gamma(t_{i+1})]$, $1 \leq i \leq k-1$. The length of this polygonal arc is

$$L(\gamma,P) := \sum_{i=1}^{k-1} ||\gamma(t_{i+1}) - \gamma(t_i)||.$$

Lemma 2.1.2. *Let $\gamma : [a,b] \to \mathbb{R}^n$ be a path and let $P := \{a = t_1 < \cdots < t_k = b\}$ be a partition of $[a,b]$. If Q is another partition of $[a,b]$ and $P \subset Q$, then $L(\gamma,P) \leq L(\gamma,Q)$.*

Proof. We can assume without any loss of generality that Q is obtained from P by adding a single point. Thus, we assume $Q = P \cup \{s\}$, where $t_j < s < t_{j+1}$. Then

$$L(\gamma,Q) := \sum_{i \neq j} ||\gamma(t_{i+1}) - \gamma(t_i)|| + \| \gamma(s) - \gamma(t_j) \| + \| \gamma(t_{j+1}) - \gamma(s) \| .$$

Since

$$||\gamma(t_{j+1}) - \gamma(t_j)|| \leq \| \gamma(s) - \gamma(t_j) \| + \| \gamma(t_{j+1}) - \gamma(s) \|,$$

the conclusion follows. □

Definition 2.1.5. A path γ is said to be *rectifiable* or of *bounded variation* if

$$\sup\{L(\gamma,P) : P \text{ a partition of } [a,b]\} < +\infty.$$

If γ is rectifiable, we will refer to this supremum as the *length* of γ, and we will denote it by $L(\gamma)$.

Let $\gamma : [a,b] \to \mathbb{R}^n$ be a path and $a < c < b$. Then γ is rectifiable if and only if $\gamma|_{[a,c]}$ and $\gamma|_{[c,b]}$ are. Moreover, in that case,

$$L(\gamma) = L(\gamma|_{[a,c]}) + L(\gamma|_{[c,b]}).$$

We leave the proof to the interested reader.

Definition 2.1.6. The *norm of a partition* $P := \{a = t_1 < \cdots < t_k = b\}$ is the length of the largest subinterval defined by that partition, that is,

$$\|P\| = \max\{|t_{i+1} - t_i| \ : \ i = 1, \ldots, k-1\}.$$

We are going to show that any path of class C^1 is rectifiable, but before we do so, we need to remind the reader of the topological concept of compactness and of some properties enjoyed by compact sets.

Definition 2.1.7. A subset K of $\mathbb{R}^n$ is called *compact* if for each family $\mathscr{F}$ of open subsets of $\mathbb{R}^n$ that cover K, in the sense that

$$K \subset \bigcup_{G \in \mathscr{F}} G,$$

there exists a finite subfamily $G_1, \ldots, G_m$ in $\mathscr{F}$ such that $K \subset \cup_{j=1}^m G_j$.

A subset M of $\mathbb{R}^n$ is called *bounded* if it is contained in an open ball centered at the origin. If K is a compact set, then it is bounded. This is an easy exercise, but far more can be said. The proof of the following characterizations of compact subsets in $\mathbb{R}^n$ can be found in almost any book on calculus of several variables.

Theorem 2.1.1. *Let K be a subset of $\mathbb{R}^n$. The following conditions are equivalent:*

1. *K is compact.*
2. *For each sequence $(x_j)_{j=1}^\infty \subset K$ there exists a subsequence $(x_{j_k})_{k=1}^\infty$ convergent to a point $x_0 \in K$.*
3. *(Heine–Borel–Lebesgue theorem) K is bounded and closed in $\mathbb{R}^n$.*

We also need the concept of *uniform* continuity.

Definition 2.1.8. Let M be a subset of $\mathbb{R}^n$. A mapping $\boldsymbol{f} : M \subset \mathbb{R}^n \to \mathbb{R}^m$ is *uniformly continuous* on M if given $\varepsilon > 0$, there exists $\delta > 0$ such that for every $\boldsymbol{x}, \boldsymbol{y} \in M$ with $\|\boldsymbol{x} - \boldsymbol{y}\| < \delta$, we have

$$\|\boldsymbol{f}(\boldsymbol{x}) - \boldsymbol{f}(\boldsymbol{y})\| < \varepsilon.$$

Uniform continuity of course implies continuity, but is, in fact, a stronger property. Nevertheless, both concepts coincide if the set M is a compact set, a result that we state here without proof.

Theorem 2.1.2 (Heine–Cantor theorem). *Every continuous mapping $\boldsymbol{f} : K \subset \mathbb{R}^n \to \mathbb{R}^m$ on a compact set K is uniformly continuous on K.*

We return to our study of paths.

Theorem 2.1.3. *Let $\boldsymbol{\gamma} : [a,b] \to \mathbb{R}^n$ be a path of class C^1. Then $\boldsymbol{\gamma}$ is rectifiable and*

$$L(\boldsymbol{\gamma}) = \int_a^b \|\boldsymbol{\gamma}'(t)\| \mathrm{d}t.$$

Proof. We define $F : [a,b]^n \to \mathbb{R}$ by

$$F(s_1,\ldots,s_n) := \sqrt{\sum_{j=1}^{n} |\boldsymbol{\gamma}_j'(s_j)|^2}.$$

Then $F(t,\ldots,t) = ||\boldsymbol{\gamma}'(t)||$. Moreover, given a partition

$$P := \{a = t_1 < \cdots < t_k = b\},$$

we can apply the mean value theorem to deduce that for every $1 \leq i \leq k-1$ and $1 \leq j \leq n$, there exists $s_{ji} \in [t_i, t_{i+1}]$ such that

$$\begin{aligned} L(\boldsymbol{\gamma},P) = \sum_{i=1}^{k-1} \|\boldsymbol{\gamma}(t_{i+1}) - \boldsymbol{\gamma}(t_i)\| &= \sum_{i=1}^{k-1} \sqrt{\sum_{j=1}^{n} \left(\boldsymbol{\gamma}_j(t_{i+1}) - \boldsymbol{\gamma}_j(t_i)\right)^2} \\ &= \sum_{i=1}^{k-1} F(s_{1i},\ldots,s_{ni})(t_{i+1}-t_i) \\ &= \sum_{i=1}^{k-1} \int_{t_i}^{t_{i+1}} F(s_{1i},\ldots,s_{ni})\mathrm{d}t. \end{aligned}$$

Since F is a continuous function on the compact set $[a,b]^n$, it follows by the Heine–Cantor theorem that F is in fact uniformly continuous on $[a,b]^n$. That is, for every $\varepsilon > 0$ there is a $\delta > 0$ such that $\boldsymbol{x},\boldsymbol{y} \in [a,b]^n$ and $\|\boldsymbol{x} - \boldsymbol{y}\| < \delta$ imply

$$|F(\boldsymbol{x}) - F(\boldsymbol{y})| < \varepsilon.$$

Let us now assume that the previous partition P satisfies $\|P\| < \frac{\delta}{\sqrt{n}}$. Then

$$||(t,\ldots,t) - (s_{1i},\ldots,s_{ni})|| < \delta,$$

whenever $t \in [t_i, t_{i+1}]$. Hence

$$\begin{aligned} \left| L(\boldsymbol{\gamma},P) - \int_a^b ||\boldsymbol{\gamma}'(t)||\mathrm{d}t \right| &\leq \sum_{i=1}^{k-1} \int_{t_i}^{t_{i+1}} |F(s_{1i},\ldots,s_{ni}) - F(t,\ldots,t)|\mathrm{d}t \\ &\leq \sum_{i=1}^{k-1} \int_{t_i}^{t_{i+1}} \varepsilon \,\mathrm{d}t = \varepsilon \sum_{i=1}^{k-1} (t_{i+1} - t_i) = \varepsilon(b-a). \end{aligned}$$

To finish the proof, we fix a partition P_0 with norm less than $\frac{\delta}{\sqrt{n}}$. For an arbitrary partition P of $[a,b]$ we then have

$$L(\boldsymbol{\gamma},P) \leq L(\boldsymbol{\gamma},P \cup P_0) \leq \varepsilon(b-a) + \int_a^b ||\boldsymbol{\gamma}'(t)||\mathrm{d}t,$$

which shows that $\boldsymbol{\gamma}$ is rectifiable and $L(\boldsymbol{\gamma}) \leq \varepsilon(b-a) + \int_a^b ||\boldsymbol{\gamma}'(t)||\mathrm{d}t$. On the other hand,

$$L(\boldsymbol{\gamma}) \geq L(\boldsymbol{\gamma}, P_0) \geq \int_a^b ||\boldsymbol{\gamma}'(t)||\mathrm{d}t - \varepsilon(b-a).$$

Taking limits as ε tends to zero, we reach the desired conclusion. □

It is worth mentioning that many texts, for instance Do Carmo [5, 1-2, p. 6] or Marsden–Tromba [14, 3.2, p. 201], define the length of a path of class C^1 as $L(\boldsymbol{\gamma}) = \int_a^b \|\boldsymbol{\gamma}'(t)\| \,\mathrm{d}t$. While this point of view is very efficient, it can be puzzling to the student as to why this should be the "correct" definition. In our opinion, the polygonal arc approach to path length is far more intuitive, and as shown by the theorem above, entirely consistent.

Corollary 2.1.1. *Let $\boldsymbol{\gamma} : [a,b] \to \mathbb{R}^n$ be a piecewise C^1 path. Then $\boldsymbol{\gamma}$ is rectifiable and $L(\boldsymbol{\gamma}) = \int_a^b \|\boldsymbol{\gamma}'(t)\|\mathrm{d}t$.*

Proof. Let $\{a = t_1 < t_2 < \cdots < t_k = b\}$ be a partition of $[a,b]$ such that $\boldsymbol{\gamma}_j := \boldsymbol{\gamma}_{|[t_j,t_{j+1}]}$ is of class C^1 for every $j = 1,\ldots,k-1$. By Theorem 2.1.3, every $\boldsymbol{\gamma}_j$ is a rectifiable path on $[t_j,t_{j+1}]$ and

$$L(\boldsymbol{\gamma}_j) = \int_{t_j}^{t_{j+1}} \|\boldsymbol{\gamma}'(t)\|\mathrm{d}t.$$

Consequently, $\boldsymbol{\gamma}$ is also rectifiable and

$$L(\boldsymbol{\gamma}) = \sum_{j=1}^{k-1} L(\boldsymbol{\gamma}_j) = \sum_{j=1}^{k-1} \int_{t_j}^{t_{j+1}} \| \boldsymbol{\gamma}'(t) \| \,\mathrm{d}t.$$

The function $t \mapsto \|\boldsymbol{\gamma}'(t)\|$ is well defined at each point of the interval $[a,b]$ except for a finite set. Moreover, its restriction to each interval $[t_j,t_{j+1}]$ is continuous, and hence Riemann integrable on $[t_j,t_{j+1}]$ for every $j = 1,\ldots,k-1$. It follows that the function is Riemann integrable on $[a,b]$. We can therefore deduce, from the properties of Riemann integrable functions, that

$$L(\boldsymbol{\gamma}) = \int_a^b \| \boldsymbol{\gamma}'(t) \| \,\mathrm{d}t.$$

□

Let $\boldsymbol{\gamma} : [0,1] \to \mathbb{R}^2$ be given by $\boldsymbol{\gamma}(0) := (0,0)$ and $\boldsymbol{\gamma}(t) := (t, t\cos\frac{\pi}{t})$ when $0 < t \leq 1$. We leave it to the reader to prove that $\boldsymbol{\gamma}$ is a continuous path but is not rectifiable.

Although we focus attention on piecewise C^1 paths throughout the text, we have found it natural to deal with the more general class of rectifiable paths for our treatment of path length. The generalization is also of benefit in the next section, where we consider work done by a vector field.

2.2 Integration of Vector Fields

The line integral was originally motivated by problems involving fluid motion and electromagnetic or other force fields.[2] Let us assume, for instance, that $\boldsymbol{\gamma} : [a,b] \to \mathbb{R}^3$ is a smooth path contained in the open set $U \subset \mathbb{R}^3$ and that there is present a force field $\boldsymbol{F} : U \to \mathbb{R}^3$. We want to evaluate the work done by the force on an object moving along the arc $\boldsymbol{\gamma}([a,b])$ from $\boldsymbol{\gamma}(a)$ to $\boldsymbol{\gamma}(b)$. We will have to take into account two basic principles:

1. The work depends only on the component of the force that is acting in the same direction as that in which the object is moving (that is, the tangent direction to the path at each point).
2. The work done by a constant field $\boldsymbol{F}_0$ to move the object through a line segment, in the same direction as $\boldsymbol{F}_0$, is the product of $\|\boldsymbol{F}_0\|$ and the length of that segment.

We recall that

$$\boldsymbol{T}(t) := \frac{\boldsymbol{\gamma}'(t)}{\|\boldsymbol{\gamma}'(t)\|}$$

is a unit vector that is tangent to the path at $\boldsymbol{\gamma}(t)$, and so the component of $\boldsymbol{F}$ that acts in the tangent direction to the path at $\boldsymbol{\gamma}(t_j)$ is

$$\langle \boldsymbol{F}(\boldsymbol{\gamma}(t_j)), \boldsymbol{T}(t_j) \rangle .$$

If we consider a short interval $[t_j, t_{j+1}]$, then the length of $\boldsymbol{\gamma}|_{[t_j, t_{j+1}]}$ is approximated by

$$\|\boldsymbol{\gamma}'(t_j)\| \cdot (t_{j+1} - t_j).$$

Hence, using the basic principles above on a very fine partition

$$P := \{a = t_1 < \cdots < t_k = b\}$$

of the interval, we see that a good approximation to the work done in moving the particle along $\boldsymbol{\gamma}$ is

$$\sum_{j=1}^{k-1} \langle \boldsymbol{F}(\boldsymbol{\gamma}(t_j)), \boldsymbol{\gamma}'(t_j) \rangle \cdot (t_{j+1} - t_j).$$

The following result tells us the limiting value of this quantity as we take finer and finer partitions (Fig. 2.1).

[2] A force field is, mathematically speaking, the same thing as a vector field. The term is often used when the field has a physical interpretation.

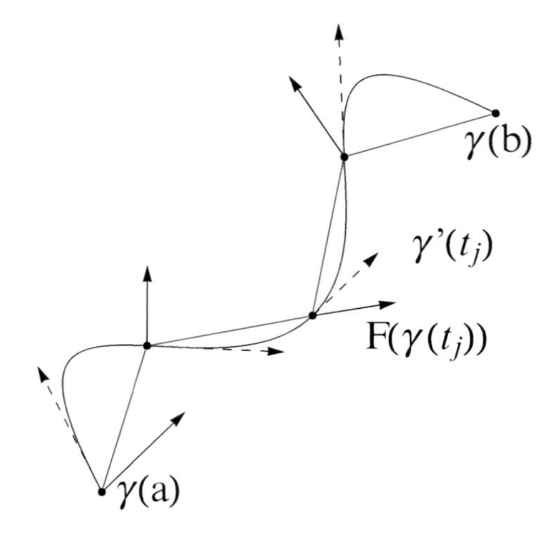

Fig. 2.1 Vector field along a path

Lemma 2.2.1. *Let $P := \{a = t_1 < t_2 < \cdots < t_k = b\}$ be a partition of the interval $[a,b]$. For any continuous function $f : [a,b] \to \mathbb{R}$ and for each selection $u_j \in [t_j, t_{j+1}]$ we have*

$$\lim_{\|P\| \to 0} \sum_{j=1}^{k-1} (t_{j+1} - t_j) \cdot f(u_j) \;=\; \int_a^b f(t)\, \mathrm{d}t.$$

Proof. By the Heine–Cantor theorem, f is uniformly continuous, which means that for every $\varepsilon > 0$ one can find $\delta > 0$ such that

$$|f(s) - f(t)| \leq \varepsilon$$

whenever $s,t \in [a,b]$ and $|s-t| \leq \delta$. Consider now a partition

$$P = \{a = t_1 < t_2 < \cdots < t_k = b\}$$

with $0 <\| P \|< \delta$. Then

$$\begin{aligned}
\left| \sum_{j=1}^{k-1} (t_{j+1} - t_j) \cdot f(u_j) - \int_a^b f(t) \right| &= \left| \sum_{j=1}^{k-1} \int_{t_j}^{t_{j+1}} \Big(f(u_j) - f(t) \Big)\, \mathrm{d}t \right| \\
&\leq \sum_{j=1}^{k-1} \int_{t_j}^{t_{j+1}} \big| f(u_j) - f(t) \big|\, \mathrm{d}t \\
&\leq \varepsilon \sum_{j=1}^{k-1} \big(t_{j+1} - t_j \big) = \varepsilon (b-a).
\end{aligned}$$

Since ε is arbitrarily small, the result follows. □

Lemma 2.2.1 and its preceding discussion show that the work done by the force field $\boldsymbol{F}$ on a particle as it moves along $\boldsymbol{\gamma}$ is

$$\int_a^b \langle \boldsymbol{F}(\boldsymbol{\gamma}(t)), \boldsymbol{\gamma}'(t) \rangle \, \mathrm{d}t.$$

This type of integral, known as a line integral, appears often in the sequel, and so we make a formal definition.

Definition 2.2.1. Let $\boldsymbol{F} : U \subset \mathbb{R}^n \to \mathbb{R}^n$ be a continuous vector field and let $\boldsymbol{\gamma}$ be a piecewise C^1 path in U. The line integral of $\boldsymbol{F}$ along $\boldsymbol{\gamma}$ is given by

$$\int_{\boldsymbol{\gamma}} \boldsymbol{F} := \int_a^b \langle \boldsymbol{F}(\boldsymbol{\gamma}(t)), \boldsymbol{\gamma}'(t) \rangle \, \mathrm{d}t.$$

2.3 Integration of Differential Forms

In order to facilitate later calculations, and also to provide a better framework for dealing with the general Stokes's theorem, it is convenient at this point to introduce the notion of differential form. A key observation is that any vector $\boldsymbol{v} \in \mathbb{R}^n$ defines a linear $\mathbb{R}$-valued mapping, that is, a *linear form*,

$$\varphi_v : \mathbb{R}^n \to \mathbb{R},$$

by

$$\varphi_v(\boldsymbol{h}) := \langle \boldsymbol{v}, \boldsymbol{h} \rangle .$$

Conversely, any linear form L on $\mathbb{R}^n$ coincides with φ_v for a unique vector $\boldsymbol{v} \in \mathbb{R}^n$. Indeed, if $L : \mathbb{R}^n \to \mathbb{R}$, the mapping

$$\boldsymbol{\varphi} : \mathbb{R}^n \to (\mathbb{R}^n)^*, \quad \boldsymbol{v} \mapsto \boldsymbol{\varphi}_v,$$

maps the vector $(L(e_1), \dots, L(e_n))$ to the linear form L. Thus $\boldsymbol{\varphi}$ is a linear isomorphism from the vectors in $\mathbb{R}^n$ to the linear forms on $\mathbb{R}^n$.

We denote by $\mathrm{d}x_j$ the linear form associated with the vector $\boldsymbol{e}_j$ of the canonical basis of $\mathbb{R}^n$. That is,

$$\mathrm{d}x_j(\boldsymbol{h}) = \langle \boldsymbol{e}_j, \boldsymbol{h} \rangle = h_j.$$

It easily follows that if $\boldsymbol{v} = (v_1, \dots, v_n)$, then φ_v is the linear form

$$\sum_{j=1}^{n} v_j \cdot \mathrm{d}x_j.$$

After identifying vectors with linear forms, it is quite natural to identify a vector field on a set U with a mapping that associates to each point of U a linear form.

Definition 2.3.1. Let $U \subset \mathbb{R}^n$ be an open set. A *differential form of degree* 1 on U, or simply a 1*-form*, is a mapping

$$\boldsymbol{\omega} : U \subset \mathbb{R}^n \to \mathscr{L}(\mathbb{R}^n, \mathbb{R}) = (\mathbb{R}^n)^*.$$

Given a 1-form $\boldsymbol{\omega}$, a vector $\boldsymbol{x} \in U$, and an integer $j \in \{1, \dots, n\}$, we will denote the scalar $\boldsymbol{\omega}(\mathbf{x})(e_j) \in \mathbb{R}$ by $f_j(\mathbf{x})$. Evidently, each f_j is a function from U to $\mathbb{R}$, and by linearity of $\boldsymbol{\omega}(x) \in (\mathbb{R}^n)^*$, for any $\boldsymbol{h} = (h_1, \dots, h_n) \in \mathbb{R}^n$ we can write

$$\boldsymbol{\omega}(\boldsymbol{x})(\boldsymbol{h}) = \sum_{j=1}^{n} \boldsymbol{\omega}(\mathbf{x})(e_j).h_j = \sum_{j=1}^{n} f_j(\boldsymbol{x}) h_j = \left(\sum_{j=1}^{n} f_j(\boldsymbol{x}) \mathrm{d}x_j \right)(\boldsymbol{h}).$$

Hence $\boldsymbol{\omega}(\boldsymbol{x}) = \sum_{j=1}^{n} f_j(\boldsymbol{x})\mathrm{d}x_j$ for all $\boldsymbol{x} \in U$, which we will abbreviate to

$$\boldsymbol{\omega} = \sum_{j=1}^{n} f_j \cdot \mathrm{d}x_j$$

and call f_j the *component* functions of $\boldsymbol{\omega}$.

The notion of differential form of degree 1 is, as we shall highlight below, a generalization of the concept of differential of a function. It is a very powerful algebraic tool for studying integration on curves or surfaces.

A 1-form $\boldsymbol{\omega}$ is continuous or of class C^q if its component functions f_j are continuous or, respectively, of class C^q. From now on *every differential form of degree* 1 *is assumed to be continuous*.

Observe that $\boldsymbol{\omega}(\boldsymbol{x})$ is the linear form associated with the vector

$$\boldsymbol{F}(\boldsymbol{x}) := (f_1(\boldsymbol{x}), \ldots, f_n(\boldsymbol{x})),$$

where, as usual, the f_j denote the component functions of $\boldsymbol{\omega}$. Consequently, the study of the 1-form $\boldsymbol{\omega}$ is essentially equivalent to the study of the vector field $\boldsymbol{F}$, and we may interpret differential forms of degree 1 and vector fields as two different ways of visualizing the same mathematical object. Vector fields provide the most appropriate point of view for formulating problems from physics or engineering, but in order to solve these problems mathematically, it is often more convenient to express them in terms of differential forms.

What we have just seen is another important example of changing the meaning, or interpretation, of a mathematical concept by renaming it. We have moved from the realm of linear algebra into that of differential geometry by viewing a linear mapping, namely the projection $\boldsymbol{h} \to h_j$, as a differential form, $\mathrm{d}x_j$.

Example 2.3.1. Let $g : U \subset \mathbb{R}^n \to \mathbb{R}$ be a function of class C^1 on the open set U. The differential of g at point $\boldsymbol{x} \in U$ is the linear mapping $\mathrm{d}g(\boldsymbol{x}) : \mathbb{R}^n \to \mathbb{R}$ given by

$$(\mathrm{d}g)(\boldsymbol{x})(\boldsymbol{h}) = \sum_{j=1}^{n} \frac{\partial g}{\partial x_j}(\boldsymbol{x})h_j = \sum_{j=1}^{n} \frac{\partial g}{\partial x_j}(\boldsymbol{x})\mathrm{d}x_j(\boldsymbol{h}).$$

Hence

$$\mathrm{d}g = \sum_{j=1}^{n} \frac{\partial g}{\partial x_j}\mathrm{d}x_j.$$

Since each partial derivative $\frac{\partial g}{\partial x_j}$ is a continuous function, we conclude that the mapping $U \to \mathscr{L}(\mathbb{R}^n, \mathbb{R})$, $\boldsymbol{x} \to (\mathrm{d}g)(\boldsymbol{x})$, is a (continuous) differential form of degree 1. We represent it by $\boldsymbol{\omega} = \mathrm{d}g$ (the differential of g).

Definition 2.3.2. Let $\boldsymbol{\omega}$ be a continuous 1-form on U and let $\boldsymbol{\gamma} : [a,b] \to U$ be a piecewise C^1 path. Then

$$\int_{\boldsymbol{\gamma}} \boldsymbol{\omega} := \int_a^b \boldsymbol{\omega}(\boldsymbol{\gamma}(t))\,(\boldsymbol{\gamma}'(t))\,\mathrm{d}t.$$

We observe that the previous function is well defined and continuous except on a finite set. Since it is bounded, it is integrable in the sense of Riemann and Lebesgue. Moreover, expressing the 1-form $\boldsymbol{\omega}$ in terms of its component functions, $\boldsymbol{\omega} = \sum_{j=1}^n f_j \mathrm{d}x_j$, we have

$$\int_{\boldsymbol{\gamma}} \boldsymbol{\omega} = \int_a^b \left(\sum_{j=1}^n f_j(\boldsymbol{\gamma}(t))\mathrm{d}x_j(\boldsymbol{\gamma}'(t)) \right) \mathrm{d}t = \int_a^b \left(\sum_{j=1}^n f_j(\boldsymbol{\gamma}(t))\boldsymbol{\gamma}_j'(t) \right) \mathrm{d}t$$
$$= \int_a^b \langle \boldsymbol{F}(\boldsymbol{\gamma}(t)),\, \boldsymbol{\gamma}'(t) \rangle \,\mathrm{d}t,$$

where $\boldsymbol{F} := (f_1, \ldots, f_n) : U \subset \mathbb{R}^n \to \mathbb{R}^n$ is the vector field associated with $\boldsymbol{\omega}$. This simple calculation highlights an important fact: when $\boldsymbol{F}$ is the vector field associated with the differential form $\boldsymbol{\omega}$ of degree 1, we have

$$\int_{\boldsymbol{\gamma}} \boldsymbol{\omega} = \int_{\boldsymbol{\gamma}} \boldsymbol{F}.$$

2.4 Parameter Changes

The line integral $\int_{\boldsymbol{\gamma}} \boldsymbol{F}$ depends on the vector field $\boldsymbol{F}$ and also on the path $\boldsymbol{\gamma}$. In this section, we plan to analyze what happens on replacing $\boldsymbol{\gamma}$ by some other path with the same trace. Further results in this direction will be obtained in the optional Sect. 2.7.

Definition 2.4.1. Let $\boldsymbol{\alpha} : [a,b] \to \mathbb{R}^n$ and $\boldsymbol{\beta} : [c,d] \to \mathbb{R}^n$ be two paths. We say that $\boldsymbol{\alpha}$ and $\boldsymbol{\beta}$ are *equivalent*, and write $\boldsymbol{\alpha} \sim \boldsymbol{\beta}$, if there is a mapping $\varphi : [a,b] \to [c,d]$, of class C^1, such that $\varphi([a,b]) = [c,d]$, $\varphi'(t) > 0$ for every $t \in [a,b]$ and $\boldsymbol{\alpha} = \boldsymbol{\beta} \circ \varphi$ (Fig. 2.2).

By the mean value theorem, the conditions on φ in this definition ensure that φ is strictly increasing, hence bijective, and that $c = \varphi(a)$ and $d = \varphi(b)$. It then follows that $\boldsymbol{\alpha} \sim \boldsymbol{\beta}$ implies $\boldsymbol{\beta} \sim \boldsymbol{\alpha}$. Indeed, $\boldsymbol{\beta} = \boldsymbol{\alpha} \circ \varphi^{-1}$, and φ^{-1} has the necessary properties for equivalence.

The next result is usually referred to as the *chain rule* or the *composite function theorem.* Its proof can be found in any text on differential calculus of several variables; for example [9, Theorem 3.1, p. 76] or [10, 4.4, p. 134].

Theorem 2.4.1. *Let U and V be open subsets of $\mathbb{R}^n$ and $\mathbb{R}^m$ respectively. If the mappings $\boldsymbol{F} : U \longrightarrow V$ and $\boldsymbol{G} : V \longrightarrow \mathbb{R}^p$ are differentiable at $\boldsymbol{a} \in U$ and $\boldsymbol{F}(\boldsymbol{a}) \in V$ respectively, then their composition $\boldsymbol{H} = \boldsymbol{G} \circ \boldsymbol{F}$ is differentiable at $\boldsymbol{a}$, and*

$$\mathrm{d}\boldsymbol{H}(\boldsymbol{a}) = \mathrm{d}\boldsymbol{G}(\boldsymbol{F}(\boldsymbol{a})) \circ \mathrm{d}\boldsymbol{F}(\boldsymbol{a}),$$

or, in terms of their associated matrices,

$$\boldsymbol{H}'(\boldsymbol{a}) = \boldsymbol{G}'(\boldsymbol{F}(\boldsymbol{a}))\boldsymbol{F}'(\boldsymbol{a}).$$

Furthermore, if $\boldsymbol{F}$ and $\boldsymbol{G}$ are of class C^q ($1 \leq q \leq \infty$) on their respective domains of definition, then $\boldsymbol{H}$ is also C^q on U.

Observe that the chain rule hypothesis involves functions defined on open sets. Nevertheless, when the first function is defined on a closed interval $[a,b]$, a variant of the result is also true.

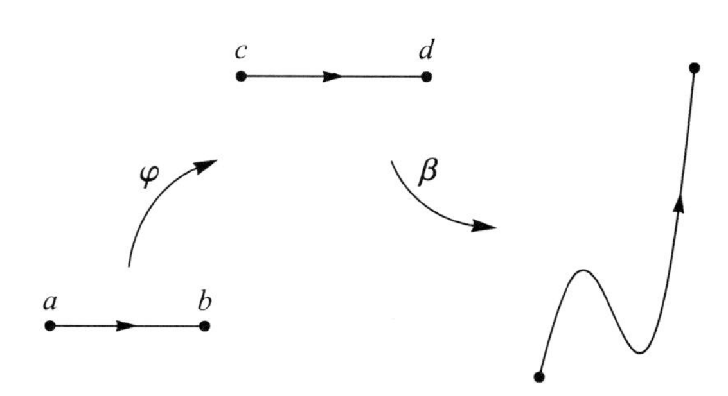

Fig. 2.2 $\boldsymbol{\beta}$ and $\boldsymbol{\beta} \circ \varphi$ are equivalent paths

Corollary 2.4.1. *Let V be an open subset of $\mathbb{R}^m$, $\boldsymbol{\gamma} : [a,b] \longrightarrow V$ a path that has a derivative at each point of $[a,b]$, and suppose $\boldsymbol{G} : V \longrightarrow \mathbb{R}^m$ is differentiable at each point of $\boldsymbol{\gamma}([a,b])$. Then $\boldsymbol{G} \circ \boldsymbol{\gamma} : [a,b] \longrightarrow \mathbb{R}^m$ has a derivative at each point of $[a,b]$ and*

$$(\boldsymbol{G} \circ \boldsymbol{\gamma})'(t) = \boldsymbol{G}'(\boldsymbol{\gamma}(t)) \cdot \boldsymbol{\gamma}'(t),$$

for every $t \in [a,b]$. Here, the derivatives at $t = a$ and $t = b$ are understood to be derivatives from the right and from the left respectively. Furthermore, if $\boldsymbol{\gamma}$ is C^q on $[a,b]$ and $\boldsymbol{G}$ is C^q on the open set V, then $\boldsymbol{G} \circ \boldsymbol{\gamma}$ is also C^q on $[a,b]$.

Proof. We write $\boldsymbol{\gamma}(t) = (\gamma_1(t), \ldots, \gamma_m(t))$, where $\gamma_j : [a,b] \longrightarrow \mathbb{R}$ has a derivative on $[a,b]$ for each $j = 1, \ldots, n$ (or C^q on $[a,b]$ respectively).

It is known that γ_j can be extended to $\tilde{\gamma}_j : \mathbb{R} \to \mathbb{R}$ with a derivative at each point of $\mathbb{R}$ (or C^q on $\mathbb{R}$ respectively). We take $\tilde{\boldsymbol{\gamma}} = (\tilde{\gamma}_1, \ldots, \tilde{\gamma}_m) : \mathbb{R} \longrightarrow \mathbb{R}^m$. Now we define $U = (\tilde{\boldsymbol{\gamma}})^{-1}(V)$. Since $\tilde{\boldsymbol{\gamma}}$ is continuous on $\mathbb{R}$, U is an open subset of $\mathbb{R}$ containing $[a,b]$. Moreover, we have $\tilde{\boldsymbol{\gamma}} : U \longrightarrow V$ and $\boldsymbol{G} : V \longrightarrow \mathbb{R}^m$ with $\tilde{\boldsymbol{\gamma}}$ and $\boldsymbol{G}$ differentiable (respectively of class C^q) on their respective domains. Now the chain rule gives the conclusion. □

The same argument clearly proves the following further variation of the chain rule.

Corollary 2.4.2. *Let $\alpha : [a,b] \longrightarrow \mathbb{R}$ be differentiable at each point of $[a,b]$ and let $\boldsymbol{\beta} : [c,d] \longrightarrow \mathbb{R}^m$ be a path that has a derivative at each point of $[c,d]$ with $\alpha([a,b]) \subset [c,d]$. Then $\boldsymbol{\beta} \circ \alpha : [a,b] \longrightarrow \mathbb{R}^m$ has a derivative at each point of $[a,b]$ and*

$$(\boldsymbol{\beta} \circ \alpha)'(t) = \boldsymbol{\beta}'(\alpha(t))\alpha'(t),$$

for every $t \in [a,b]$. Furthermore, if α is C^q on $[a,b]$ and $\boldsymbol{\beta}$ is C^q on $[c,d]$, then $\boldsymbol{\beta} \circ \alpha$ is also C^q on $[a,b]$.

Proposition 2.4.1. *If $\boldsymbol{\alpha}$ and $\boldsymbol{\beta}$ are two equivalent paths and one of them, say $\boldsymbol{\beta}$, is piecewise C^1, then $\boldsymbol{\alpha}$ is piecewise C^1 and $L(\boldsymbol{\alpha}) = L(\boldsymbol{\beta})$.*

Proof. Write $\boldsymbol{\alpha} = \boldsymbol{\beta} \circ \varphi$, where φ is as in Definition 2.4.1 and let $Q = \{c = u_1 < \cdots < u_k = d\}$ be a partition of $[c,d]$ such that $\boldsymbol{\beta}_{|[u_j,u_{j+1}]}$ is C^1 on $[u_j, u_{j+1}]$. If we take $t_j := \varphi^{-1}(u_j)$, then $P = \{a = t_1 < \cdots < t_k = b\}$ is a partition of $[a,b]$.

Since $\boldsymbol{\alpha}|_{[t_j,t_{j+1}]} = (\boldsymbol{\beta} \circ \varphi)|_{[t_j,t_{j+1}]} = (\boldsymbol{\beta}|_{[u_j,u_{j+1}]}) \circ \varphi|_{[t_j,t_{j+1}]}$ is the composition of two C^1 mappings (on closed intervals), it is itself C^1 on $[t_j, t_{j+1}]$, and thus $\boldsymbol{\alpha}$ is piecewise C^1. Now,

$$L(\boldsymbol{\beta}) = \int_{c=\varphi(a)}^{d=\varphi(b)} ||\boldsymbol{\beta}'(u)||\mathrm{d}u = \sum_{j=1}^{k-1} \int_{u_j}^{u_{j+1}} ||\boldsymbol{\beta}'(u)||\mathrm{d}u.$$

For each $1 \le j \le k-1$ we can apply the change of variable $u = \varphi(t)$ to obtain

$$\int_{u_j}^{u_{j+1}} ||\boldsymbol{\beta}'(u)||\mathrm{d}u = \int_{t_j}^{t_{j+1}} ||\boldsymbol{\beta}'(\varphi(t))||\varphi'(t)\mathrm{d}t$$

$$= \int_{t_j}^{t_{j+1}} ||\boldsymbol{\beta}'(\varphi(t))\varphi'(t)||\mathrm{d}t = \int_{t_j}^{t_{j+1}} ||(\boldsymbol{\beta}\circ\varphi)'(t)||\mathrm{d}t$$

$$= \int_{t_j}^{t_{j+1}} ||\boldsymbol{\alpha}'(t)||\mathrm{d}t.$$

Summing for all values of $1 \leq j \leq k-1$, we conclude that

$$L(\boldsymbol{\beta}) = \int_a^b ||\boldsymbol{\alpha}'(t)||\mathrm{d}t = L(\boldsymbol{\alpha}).$$

□

Proposition 2.4.2. *Let* $\boldsymbol{\omega}$ *be a continuous* 1*-form on* U*, and let* $\boldsymbol{\alpha}$ *and* $\boldsymbol{\beta}$ *be two piecewise* C^1 *paths in* U *with* $\boldsymbol{\alpha} \sim \boldsymbol{\beta}$*. Then*

$$\int_{\boldsymbol{\alpha}} \boldsymbol{\omega} = \int_{\boldsymbol{\beta}} \boldsymbol{\omega}.$$

Proof. Let us first assume that $\boldsymbol{\alpha}$ and $\boldsymbol{\beta}$ are paths of class C^1. Then, employing the same notation as the previous proof, we have

$$\int_{\boldsymbol{\beta}} \boldsymbol{\omega} = \int_{\varphi(a)}^{\varphi(b)} \boldsymbol{\omega}(\boldsymbol{\beta}(u))\left(\boldsymbol{\beta}'(u)\right)\mathrm{d}u$$

$$= \int_a^b \boldsymbol{\omega}(\boldsymbol{\beta}(\varphi(t)))\left(\boldsymbol{\beta}'(\varphi(t))\right)\varphi'(t)\mathrm{d}t$$

$$= \int_a^b \boldsymbol{\omega}(\boldsymbol{\beta}(\varphi(t)))\left((\boldsymbol{\beta}\circ\varphi)'(t)\right)\mathrm{d}t$$

$$= \int_a^b \boldsymbol{\omega}(\boldsymbol{\alpha}(t))(\boldsymbol{\alpha}'(t))\mathrm{d}t = \int_{\boldsymbol{\alpha}} \boldsymbol{\omega}.$$

Again we are using the chain rule theorem for extensions of $\boldsymbol{\beta}$ and φ to the whole real line. In the general case, we consider a partition

$$P := \{a = t_1 < t_2 < \cdots < t_k = b\}$$

of the interval $[a,b]$ for which $\boldsymbol{\beta}$ is of class C^1 on each subinterval $[u_j, u_{j+1}] = [\varphi(t_j), \varphi(t_{j+1})]$. Then $\boldsymbol{\alpha} = \boldsymbol{\beta}\circ\varphi$ is also of class C^1 on each subinterval $[t_j, t_{j+1}]$. We therefore have from above that

$$\int_{\boldsymbol{\beta}_j} \boldsymbol{\omega} = \int_{\boldsymbol{\alpha}_j} \boldsymbol{\omega},$$

Fig. 2.3 $\boldsymbol{\gamma}$ and $-\boldsymbol{\gamma}$ are opposite paths

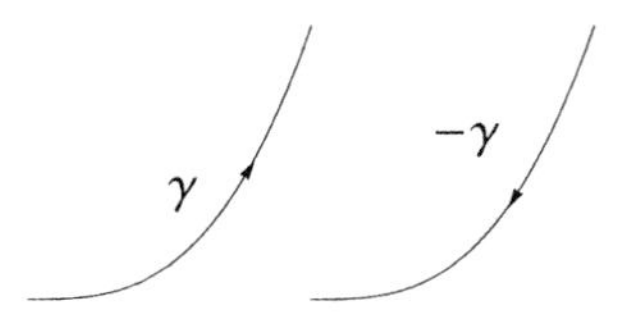

where $\boldsymbol{\beta}_j = \boldsymbol{\beta}_{|[u_j,u_{j+1}]}$ and $\boldsymbol{\alpha}_j = \boldsymbol{\alpha}_{|[t_j,t_{j+1}]}$. Summing this identity for all values of j from 1 to $k-1$, we get

$$\int_{\boldsymbol{\beta}} \omega = \int_{\boldsymbol{\alpha}} \omega .$$

□

Notice that in the above proof neither $\varphi'(t_j)$ nor $(\boldsymbol{\beta} \circ \varphi)'(t_j)$ is necessarily defined for $j = 1, \ldots, k-1$.

Definition 2.4.2. Let $\boldsymbol{\gamma} : [a,b] \to \mathbb{R}^n$ be a piecewise C^1 path. The *opposite* path (Fig. 2.3) is defined by $(-\boldsymbol{\gamma}) : [-b,-a] \to \mathbb{R}^n$, $(-\boldsymbol{\gamma})(t) := \boldsymbol{\gamma}(-t)$.

The opposite path $-\boldsymbol{\gamma}$ moves along $\boldsymbol{\gamma}([a,b])$ in the opposite direction to that given by the path $\boldsymbol{\gamma}$. The initial point of $-\boldsymbol{\gamma}$ is the final point of $\boldsymbol{\gamma}$ and vice versa. The reader will note the difference between $(-\boldsymbol{\gamma})(t)$ and $-(\boldsymbol{\gamma}(t))$.

Definition 2.4.3. Let $\boldsymbol{\alpha} : [a,b] \to \mathbb{R}^n$ and $\boldsymbol{\beta} : [c,d] \to \mathbb{R}^n$ be two piecewise C^1 paths such that $\boldsymbol{\alpha}(b) = \boldsymbol{\beta}(c)$. By a *union of the paths* $\boldsymbol{\alpha}$ and $\boldsymbol{\beta}$, denoted by $\boldsymbol{\alpha} \cup \boldsymbol{\beta}$, we mean any piecewise C^1 path $\boldsymbol{\xi} : [e,f] \to \mathbb{R}^n$ with the property that for some $e < r < f$,

$$\boldsymbol{\xi}_{|[e,r]} \sim \boldsymbol{\alpha} \quad \text{and} \quad \boldsymbol{\xi}_{|[r,f]} \sim \boldsymbol{\beta}.$$

It is obvious that[3] the trace of $\boldsymbol{\alpha} \cup \boldsymbol{\beta}$ is $\boldsymbol{\alpha}([a,b]) \cup \boldsymbol{\beta}([c,d])$ and $\boldsymbol{\alpha} \cup \boldsymbol{\beta}$ consists of tracing over $\boldsymbol{\alpha}$ first and then over $\boldsymbol{\beta}$. Such a union of paths always exists, a concrete example being given by $\boldsymbol{\xi} : [0,1] \to \mathbb{R}^n$, where

$$\boldsymbol{\xi}(t) = \begin{cases} \boldsymbol{\alpha}\big(2t(b-a)+a\big), & 0 \le t \le \frac{1}{2}, \\ \boldsymbol{\beta}\big((2t-1)(d-c)+c\big), & \frac{1}{2} \le t \le 1. \end{cases}$$

Proposition 2.4.3. *Let ω be a continuous 1-form on an open subset U of $\mathbb{R}^n$ and let $\boldsymbol{\alpha}, \boldsymbol{\beta}, \boldsymbol{\gamma}$ be three piecewise C^1 paths in U with the final point of $\boldsymbol{\alpha}$ coinciding with the initial point of $\boldsymbol{\beta}$. Then*

(1) $\displaystyle\int_{-\boldsymbol{\gamma}} \omega = -\int_{\boldsymbol{\gamma}} \omega,$

(2) $\displaystyle\int_{\boldsymbol{\alpha} \cup \boldsymbol{\beta}} \omega = \int_{\boldsymbol{\alpha}} \omega + \int_{\boldsymbol{\beta}} \omega.$

---

[3]The right-hand side here is a normal set union. Note that unlike set union, path union is not a symmetric operation.

Proof. Since $(-\gamma)'(t) = -\gamma'(-t)$, the substitution $u = -t$ gives

$$\int_{-\boldsymbol{\gamma}} \boldsymbol{\omega} = \int_{-b}^{-a} \boldsymbol{\omega}(\boldsymbol{\gamma}(-t))\left(-\boldsymbol{\gamma}'(-t)\right) \mathrm{d}t$$
$$= \int_{a}^{b} -\boldsymbol{\omega}(\boldsymbol{\gamma}(u))\left(\boldsymbol{\gamma}'(u)\right) \mathrm{d}u = -\int_{\boldsymbol{\gamma}} \boldsymbol{\omega}.$$

Also

$$\int_{\boldsymbol{\alpha}\cup\boldsymbol{\beta}} \boldsymbol{\omega} = \int_{\mathrm{e}}^{r} \boldsymbol{\omega}(\boldsymbol{\xi}(t))\left(\boldsymbol{\xi}'(t)\right) \mathrm{d}t + \int_{r}^{f} \boldsymbol{\omega}(\boldsymbol{\xi}(t))\left(\boldsymbol{\xi}'(t)\right) \mathrm{d}t$$
$$= \int_{\boldsymbol{\xi}_{|[e,r]}} \boldsymbol{\omega} + \int_{\boldsymbol{\xi}_{|[r,f]}} \boldsymbol{\omega} = \int_{\boldsymbol{\alpha}} \boldsymbol{\omega} + \int_{\boldsymbol{\beta}} \boldsymbol{\omega}.$$

□

2.5 Conservative Fields: Exact Differential Forms

We begin this section with an example of a vector field for which the line integral over two different paths from $(0,0)$ to $(1,1)$ is the same.

Example 2.5.1. Let $\boldsymbol{F} : \mathbb{R}^2 \to \mathbb{R}^2$ be the vector field defined by $\boldsymbol{F}(x,y) = (x,y)$ and let us consider the paths (Fig. 2.4)

$$\boldsymbol{\gamma}_1 : [0,1] \to \mathbb{R}^2 \text{ and } \boldsymbol{\gamma}_2 : [0,1] \to \mathbb{R}^2$$

defined by $\boldsymbol{\gamma}_1(t) = (t,t)$ and $\boldsymbol{\gamma}_2(t) = (t,t^2)$. Then

$$\int_{\boldsymbol{\gamma}_1} \boldsymbol{F} = \int_0^1 \langle (t,t),(1,1)\rangle \, \mathrm{d}t = \int_0^1 (t+t)\, \mathrm{d}t = 1.$$

Also

$$\int_{\boldsymbol{\gamma}_2} \boldsymbol{F} = \int_0^1 \langle (t,t^2),(1,2t)\rangle \, \mathrm{d}t = \int_0^1 (t+2t^3)\, \mathrm{d}t = 1.$$

The fact that the two line integrals here take the same value is not surprising if one remembers that a line integral represents the work done by a force field in moving a particle along a path, while it is well known from physics that under the action of a gravitational field,[4] the work done to move an object between two different points depends only on the difference of the potential energies at these points, and is therefore independent of the path taken (see Example 2.5.4). Our next example, a vector field for which the line integral over two different paths from $(0,0)$ to $(1,1)$ is different, shows that this behavior is not universal.

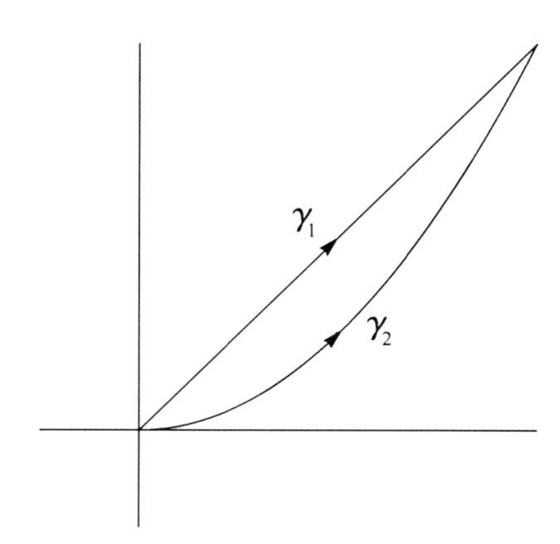

Fig. 2.4 The paths of Example 2.5.1

[4] Although the vector field in our example is not the gravitational field of Example 1.2.2, it is similar enough, in a sense yet to be defined, that the physical argument remains valid.

Example 2.5.2. Let $\boldsymbol{\gamma}_1$ and $\boldsymbol{\gamma}_2$ be the paths given in Example 2.5.1 but let us consider instead the vector field

$$\boldsymbol{F}(x,y) = \left(-y+\frac{3}{8}, x-\frac{1}{2}\right).$$

Then

$$\int_{\boldsymbol{\gamma}_1} \boldsymbol{F} = \int_0^1 \left\langle \left(-t+\frac{3}{8}, t-\frac{1}{2}\right), (1,1) \right\rangle \,\mathrm{d}t = \int_0^1 -\frac{1}{8}\mathrm{d}t = -\frac{1}{8},$$

while

$$\int_{\boldsymbol{\gamma}_2} \boldsymbol{F} = \int_0^1 \left\langle \left(-t^2+\frac{3}{8}, t-\frac{1}{2}\right), (1,2t) \right\rangle \,\mathrm{d}t = \int_0^1 \left(t^2-t+\frac{3}{8}\right)\mathrm{d}t = \frac{1}{3}-\frac{1}{8}.$$

Definition 2.5.1. Let $\boldsymbol{F} = (f_1, \ldots, f_n)$ be a vector field on an open subset U of $\mathbb{R}^n$ with associated 1-form $\boldsymbol{\omega} = f_1 \cdot \mathrm{d}x_1 + \cdots + f_n \cdot \mathrm{d}x_n$. If there is a function of class C^1, $f : U \subset \mathbb{R}^n \to \mathbb{R}$, such that $\nabla f = \boldsymbol{F}$ (or equivalently, $\mathrm{d}f = \boldsymbol{\omega}$) on U, then the vector field $\boldsymbol{F}$ is said to be *conservative* and the 1-form $\boldsymbol{\omega}$ is said to be *exact*. The scalar function f is called a *potential* of the conservative vector field $\boldsymbol{F}$.

Conservative fields have a similar behavior to that of Example 2.5.1 in that their line integrals are independent of path. This is an immediate consequence of the following result, which is a generalization of the fundamental theorem of calculus.

Theorem 2.5.1. *Let $g : U \subset \mathbb{R}^n \to \mathbb{R}$ be a function of class C^1 on an open set U and $\boldsymbol{\gamma} : [a,b] \to U$ a piecewise C^1 path. Then*

$$\int_{\boldsymbol{\gamma}} \boldsymbol{\nabla} g = \int_{\boldsymbol{\gamma}} \mathrm{d}g = g(\boldsymbol{\gamma}(b)) - g(\boldsymbol{\gamma}(a)).$$

Proof. Since $\mathrm{d}g$ is the 1-form associated to the vector field $\boldsymbol{\nabla} g$, the first equation follows from our observation immediately preceding Sect. 2.4. For the second equation, we take the usual expansion of the 1-form $\mathrm{d}g$,

$$\mathrm{d}g = \sum_{j=1}^{n} \frac{\partial g}{\partial x_j} \cdot \mathrm{d}x_j$$

and apply the chain rule,

$$\sum_{j=1}^{n} \frac{\partial g}{\partial x_j}(\boldsymbol{\gamma}(t))\boldsymbol{\gamma}_j'(t) = (g \circ \boldsymbol{\gamma})'(t),$$

which is valid for all but a finite set of t in $[a,b]$, to obtain

$$\int_{\boldsymbol{\gamma}} \mathrm{d}g = \int_a^b \left(\sum_{j=1}^{n} \frac{\partial g}{\partial x_j}(\boldsymbol{\gamma}(t))\boldsymbol{\gamma}_j'(t) \right) \mathrm{d}t = \int_a^b (g \circ \boldsymbol{\gamma})'(t)\mathrm{d}t.$$

Let $P := \{a = t_1 < \cdots < t_k = b\}$ be a partition such that $\boldsymbol{\gamma}|_{[t_i,t_{i+1}]}$ is of class C^1 for every $1 \le i \le k-1$. The fundamental theorem of calculus gives

$$\begin{aligned}\int_{\boldsymbol{\gamma}} \mathrm{d}g &= \sum_{i=1}^{k-1} \int_{t_i}^{t_{i+1}} (g \circ \boldsymbol{\gamma})'(t)\mathrm{d}t \\ &= \sum_{i=1}^{k-1} \big((g \circ \boldsymbol{\gamma})(t_{i+1}) - (g \circ \boldsymbol{\gamma})(t_i)\big) \\ &= g(\boldsymbol{\gamma}(b)) - g(\boldsymbol{\gamma}(a)).\end{aligned}$$

□

For our next result we recall the concepts of connected set and path-connected set.

Definition 2.5.2. Let C be a subset of $\mathbb{R}^n$.

1. C is said to be *connected* if there do not exist two open sets V and W in $\mathbb{R}^n$ such that

   a. $C \subset V \cup W$;
   b. $C \cap V \neq \varnothing$ and $C \cap W \neq \varnothing$;
   c. $C \cap V \cap W = \varnothing$.

   In other words, a set C is connected if and only if with the topology induced on C by $\mathbb{R}^n$, the only subsets of C that are both open and closed are C itself and the empty set $\varnothing$.
2. C is said to be *path connected* if given $\boldsymbol{x}, \boldsymbol{y} \in C$, there exists a path $\boldsymbol{\alpha} : [a,b] \to C$ such that $\boldsymbol{\alpha}(a) = \boldsymbol{x}$ and $\boldsymbol{\alpha}(b) = \boldsymbol{y}$. If the connecting path can be chosen to be polygonal, then the set is called *polygonally connected*.

The continuous image of a connected set is connected, from which it follows that every path-connected set is connected. For an open subset of $\mathbb{R}^n$, the converse also holds, and the two concepts coincide. This result can be found, for example, in [1, Theorem 4.43].

Theorem 2.5.2. *Let U be a connected open subset of $\mathbb{R}^n$. Then U is polygonally connected.*

Theorem 2.5.3. *Let $\boldsymbol{F} : U \subset \mathbb{R}^n \to \mathbb{R}^n$ be a continuous vector field on an open set U. The following conditions are equivalent:*

(1) $\boldsymbol{F}$ is a conservative vector field.
(2) If $\boldsymbol{\gamma} : [a,b] \to U \subset \mathbb{R}^n$ is a piecewise C^1 path that is closed ($\boldsymbol{\gamma}(a) = \boldsymbol{\gamma}(b)$), then

$$\int_{\boldsymbol{\gamma}} \boldsymbol{F} = 0.$$

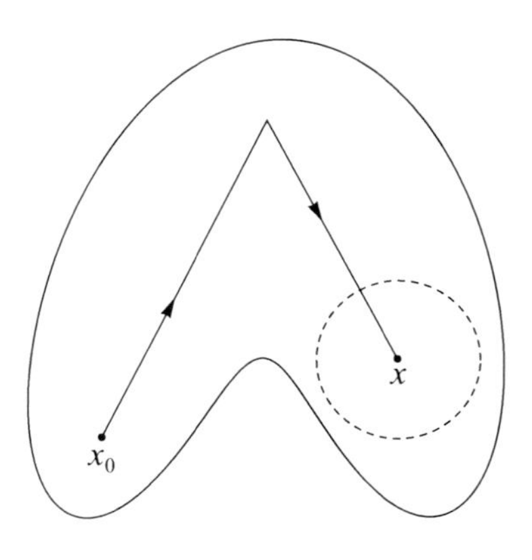

Fig. 2.5 A polygonal path in U from $\mathbf{x}_0$ to $\boldsymbol{x}$

(3) *If $\boldsymbol{\gamma}_1$ and $\boldsymbol{\gamma}_2$ are two piecewise C^1 paths with the same initial and final points, then*

$$\int_{\boldsymbol{\gamma}_1} \boldsymbol{F} = \int_{\boldsymbol{\gamma}_1} \boldsymbol{F}.$$

Proof. (1) $\Rightarrow$ (2). By hypothesis, there is a C^1 function f on U,

$$f : U \subset \mathbb{R}^n \to \mathbb{R},$$

with $\boldsymbol{\nabla} f = \boldsymbol{F}$. Let $\boldsymbol{\gamma} : [a,b] \to U \subset \mathbb{R}^n$ be a closed piecewise C^1 path. According to Theorem 2.5.1,

$$\int_{\boldsymbol{\gamma}} \boldsymbol{F} = f(\boldsymbol{\gamma}(b)) - f(\boldsymbol{\gamma}(a)) = 0.$$

(2) $\Rightarrow$ (3). If $\boldsymbol{\gamma}_1$ and $\boldsymbol{\gamma}_2$ have the same initial point and also the same final point, then

$$\boldsymbol{\gamma} := \boldsymbol{\gamma}_1 \cup (-\boldsymbol{\gamma}_2)$$

is a closed piecewise C^1 path in U. Consequently,

$$\int_{\boldsymbol{\gamma}_1} \boldsymbol{F} - \int_{\boldsymbol{\gamma}_2} \boldsymbol{F} = \int_{\boldsymbol{\gamma}} \boldsymbol{F} = 0.$$

(3) $\Rightarrow$ (1). Let us first assume that U is an open and connected set in $\mathbb{R}^n$. Then U is also path connected and polygonally connected. We fix a point $\boldsymbol{x}_0 \in U$ and define the potential function $f : U \subset \mathbb{R}^n \to \mathbb{R}$ as follows (Fig. 2.5): For every $\boldsymbol{x} \in U$ let $\boldsymbol{\gamma}_x$ be any polygonal path in U from $\mathbf{x}_0$ to $\boldsymbol{x}$ and let

$$f(\boldsymbol{x}) := \int_{\boldsymbol{\gamma}_x} \boldsymbol{F}.$$

The hypothesis of (3) implies that $f(\boldsymbol{x})$ does not depend on the particular polygonal path $\boldsymbol{\gamma}_x$ chosen, and thus the definition is unambiguous. To prove the result, we will show that f is a function of class C^1 whose gradient coincides with the vector field $\boldsymbol{F}$.

Given $\boldsymbol{x} \in U$, we can find $\delta > 0$ such that the open ball $B(\boldsymbol{x}, \delta)$ centered at $\boldsymbol{x}$ and with radius δ is contained in U. Observe that for every $1 \le j \le n$ and $0 < |t| < \delta$, if $\boldsymbol{\gamma}_x$ is a polygonal path in U from point $\mathbf{x}_0$ to point $\boldsymbol{x}$, then $\boldsymbol{\gamma}_x \cup [\boldsymbol{x}, \boldsymbol{x} + t\mathbf{e}_j]$ is a polygonal path in U from $\mathbf{x}_0$ to $\boldsymbol{x} + t\mathbf{e}_j$. Hence

$$f(\boldsymbol{x} + t\mathbf{e}_j) - f(\boldsymbol{x}) = \int_{\boldsymbol{\gamma}_x \cup [\boldsymbol{x}, \boldsymbol{x} + t\mathbf{e}_j]} \boldsymbol{F} - \int_{\boldsymbol{\gamma}_x} \boldsymbol{F} = \int_{[\boldsymbol{x}, \boldsymbol{x} + t\mathbf{e}_j]} \boldsymbol{F}.$$

Let us assume for simplicity that $t > 0$ and parameterize the line segment by $\boldsymbol{\gamma}(s) = \boldsymbol{x} + s\mathbf{e}_j$, $0 \le s \le t$. We also write $\boldsymbol{F}$ in terms of component functions, $\boldsymbol{F} = (f_1, \ldots, f_n)$. Then for each $1 \le j \le n$,

$$f(\boldsymbol{x} + t\mathbf{e}_j) - f(\boldsymbol{x}) = \int_0^t \langle \boldsymbol{F}(\boldsymbol{x} + s\mathbf{e}_j), \mathbf{e}_j \rangle \, \mathrm{d}s = \int_0^t f_j(\boldsymbol{x} + s\mathbf{e}_j) \, \mathrm{d}s.$$

It follows that

$$\begin{aligned} \left| \frac{f(\boldsymbol{x} + t\mathbf{e}_j) - f(\boldsymbol{x})}{t} - f_j(\mathbf{x}) \right| &= \left| \frac{1}{t} \int_0^t \left(f_j(\boldsymbol{x} + s\mathbf{e}_j) - f_j(\boldsymbol{x}) \right) \, \mathrm{d}s \right| \\ &\le \frac{1}{t} \int_0^t |f_j(\boldsymbol{x} + s\mathbf{e}_j) - f_j(\boldsymbol{x})| \mathrm{d}s \\ &\le \max_{0 \le s \le t} |f_j(\boldsymbol{x} + s\mathbf{e}_j) - f_j(\boldsymbol{x})|. \end{aligned}$$

Since each f_j is a continuous function, the above expression tends to zero as t tends to zero. As a consequence,

$$\frac{\partial f}{\partial x_j}(\boldsymbol{x}) = f_j(\boldsymbol{x}).$$

In particular, f is a function of class C^1 and $\boldsymbol{\nabla} f = \boldsymbol{F}$ on U. In the general case in which U is not connected, the above argument can be applied to each connected component of U in order to construct a suitable potential function (see, for instance, Burkill [4, p. 60]). $\square$

Definition 2.5.3. An open set $U \subset \mathbb{R}^n$ is said to be *starlike* with respect to the point $\boldsymbol{a} \in U$ if the segment $[\boldsymbol{a}, \boldsymbol{x}]$ is contained in U for every $\boldsymbol{x} \in U$.

We recall that the triangle with vertices $\boldsymbol{a}$, $\boldsymbol{x}$, and $\boldsymbol{y}$ is the set

$$\{\alpha \boldsymbol{a} + \beta \boldsymbol{x} + \gamma \boldsymbol{y} \, : \, 0 \le \alpha, \beta, \gamma \text{ and } \alpha + \beta + \gamma = 1\}.$$

By its boundary we understand the closed polygonal arc $[\boldsymbol{a}, \boldsymbol{x}] \cup [\boldsymbol{x}, \boldsymbol{y}] \cup [\boldsymbol{y}, \boldsymbol{a}]$.

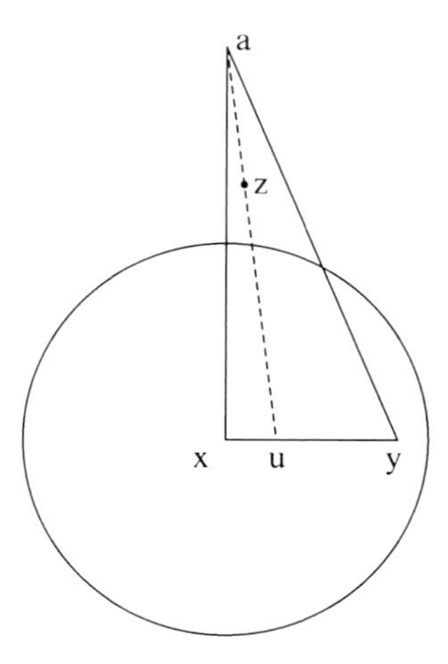

Fig. 2.6 Triangle of Lemma 2.5.1

Lemma 2.5.1. *Let U be starlike with respect to the point $\boldsymbol{a} \in U$. If $\boldsymbol{y} \in B(\boldsymbol{x}, R) \subset U$, then the triangle (Fig. 2.6) with vertex $\{\boldsymbol{a}, \boldsymbol{x}, \boldsymbol{y}\}$ is contained in U.*

Proof. Let $\mathbf{z} = \alpha\boldsymbol{a} + \beta\boldsymbol{x} + \gamma\boldsymbol{y}$ be given, where $\alpha, \beta, \gamma \geq 0$ and $\alpha + \beta + \gamma = 1$. If $\alpha = 1$, then $z = a \in U$. In other case, we put

$$\boldsymbol{u} = \frac{\beta}{1-\alpha}\boldsymbol{x} + \frac{\gamma}{1-\alpha}\boldsymbol{y}.$$

Since

$$\frac{\beta}{1-\alpha} + \frac{\gamma}{1-\alpha} = 1,$$

we obtain that $\boldsymbol{u} \in [\boldsymbol{x}, \boldsymbol{y}] \subset B(\boldsymbol{x}, R) \subset U$. Finally, from $z = \alpha\boldsymbol{a} + (1-\alpha)\boldsymbol{u}$ we conclude that

$$z \in [\boldsymbol{a}, \boldsymbol{u}] \subset U.$$

□

Theorem 2.5.4. *If the open set U is starlike with respect to the point $\boldsymbol{a} \in U$, then the conditions of Theorem 2.5.3 are equivalent to the following:*

(4) *If $\boldsymbol{\gamma}$ is the boundary of a triangle contained in U, then*

$$\int_{\boldsymbol{\gamma}} \omega = 0.$$

Proof. Since condition (2) clearly implies (4), it is enough to prove that (4) implies (1). We do this by arguing that the function f defined by

$$f(\boldsymbol{x}) := \int_{[\boldsymbol{a}, \boldsymbol{x}]} \boldsymbol{F}$$

is, in fact, a potential for the vector field $\boldsymbol{F}$. For every $\boldsymbol{x} \in U$ we choose $R_{\boldsymbol{x}} > 0$ with $B(\boldsymbol{x}, R_{\boldsymbol{x}}) \subset U$. According to the previous lemma, for each $1 \leq j \leq n$ and for all $t \in \mathbb{R}$ with $|t| < R_{\boldsymbol{x}}$, the triangle with vertices $\boldsymbol{a}$, $\boldsymbol{x}$, and $\boldsymbol{x} + t\mathbf{e}_j$ is contained in U. Hence, we deduce from condition (4) that

$$f(\boldsymbol{x} + t\mathbf{e}_j) - f(\boldsymbol{x}) = \int_{[\boldsymbol{a}, \boldsymbol{x}+t\mathbf{e}_j]} \boldsymbol{F} - \int_{[\boldsymbol{a}, \boldsymbol{x}]} \boldsymbol{F} = \int_{[\boldsymbol{x}, \boldsymbol{x}+t\mathbf{e}_j]} \boldsymbol{F}.$$

The argument now proceeds as in the proof of (3) $\Rightarrow$ (1) in Theorem 2.5.3. □

Of course, Theorem 2.5.3 can also be interpreted as a characterization of those differential forms of degree 1 that are exact.

Example 2.5.3. Let

$$\omega = -\frac{y}{x^2 + y^2} \cdot \mathrm{d}x + \frac{x}{x^2 + y^2} \cdot \mathrm{d}y,$$

which is a continuous 1-form in $U := \mathbb{R}^2 \setminus \{(0,0)\}$. Then ω is not exact, since

$$\boldsymbol{\gamma} : [0, 2\pi] \to \mathbb{R}^2, \ \boldsymbol{\gamma}(t) := (\cos(t), \sin(t)),$$

defines a closed path contained in U for which $\displaystyle\int_{\boldsymbol{\gamma}} \omega = 2\pi \neq 0$.

However, if we consider $V := \mathbb{R}^2 \setminus \{(0,y) : y \in \mathbb{R}\}$ and $g : V \to (-\frac{\pi}{2}, \frac{\pi}{2})$ defined by

$$g(x,y) := \arctan \frac{y}{x},$$

then $\omega = \mathrm{d}g$ on V.

Thus, the fact that a vector field is conservative depends not only on the expression of the field but also on the region U we are dealing with. That is, a vector field admitting a potential function on a given open set may not be conservative on some larger set.

Example 2.5.4. The gravitational field is conservative.

Let us consider a particle of mass M located at the origin. The force of attraction exerted on a particle of unit mass located at point $(x,y,z) \in \mathbb{R}^3 \setminus \{(0,0,0)\}$ is

$$\boldsymbol{F}(x,y,z) = -\frac{GM}{(x^2 + y^2 + z^2)^{\frac{3}{2}}} (x,y,z),$$

where G is the gravitational constant. Since the magnitude of the force is the same at all points equidistant from the origin, it seems reasonable to expect that this is also true for the potential function.

Consequently, we look for a function of $r = \sqrt{x^2+y^2+z^2}$ whose derivative is $-\frac{GM}{r^2}$. An example of such a function is $\frac{GM}{r}$, and it is easy to check that

$$f(x,y,z) := \frac{GM}{\sqrt{x^2+y^2+z^2}}$$

satisfies $\nabla f = \boldsymbol{F}$. That is, f is a potential function for the gravitational field.

One should remark that what is called the gravitational potential in physics is the function $V := -f$. Hence, the work done by the field to move a particle from point $\boldsymbol{A}$ to point $\boldsymbol{B}$ is independent of the trajectory of the particle, and its value is the difference of potentials $V(\boldsymbol{A}) - V(\boldsymbol{B})$.

We now obtain a necessary condition for a vector field to be conservative.

Theorem 2.5.5. *Let $\boldsymbol{F} : U \subset \mathbb{R}^n \to \mathbb{R}^n$ be a conservative vector field of class C^1 on an open set U and write $\boldsymbol{F} := (f_1, f_2, \ldots, f_n)$. Then*

$$\frac{\partial f_j}{\partial x_k}(\boldsymbol{x}) = \frac{\partial f_k}{\partial x_j}(\boldsymbol{x})$$

for every $j,k \in \{1,\ldots,n\}$ and every $\boldsymbol{x} \in U$.

Proof. By hypothesis, there is a function $g : U \subset \mathbb{R}^n \to \mathbb{R}$ such that $\nabla g = \boldsymbol{F}$. Therefore $f_j = \frac{\partial g}{\partial x_j}$, and we can write

$$\frac{\partial f_j}{\partial x_k}(\boldsymbol{x}) = \frac{\partial^2 g}{\partial x_k\, \partial x_j}(\boldsymbol{x})$$

and similarly

$$\frac{\partial f_k}{\partial x_j}(\boldsymbol{x}) = \frac{\partial^2 g}{\partial x_j\, \partial x_k}(\boldsymbol{x})$$

for all $\boldsymbol{x} \in U$. Since $\boldsymbol{F}$ is a function of class C^1 on U, it follows that g is of class C^2, and Schwarz's theorem concerning the symmetry of second-order partial derivatives gives our result. □

In particular, if $\boldsymbol{F} = (P,Q)$ is a conservative vector field on $U \subset \mathbb{R}^2$ of class C^1, then for every $(x,y) \in U$,

$$\frac{\partial Q}{\partial x}(x,y) = \frac{\partial P}{\partial y}(x,y).$$

It also follows from Definition 1.2.11 and Theorem 2.5.5 that every conservative vector field $\boldsymbol{F} = (f_1, f_2, f_3)$ on the open set $U \subset \mathbb{R}^3$ of class C^1 satisfies

$$\begin{aligned}\text{Curl}\, \boldsymbol{F}(x,y,z) &= \left(\tfrac{\partial f_3}{\partial y} - \tfrac{\partial f_2}{\partial z}, \tfrac{\partial f_1}{\partial z} - \tfrac{\partial f_3}{\partial x}, \tfrac{\partial f_2}{\partial x} - \tfrac{\partial f_1}{\partial y}\right) \\ &= 0\end{aligned}$$

for all $(x,y,z) \in U$.

In Example 2.5.3 we proved that the vector field

$$\boldsymbol{F} = (P,Q) : U \subset \mathbb{R}^2 \to \mathbb{R}^2$$

defined on $U = \mathbb{R}^2 \setminus \{(0,0)\}$ by

$$P(x,y) = -\frac{y}{x^2+y^2}, \; Q(x,y) = \frac{x}{x^2+y^2}$$

is not conservative. However,

$$\frac{\partial Q}{\partial x}(x,y) = \frac{\partial P}{\partial y}(x,y) = \frac{y^2-x^2}{(x^2+y^2)^2}.$$

Thus, the condition in Theorem 2.5.5 is not sufficient, in general, for a vector field to be conservative. However, if we impose certain geometric conditions on the domain U, then the condition of Theorem 2.5.5 is sufficient, as the next theorem shows. Before stating the result, we need to recall the following fact about the derivative of an integral (also called a parametric derivative).

Proposition 2.5.1. *Suppose $f : [a,b] \times [c,d] \to \mathbb{R}$ is a function of class C^1 and $g : [a,b] \to \mathbb{R}$ is defined by*

$$g(x) = \int_c^d f(x,t)\mathrm{d}t.$$

Then g is of class C^1 on $[a,b]$ and

$$g'(x) = \int_c^d \frac{\partial f}{\partial x}(x,t)\mathrm{d}t$$

for all $x \in [a,b]$.

The proof of this statement follows the proof of (3) $\Rightarrow$ (1) in Theorem 2.5.3, but the interested reader can find stronger results concerning the derivative of an integral, for example in [15, Theorem 9.42, p. 236].

Theorem 2.5.6 (Poincaré's lemma). *Let $\boldsymbol{F} : U \subset \mathbb{R}^n \to \mathbb{R}^n$ be a vector field of class C^1, $\boldsymbol{F} := (f_1, f_2, \ldots, f_n)$, on an open set U that is starlike with respect to the point $\boldsymbol{a}$. If*

$$\frac{\partial f_j}{\partial x_k}(\boldsymbol{x}) = \frac{\partial f_k}{\partial x_j}(\boldsymbol{x})$$

for every choice of the indices $1 \le j,k \le n$ and for all $\boldsymbol{x} \in U$, then $\boldsymbol{F}$ is conservative.

Proof. We define $g : U \subset \mathbb{R}^n \to \mathbb{R}$ by

$$g(\boldsymbol{x}) := \int_0^1 \sum_{j=1}^n f_j(\boldsymbol{a} + t(\boldsymbol{x} - \boldsymbol{a})) \cdot (x_j - a_j)\,\mathrm{d}t.$$

It follows that g is a function of class C^1 on U and that

$$
\begin{aligned}
\frac{\partial g}{\partial x_k}(\boldsymbol{x}) &= \sum_{j\neq k}(x_j - a_j)\cdot\int_0^1 \frac{\partial f_j}{\partial x_k}(\boldsymbol{a}+t(\boldsymbol{x}-\boldsymbol{a}))t\,\mathrm{d}t \\
&\quad + \int_0^1 \left\{\frac{\partial f_k}{\partial x_k}(\boldsymbol{a}+t(\boldsymbol{x}-\boldsymbol{a}))t(x_k - a_k) + f_k(\boldsymbol{a}+t(\boldsymbol{x}-\boldsymbol{a}))\right\}\mathrm{d}t \\
&= \int_0^1 \left[\sum_{j=1}^{n}(x_j - a_j)\frac{\partial f_k}{\partial x_j}(\boldsymbol{a}+t(\boldsymbol{x}-\boldsymbol{a}))t + f_k(\boldsymbol{a}+t(\boldsymbol{x}-\boldsymbol{a}))\right]\mathrm{d}t \\
&= \int_0^1 \frac{d}{\mathrm{d}t}\left(f_k(\boldsymbol{a}+t(\boldsymbol{x}-\boldsymbol{a}))t\right)\,\mathrm{d}t = f_k(\boldsymbol{x}),
\end{aligned}
$$

for every $\boldsymbol{x} \in U$. This proves that $\boldsymbol{\nabla} g = \boldsymbol{F}$ and $\boldsymbol{F}$ is a conservative vector field on U. □

Corollary 2.5.1. *Let $\boldsymbol{F} : U \subset \mathbb{R}^3 \to \mathbb{R}^3$ be a vector field of class C^1 on an open starlike set U. Then $\boldsymbol{F}$ is conservative on U if and only if* Curl $\boldsymbol{F} = 0$.

2.6 Green's Theorem

This section constitutes a first approach to studying the result discovered in 1828 by George Green. This result is now known as Green's theorem and can be viewed as a generalization of the fundamental theorem of calculus. It states that the value of a double integral over the region bounded by a path is determined by the value of a line integral over that path. We will encounter Green's theorem again in Chap. 9, where it will appear as a particular case of the general Stokes's theorem. The results of Chap. 9 will allow us to apply Green's theorem to more general regions than those considered in this section, where we restrict attention to regions of type I (vertically simple) and type II (horizontally simple) (see Marsden–Tromba [14, 8.1, p. 494]).

For each $\varepsilon > 0$ let us consider the paths

$$\boldsymbol{\gamma}_\varepsilon^j : \left[-\frac{\varepsilon}{2}, \frac{\varepsilon}{2}\right] \longrightarrow \mathbb{R}^2$$

defined by

$$\boldsymbol{\gamma}_\varepsilon^1(t) := \left(t, -\frac{\varepsilon}{2}\right), \; \boldsymbol{\gamma}_\varepsilon^2(t) := \left(\frac{\varepsilon}{2}, t\right), \; \boldsymbol{\gamma}_\varepsilon^3(t) := \left(t, \frac{\varepsilon}{2}\right), \; \boldsymbol{\gamma}_\varepsilon^4(t) := \left(-\frac{\varepsilon}{2}, t\right).$$

Then

$$\boldsymbol{\gamma}_\varepsilon = \boldsymbol{\gamma}_\varepsilon^1 \cup \boldsymbol{\gamma}_\varepsilon^2 \cup (-\boldsymbol{\gamma}_\varepsilon^3) \cup (-\boldsymbol{\gamma}_\varepsilon^4)$$

represents the boundary of a square oriented counterclockwise (Fig. 2.7).

With this notation we have the following.

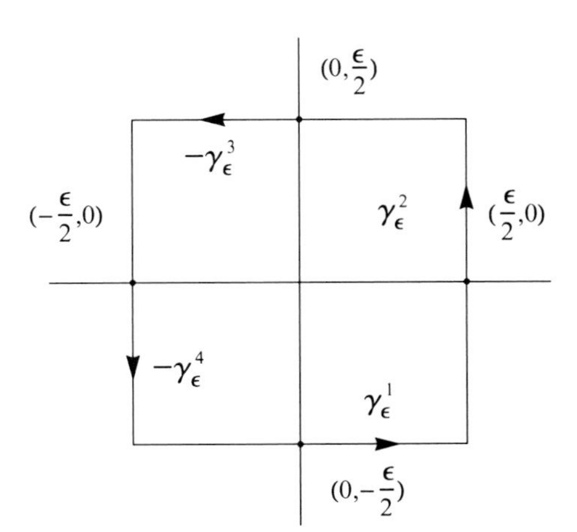

Fig. 2.7 The paths $\boldsymbol{\gamma}_\varepsilon^j$

Proposition 2.6.1. *Let $\boldsymbol{F} : U \subset \mathbb{R}^2 \to \mathbb{R}^2$ be a vector field of class C^1 on the open set U with components $\boldsymbol{F} = (P,Q)$. Then for every $(x_0,y_0) \in U$ we have*

$$\lim_{\varepsilon\to 0} \frac{1}{\varepsilon^2} \int_{(x_0,y_0)+\boldsymbol{\gamma}_\varepsilon} \boldsymbol{F} \quad = \quad \left(\frac{\partial Q}{\partial x} - \frac{\partial P}{\partial y}\right)(x_0,y_0).$$

Proof. The line integral $\int_{(x_0,y_0)+\boldsymbol{\gamma}_\varepsilon} \boldsymbol{F}$ can be written as the difference between

$$\int_{-\frac{\varepsilon}{2}}^{\frac{\varepsilon}{2}} \left(Q\left(x_0 + \frac{\varepsilon}{2}, y_0 + t\right) - Q\left(x_0 - \frac{\varepsilon}{2}, y_0 + t\right)\right) \mathrm{d}t$$

and

$$\int_{-\frac{\varepsilon}{2}}^{\frac{\varepsilon}{2}} \left(P\left(x_0 + t, y_0 + \frac{\varepsilon}{2}\right) - P\left(x_0 + t, y_0 - \frac{\varepsilon}{2}\right)\right) \mathrm{d}t.$$

For every $t \in (-\frac{\varepsilon}{2}, \frac{\varepsilon}{2})$, we apply the mean value theorem to the functions $Q(\cdot, y_0 + t)$ and $P(x_0 + t, \cdot)$ to obtain points x_ε, y_ε (which depend on ε and also on t) such that $|x_0 - x_\varepsilon| < \frac{\varepsilon}{2}$, $|y_0 - y_\varepsilon| < \frac{\varepsilon}{2}$, and

$$\frac{1}{\varepsilon^2} \int_{(x_0,y_0)+\boldsymbol{\gamma}_\varepsilon} \boldsymbol{F} = \frac{1}{\varepsilon} \int_{-\frac{\varepsilon}{2}}^{\frac{\varepsilon}{2}} \left(\frac{\partial Q}{\partial x}(x_\varepsilon, y_0 + t) - \frac{\partial P}{\partial y}(x_0 + t, y_\varepsilon)\right) \mathrm{d}t.$$

Since $\frac{\partial Q}{\partial x}$ is continuous at (x_0,y_0), for every $r > 0$ there exist $\delta > 0$ such that $(x,y) \in U$ and

$$\left|\frac{\partial Q}{\partial x}(x,y) - \frac{\partial Q}{\partial x}(x_0,y_0)\right| \le r,$$

whenever $|x - x_0| < \delta$ and $|y - y_0| < \delta$. Then for each $0 < \varepsilon < \delta$ and $t \in (-\frac{\varepsilon}{2}, \frac{\varepsilon}{2})$, we obtain

$$\left|\frac{\partial Q}{\partial x}(x_\varepsilon, y_0 + t) - \frac{\partial Q}{\partial x}(x_0,y_0)\right| \le r.$$

Now we can write

$$\left|\frac{1}{\varepsilon} \int_{-\frac{\varepsilon}{2}}^{\frac{\varepsilon}{2}} \frac{\partial Q}{\partial x}(x_\varepsilon, y_0 + t)\, \mathrm{d}t - \frac{\partial Q}{\partial x}(x_0,y_0)\right|$$
$$= \left|\frac{1}{\varepsilon} \int_{-\frac{\varepsilon}{2}}^{\frac{\varepsilon}{2}} \left(\frac{\partial Q}{\partial x}(x_\varepsilon, y_0 + t) - \frac{\partial Q}{\partial x}(x_0,y_0)\right) \mathrm{d}t\right| \le \frac{1}{\varepsilon} \int_{-\frac{\varepsilon}{2}}^{\frac{\varepsilon}{2}} r\, \mathrm{d}t = r.$$

We have used here the fact that $\left|\int_a^b f(t)\,\mathrm{d}t\right| \le \int_a^b |f(t)|\,\mathrm{d}t$. This proves that

$$\lim_{\varepsilon\to 0}\frac{1}{\varepsilon}\int_{-\frac{\varepsilon}{2}}^{\frac{\varepsilon}{2}} \frac{\partial Q}{\partial x}(x_\varepsilon, y_0+t)\,\mathrm{d}t = \frac{\partial Q}{\partial x}(x_0,y_0).$$

A similar argument gives

$$\lim_{\varepsilon\to 0}\frac{1}{\varepsilon}\int_{-\frac{\varepsilon}{2}}^{\frac{\varepsilon}{2}} \frac{\partial P}{\partial y}(x_0+t, y_\varepsilon)\,\mathrm{d}t = \frac{\partial P}{\partial y}(x_0,y_0),$$

from which the conclusion follows. □

In some texts, the line integral $\int_{\boldsymbol{\alpha}} \boldsymbol{F}$ is called the *circulation* of the vector field $\boldsymbol{F}$ along the path $\boldsymbol{\alpha}$, and the limit

$$\lim_{\varepsilon\to 0}\frac{1}{\varepsilon^2}\int_{(x_0,y_0)+\boldsymbol{\gamma}_\varepsilon} \boldsymbol{F}$$

is called the *rate of circulation* of the vector field $\boldsymbol{F}$ at the point (x_0,y_0). We refer to the comments after Corollary 9.4.1 for the physical interpretation of this expression.

If $f(x,y)$ denotes the density (mass per unit area) of a planar object U, then one can evaluate its total mass as

$$\iint_U f(x,y)\,\mathrm{d}(x,y).$$

By analogy, it seems reasonable to expect that the circulation of the vector field $\boldsymbol{F}$ along a path bounding a region U can be obtained as a double integral over U of the rate of circulation, that is, of the function

$$\frac{\partial Q}{\partial x} - \frac{\partial P}{\partial y}.$$

Green's theorem shows that this intuition is correct in some cases.

Up to this point, we have dealt with the integral only of continuous functions on an interval $[a,b]$, and so we have required only the elementary properties of the Riemann integral. Henceforth, however, we will need tools from the theory of integration in several variables. In that setting, one has to choose between two possibilities: the Riemann integral and the Lebesgue integral. The Riemann integral on an n-rectangle in $\mathbb{R}^n$, and its extension to bounded subsets with Jordan content, usually called Jordan measurable sets, is by far the more intuitive. On the other hand, the Lebesgue integral on $\mathbb{R}^n$, or specifically, on a measurable subset of $\mathbb{R}^n$, is the more powerful theory, even if less intuitive. We have decided to use the Lebesgue integral. Why? A key point is that any open or any closed subset of $\mathbb{R}^n$

is Lebesgue measurable but, in general, not Jordan measurable. Moreover, every continuous function on a compact set is Lebesgue integrable, and two of the most important theorems of integral calculus, namely Fubini's theorem and the change of variable theorem, hold in far more generality for the Lebesgue than for Riemann integral.

Notwithstanding the differences between the two theories of integration, and our preference for the theory of Lebesgue, the reader should be aware that for our purposes, and with enough hard toil, it is possible to show that most of the situations we encounter in the text actually fit into the framework of the Riemann integral and Jordan content, and so the reader may continue to think and visualize in terms of the Riemann integral. But we are not going to bridge that gap explicitly here. A book in which both theories appear and can be compared is that of Apostol [1].

Definition 2.6.1. (i) A set $R = \Pi_{j=1}^{n} I_j$ is called an *n-rectangle* in $\mathbb{R}^n$ if each I_j is a bounded interval in $\mathbb{R}$, i.e., if there exist $a_j \leq b_j$ real numbers such that $(a_j, b_j) \subseteq I_j \subseteq [a_j, b_j]$ for every $j = 1, \ldots, n$. In that case, the Lebesgue measure of R is defined to be

$$m(R) = \Pi_{j=1}^{n}(b_j - a_j).$$

(ii) A subset N of $\mathbb{R}^n$ is called a *null set* if given $\varepsilon > 0$, there exists a sequence of n-rectangles $(R_k)_{k=1}^{\infty}$ such that $N \subset \cup_{k=1}^{\infty} R_k$ and

$$\sum_{k=1}^{\infty} m(R_k) < \varepsilon.$$

(iii) A property $P(x)$, where $x \in \mathbb{R}^n$, is said to be true *almost everywhere* if there exists a null set $N \subset \mathbb{R}^n$ such that the property $P(x)$ holds for every $x \in \mathbb{R}^n \setminus N$.

(iv) Given a subset K of $\mathbb{R}^n$, the *characteristic function* of K, denoted by χ_K, is defined by

$$\chi_K(\boldsymbol{x}) = \begin{cases} 1 & \text{if } x \in K, \\ 0 & \text{if } x \notin K. \end{cases}$$

(v) A function $\varphi : \mathbb{R}^n \to \mathbb{R}$ is called a *step function* if φ is a linear combination of characteristic functions of n-rectangles, i.e., if

$$\varphi = \sum_{k=1}^{m} c_j \chi_{R_j},$$

where $c_j \in \mathbb{R}$ and R_j is an n-rectangle.

(vi) A subset $M \subset \mathbb{R}^n$ is said to be *measurable* if there exists a sequence $(\varphi_h)_{h=1}^{\infty}$ of step functions and a null subset N of $\mathbb{R}^n$ such that the sequence of real numbers $(\varphi_h(x))$ converges to $\chi_M(x)$ for every $x \in \mathbb{R}^n \setminus N$, i.e., if the sequence $(\varphi_h)_{h=1}^{\infty}$ converges pointwise almost everywhere to χ_M.

It can be proved that for every $\varepsilon > 0$, a null set can be covered by a sequence (R_k) of n-cubes, that is, n-rectangles with all sides of equal length, such that $\sum_{k=1}^{\infty} m(R_k) < \varepsilon$.

The family $\mathscr{M}$ of Lebesgue measurable sets in $\mathbb{R}^n$ is going to be very important throughout, but we need only the following very basic facts about that family:

1. Every open subset and every closed subset of $\mathbb{R}^n$ is measurable.
2. If M is a measurable set, then $\mathbb{R}^n \setminus M$ is also measurable.
3. If $(M_k)_{k=1}^{\infty}$ is a (countable) family of measurable sets, then $\cap_{k=1}^{\infty} M_k$ and $\cup_{k=1}^{\infty} M_k$ are measurable sets too.
4. Every null set is Lebesgue measurable.

We will need to use Fubini's theorem, and so we state it. We follow the notation of Stromberg [18, Theorem 6.121, p. 352] and refer to that book for the definition of the Lebesgue integral in $\mathbb{R}^n$. The space of Lebesgue integrable functions on $\mathbb{R}^n$ is denoted by $L(\mathbb{R}^n)$. If A is a measurable subset of $\mathbb{R}^n$, we will say that the function $f : A \to \mathbb{R}$ is Lebesgue integrable on A if extended as 0 outside A and denoting that extension by $f\chi_A$, then the function $f\chi_A$ is Lebesgue integrable on $\mathbb{R}^n$. In that case, by definition,

$$\int_A f := \int_A f(\boldsymbol{x})\mathrm{d}\boldsymbol{x} := \int_{\mathbb{R}^n} f\chi_A(\boldsymbol{x})\mathrm{d}\boldsymbol{x}.$$

If $A = [a,b]$, we keep the classical notation and write $\int_{[a,b]} f(x)\mathrm{d}x = \int_a^b f(x)\mathrm{d}x$. In the case that f has two or three variables, we will write

$$\iint_A f(x,y)\ \mathrm{d}(x,y) \text{ or } \iiint_A f(x,y,z)\ \mathrm{d}(x,y,z).$$

We prefer to write $\mathrm{d}(x,y)$ instead of $\mathrm{d}x\mathrm{d}y$ to avoid confusion with the exterior product $\mathrm{d}x \wedge \mathrm{d}y$ (see Chap. 6).

Theorem 2.6.1 (Fubini's theorem). *Let $f \in L(\mathbb{R}^{n+p})$. Then*

(i) $f_{\boldsymbol{x}}(\boldsymbol{y}) = f(\boldsymbol{x},\boldsymbol{y})$ *belongs to* $L(\mathbb{R}^p)$ *almost everywhere in* $\mathbb{R}^n$.
(ii) $f_{\boldsymbol{y}}(\boldsymbol{x}) = f(\boldsymbol{x},\boldsymbol{y})$ *belongs to* $L(\mathbb{R}^n)$ *almost everywhere in* $\mathbb{R}^p$.
(iii) $F(\boldsymbol{x}) := \int_{\mathbb{R}^p} f_{\boldsymbol{x}}(\boldsymbol{y})\mathrm{d}\boldsymbol{y}$ *is defined (that is, the integral exists) almost everywhere in* $\mathbb{R}^n$ *and is Lebesgue integrable in* $\mathbb{R}^n$. *Also* $G(\boldsymbol{y}) := \int_{\mathbb{R}^n} f_{\boldsymbol{y}}(\boldsymbol{x})\mathrm{d}\boldsymbol{x}$ *is defined almost everywhere in* $\mathbb{R}^p$*, and moreover,* G *belongs to* $L(\mathbb{R}^p)$.
(iv)

$$\int_{\mathbb{R}^n}\Big(\int_{\mathbb{R}^p} f(\boldsymbol{x},\boldsymbol{y})\mathrm{d}\boldsymbol{y}\Big)\mathrm{d}\boldsymbol{x} = \int_{\mathbb{R}^{n+p}} f(\boldsymbol{x},\boldsymbol{y})\mathrm{d}(\boldsymbol{x},\boldsymbol{y}) = \int_{\mathbb{R}^p}\Big(\int_{\mathbb{R}^n} f(\boldsymbol{x},\boldsymbol{y})\mathrm{d}\boldsymbol{x}\Big)\mathrm{d}\boldsymbol{y}.$$

A variant of Fubini's theorem particularly useful to us is the following. Let $A \subset [a,b] \times [c,d]$ be a measurable set and $f : A \to \mathbb{R}$ a Lebesgue integrable function on A. For $x \in [a,b]$, let $A_x = \{y \in \mathbb{R} : (x,y) \in A\}$. Then

$$\iint_A f(x,y)\mathrm{d}(x,y) = \int_a^b \Big(\int_{A_x} f(x,y)\mathrm{d}y\Big)\mathrm{d}x.$$

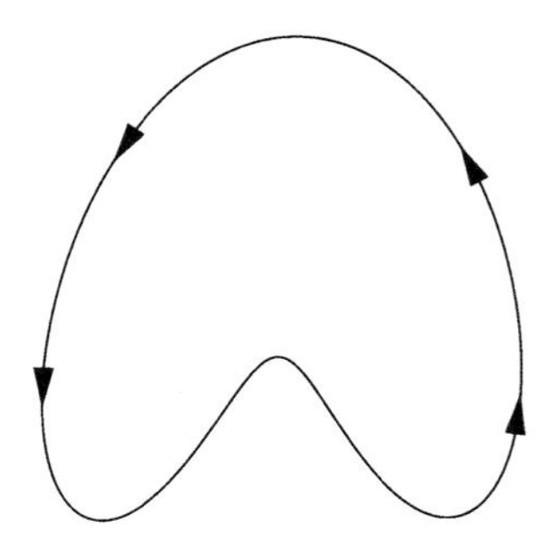

Fig. 2.8 A compact set with positively oriented boundary

The following concept will be discussed more precisely in Sect. 8.5.

Definition 2.6.2. Let $K \subset \mathbb{R}^2$ be a compact set whose boundary ∂K is equal to $\boldsymbol{\gamma}([a,b])$, where $\boldsymbol{\gamma} : [a,b] \to \mathbb{R}^2$ is a closed piecewise C^1 path. We say that $\boldsymbol{\gamma}$ is positively oriented if γ traverses ∂K once in such a way that the region K always lies to the left (Fig. 2.8). To be rigorous one must to assume that the compact set k satisfies condition (8.10), see for instance Cartan [7, 4.4.1]

Theorem 2.6.2 (Green's theorem). *Let $K \subset \mathbb{R}^2$ be a compact set with positively oriented boundary parameterized by γ (as in Definition 2.6.2). Let $\boldsymbol{\omega} = P\mathrm{d}x + Q\mathrm{d}y$ be a 1-form of class C^1 on an open set $U \subset \mathbb{R}^2$ such that $K \subset U$. Then*

$$\int_{\boldsymbol{\gamma}} P\mathrm{d}x + Q\mathrm{d}y = \iint_K \left(\frac{\partial Q}{\partial x} - \frac{\partial P}{\partial y} \right) \mathrm{d}(x,y).$$

Using the notation of vector fields, this can be written as

$$\int_{\boldsymbol{\gamma}} (P,Q) = \iint_K \left(\frac{\partial Q}{\partial x} - \frac{\partial P}{\partial y} \right) \mathrm{d}(x,y).$$

The reader may feel ill at ease with Definition 2.6.2 and Theorem 2.6.2. This is to be expected and welcomed. We have encountered for the first time a certain dichotomy that is often overlooked or brushed aside, but which we must specifically address in this book. Definition 2.6.2 and Theorem 2.6.2 have both an *intuitive* and a *mathematically rigorous* formulation, and these are not easily reconciled. If we look at Green's theorem from an intuitive point of view (typically, this is the case within the context of vector calculus), then there is generally no real difficulty. For example, if we want to apply it to situations arising in physical problems, then it is obvious what is meant by left and right, and we do not expect any trouble in deciding whether a region K lies to our left. However, from a purely mathematical point of view (i.e., within the context of vector analysis), the interpretation of "left" and "right" is not clear at all, and will demand from us much effort in developing adequate machinery to deal with the concept. Later, in Chaps. 5 and 9, we will present a

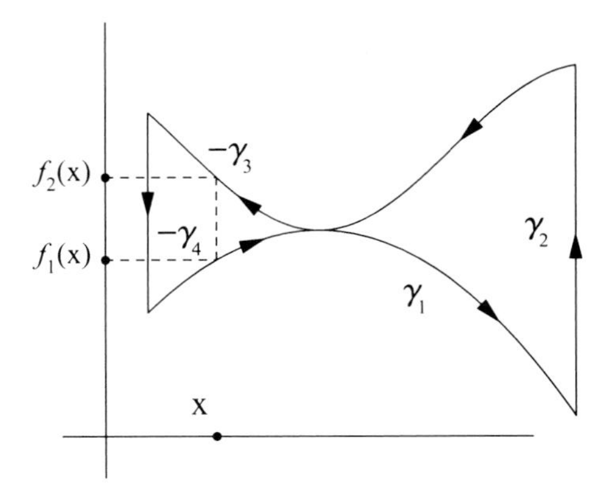

Fig. 2.9 A region of type I and its positively oriented boundary

rigorous definition of orientation and a general version of Green's theorem. For now, we content ourselves with working toward a proof of the theorem in some particular regions K for which the orientation can be very easily defined, and for which that definition of orientation coincides with our physical intuition.

Definition 2.6.3. A compact set $K \subset \mathbb{R}^2$ is said to be a region of type I (Fig. 2.9) if it can be described as

$$K = \{(x,y) \in \mathbb{R}^2 \ : \ a \le x \le b \ , \ f_1(x) \le y \le f_2(x)\},$$

where $f_1, f_2 : [a,b] \to \mathbb{R}$ are piecewise C^1 functions, $f_1 \le f_2$.

The positively oriented boundary of K is the path union

$$\gamma = \gamma_1 \cup \gamma_2 \cup (-\gamma_3) \cup (-\gamma_4),$$

where

$$\begin{aligned} \gamma_1 : [a,b] \to \mathbb{R}^2 \ , \ \gamma_1(t) &:= (t, f_1(t)); \\ \gamma_2 : [f_1(b), f_2(b)] \to \mathbb{R}^2 \ , \ \gamma_2(t) &:= (b,t); \\ \gamma_3 : [a,b] \to \mathbb{R}^2 \ , \ \gamma_3(t) &:= (t, f_2(t)); \\ \gamma_4 : [f_1(a), f_2(a)] \to \mathbb{R}^2 \ , \ \gamma_4(t) &:= (a,t). \end{aligned}$$

Lemma 2.6.1. *Let K be a compact set that is a region of type I and let P be a continuous function on K admitting continuous partial derivative $\frac{\partial P}{\partial y}$ on a neighborhood of K. Then*

$$\int_{\gamma} P\mathrm{d}x = -\iint_K \frac{\partial P}{\partial y} \mathrm{d}(x,y).$$

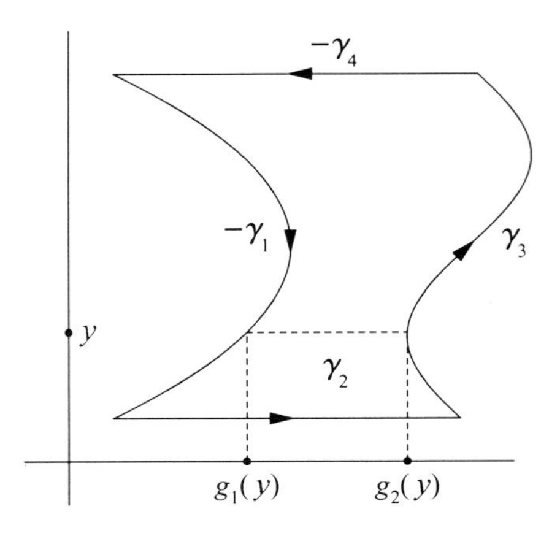

Fig. 2.10 A region of type II and its positively oriented boundary

Proof. Since $\boldsymbol{\gamma}_2'(t) = \boldsymbol{\gamma}_4'(t) = (0,1), \boldsymbol{\gamma}_1'(t) = (1, f_1'(t))$, and $\boldsymbol{\gamma}_3'(t) = (1, f_2'(t))$ for all t, we have

$$\int_{\boldsymbol{\gamma}} P\mathrm{d}x = \int_{\boldsymbol{\gamma}_1} P\mathrm{d}x - \int_{\boldsymbol{\gamma}_3} P\mathrm{d}x = \int_a^b P(t, f_1(t))\mathrm{d}t - \int_a^b P(t, f_2(t))\mathrm{d}t.$$

On the other hand, every continuous function on a compact set K is integrable, and according to Fubini's theorem,

$$\iint_K \frac{\partial P}{\partial y}\mathrm{d}(x,y) = \int_a^b \left(\int_{f_1(x)}^{f_2(x)} \frac{\partial P}{\partial y}(x,y)\mathrm{d}y \right) \mathrm{d}x = \int_a^b \left(P(x, f_2(x)) - P(x, f_1(x))\right) \mathrm{d}x.$$

Comparing these two equations, we have

$$\int_{\boldsymbol{\gamma}} P\mathrm{d}x = -\iint_K \frac{\partial P}{\partial y}\mathrm{d}(x,y).$$

□

Definition 2.6.4. The compact set K is said to be a region of type II (Fig. 2.10) if it can be analytically described as

$$K = \{(x,y) \;:\; c \le y \le d; g_1(y) \le x \le g_2(y)\},$$

where $g_1, g_2 : [c,d] \to \mathbb{R}$ are piecewise C^1 functions, $g_1 \le g_2$.

The positively oriented boundary of K is the path union

$$\boldsymbol{\gamma} = (-\boldsymbol{\gamma}_1) \cup \boldsymbol{\gamma}_2 \cup \boldsymbol{\gamma}_3 \cup (-\boldsymbol{\gamma}_4),$$

where

$$\boldsymbol{\gamma}_1 : [c,d] \to \mathbb{R}^2 \;,\; \boldsymbol{\gamma}_1(t) := (g_1(t), t);$$

$$\gamma_2 : [g_1(c), g_2(c)] \to \mathbb{R}^2 \ , \ \gamma_2(t) := (t, c);$$

$$\gamma_3 : [c, d] \to \mathbb{R}^2 \ , \ \gamma_3(t) := (g_2(t), t);$$

$$\gamma_4 : [g_1(d), g_2(d)] \to \mathbb{R}^2 \ , \ \gamma_4(t) := (t, d).$$

We have the following analogue of Lemma 2.6.1.

Lemma 2.6.2. *Let K be a compact set that is a region of type II and let Q be a continuous function on K admitting continuous partial derivative $\frac{\partial Q}{\partial x}$ on a neighborhood of K. Then*

$$\int_\gamma Q \mathrm{d}y = \iint_K \frac{\partial Q}{\partial x} \mathrm{d}(x,y).$$

Theorem 2.6.3. *Let K be a region of type I or a region of type II and let $\omega = P\mathrm{d}x + Q\mathrm{d}y$ be a 1-form of class C^1 on some open rectangle containing K. Then*

$$\int_\gamma P\mathrm{d}x + Q\mathrm{d}y = \iint_K \left(\frac{\partial Q}{\partial x} - \frac{\partial P}{\partial y} \right) \mathrm{d}(x,y).$$

Proof. We choose a, b, c, and d such that K is contained in the rectangle $R = [a,b] \times [c,d]$ and we assume that the 1-form ω is defined and is of class C^1 on the open rectangle T containing R. We will present the proof in the case that K is a region of type I. In the case that K is a region of type II, a similar argument does the job.

We have already proved that

$$\int_\gamma P\mathrm{d}x = -\iint_K \frac{\partial P}{\partial y} \mathrm{d}(x,y),$$

and so we need only to obtain

$$\int_\gamma Q\mathrm{d}y = \iint_K \frac{\partial Q}{\partial x} \mathrm{d}(x,y).$$

Notice that we cannot apply the previous lemma, since K is not necessarily a region of type II. We proceed as follows.

For $(x,y) \in T$, we define $V(x,y) := \int_c^y Q(x,t)\mathrm{d}t$. The resultant function V has the property that

$$\mathrm{d}V(x,y) = F(x,y)\ \mathrm{d}x + Q(x,y)\ \mathrm{d}y,$$

where

$$F(x,y) = \int_c^y \frac{\partial Q}{\partial x}(x,t)\mathrm{d}t.$$

The expression for F is obtained after applying Proposition 2.5.1 (see also, Apostol [1, Theorem 10.39]). Since by Theorem 2.5.1,

$$\int_{\gamma} \mathrm{d}V = 0,$$

we have

$$\int_{\gamma} Q(x,y)\, \mathrm{d}y = -\int_{\gamma} F(x,y)\, \mathrm{d}x.$$

Moreover, K is a region of type I, and on a neighborhood of K,

$$\frac{\partial F}{\partial y}(x,y) = \frac{\partial Q}{\partial x}(x,y),$$

because for each fixed x the function $y \mapsto F(x,y)$ is a primitive of $y \mapsto \frac{\partial Q}{\partial x}(x,y)$. An application of Lemma 2.6.1 to $F(x,y)\, \mathrm{d}x$ gives

$$-\int_{\gamma} F(x,y)\, \mathrm{d}x = \iint_K \frac{\partial F}{\partial y}\, \mathrm{d}(x,y).$$

Finally,

$$\int_{\gamma} Q(x,y)\, \mathrm{d}y = \iint_K \frac{\partial Q}{\partial x}(x,y)\, \mathrm{d}(x,y)$$

and

$$\int_{\gamma} \omega = \iint_K \left(\frac{\partial Q}{\partial x} - \frac{\partial P}{\partial y} \right) \mathrm{d}(x,y).$$

□

Definition 2.6.5. A compact set K is said to be a *simple region* if K is a region of type I and also a region of type II.

The hypothesis for Green's theorem on simple regions can be relaxed somewhat, since it is enough that the 1-form be defined on a neighborhood of K. Before providing the proof, let us establish a notational convention that is valid by the following fact (see Proposition 2.7.2). Suppose K is a simple region and $\gamma_1 : [a,b] \to \mathbb{R}^2$, $\gamma_2 : [c,d] \to \mathbb{R}^2$ are two simple, closed, and piecewise smooth paths with

$$\gamma_1([a,b]) = \gamma_2([c,d]) = \partial K$$

and such that γ_1 and γ_2 *induce the same orientation* on the boundary ∂K. Then according to Proposition 2.7.2,

$$\int_{\gamma_1} \omega = \int_{\gamma_2} \omega.$$

Consequently, it makes sense to write

$$\int_{\partial K} \omega$$

if we interpret the integration as being done along any *simple*, (Definition 2.7.1) closed, and piecewise C^1 path that gives a parameterization of ∂K such that the region K always lies to the left.

Theorem 2.6.4. *Let $K \subset \mathbb{R}^2$ be a simple region and let $\omega = P\mathrm{d}x + Q\mathrm{d}y$ be a 1-form of class C^1 on an open neighborhood of K. Then*

$$\int_{\partial K} \omega = \iint_K \left(\frac{\partial Q}{\partial x} - \frac{\partial P}{\partial y} \right) \mathrm{d}(x,y).$$

Proof. Since K is a region of type I, we have

$$\int_{\partial K} P\mathrm{d}x = -\iint_K \frac{\partial P}{\partial y} \mathrm{d}(x,y),$$

and since it is also a region of type II, we obtain

$$\int_{\partial K} Q\mathrm{d}y = \iint_K \frac{\partial Q}{\partial x} \mathrm{d}(x,y).$$

It is enough to sum these two identities. □

Green's Formula[5] allows us to give a direct proof of the sufficiency condition for a vector field to be conservative (Theorem 2.5.6) on an open and starlike set in the plane.

Theorem 2.6.5 (Poincaré's lemma). *Let $\boldsymbol{F} = (P,Q)$ be a vector field of class C^1 on the open and starlike set $U \in \mathbb{R}^2$ and suppose that*

$$\frac{\partial Q}{\partial x}(x,y) = \frac{\partial P}{\partial y}(x,y)$$

for every $(x,y) \in U$. Then $\boldsymbol{F}$ is conservative.

Proof. Let $\omega = P\mathrm{d}x + Q\mathrm{d}y$ be the 1-form associated with the vector field $\boldsymbol{F}$. It is enough to prove that

[5] Green's theorem is essentially a number of different results, which hold in various domains with different hypotheses, that establish a common integral identity. We shall refer to that integral identity as Green's formula.

$$\int_{\boldsymbol{\gamma}} \omega = 0$$

whenever $\boldsymbol{\gamma}$ is the boundary of some triangle K contained in U. Since K is a simple region, it follows from Green's theorem that

$$\int_{\boldsymbol{\gamma}} \omega = \iint_K \left(\frac{\partial Q}{\partial x} - \frac{\partial P}{\partial y} \right) \mathrm{d}(x,y) = 0.$$

□

A general version of Poincaré's lemma on $\mathbb{R}^n$ will be obtained in Chap. 6 (see Sect. 6.6) after we have completed our study of differential forms.

2.7 Appendix: Comments on Parameterization

If we ask a student to evaluate the integral of a vector field along the unit circle oriented counterclockwise, and we don't specify a parameterization of the circle, the student will probably use the "natural" parameterization $\boldsymbol{\gamma} : [0, 2\pi] \to \mathbb{R}^2$ given by $\boldsymbol{\gamma}(t) = (\cos(t), \sin(t))$. However, this is not the only parameterization available, and it is quite natural to ask how this freedom in the choice of parameterization might affect the solution to the problem posed. That is, suppose that we have another path $\boldsymbol{\alpha}$ whose trajectory is also the unit circle and with the property that we move along the circle counterclockwise, but without any other obvious relation to the original path $\boldsymbol{\gamma}$. Can we be sure that

$$\int_{\boldsymbol{\alpha}} \boldsymbol{F} = \int_{\boldsymbol{\gamma}} \boldsymbol{F} \,?$$

In this section, given from the point of view of vector analysis, we give a rigorous answer to this question (Proposition 2.7.2), thereby showing why so much freedom is typically allowed in parameterizing a curve.

Lemma 2.7.1. *Let $\boldsymbol{\varphi} : [c,d] \to \mathbb{R}^n$ and $\boldsymbol{\gamma} : [a,b] \to \mathbb{R}^n$ be two injective paths of class C^1 such that*

$$\boldsymbol{\gamma}([a,b]) \subset \boldsymbol{\varphi}([c,d]).$$

Let us assume that there is $1 \leq k \leq n$ with $\varphi_k'(s) \neq 0$ for all $s \in [c,d]$. Then

$$\boldsymbol{\varphi}^{-1} \circ \boldsymbol{\gamma} : [a,b] \to [c,d]$$

is a function of class C^1.

Proof. Since $\boldsymbol{\varphi}$ is C^1, the intermediate value theorem implies that $\varphi_k'(s) > 0$ for all $s \in [c,d]$ or $\varphi_k'(s) < 0$ for all $s \in [c,d]$. Without loss of generality we will assume the former. Since φ_k can be extended to a function of class C^1 on the whole real line and $\varphi_k'(s) > 0$ for all $s \in [c,d]$, we have that φ_k is a strictly increasing bijection between $[c,d]$ and $[e,f] := \varphi_k([c,d])$ and also that its inverse $h := \varphi_k^{-1}$ is a function of class C^1 on $[e,f]$. Hence $\boldsymbol{\varphi}$ is bijective onto its image. Let

$$\pi_k : \mathbb{R}^n \to \mathbb{R}$$

denote the projection onto the kth coordinate. From the fact that the mapping $\varphi_k \circ \boldsymbol{\varphi}^{-1}$ associates to each point in the range of $\boldsymbol{\varphi}$ its kth coordinate and $\boldsymbol{\gamma}([a,b]) \subset \boldsymbol{\varphi}([c,d])$ we can deduce $\varphi_k \circ \boldsymbol{\varphi}^{-1} \circ \boldsymbol{\gamma} = \pi_k \circ \boldsymbol{\gamma}$. That is,

$$\boldsymbol{\varphi}^{-1} \circ \boldsymbol{\gamma} = h \circ \pi_k \circ \boldsymbol{\gamma}.$$

Now the conclusion follows by the chain rule. □

Proposition 2.7.1. *Let $\varphi : [c,d] \to \mathbb{R}^n$ and $\gamma : [a,b] \to \mathbb{R}^n$ be two injective smooth paths. If $\gamma([a,b]) = \varphi([c,d])$ and both paths define the same "orientation" (that is, $\varphi(s) = \gamma(t)$ implies that $\varphi'(s)$ is a positive multiple of $\gamma'(t)$), then φ and γ are equivalent paths.*

Proof. The mapping $\varphi : [c,d] \to \varphi([c,d])$ is a homeomorphism because it is a continuous bijection between two compact sets. Hence, the mapping

$$\theta := \varphi^{-1} \circ \gamma : [a,b] \to [c,d]$$

is a continuous bijection and $\varphi \circ \theta = \gamma$. We have to prove that θ is a mapping of class C^1 and $\theta'(t) > 0$ for all $t \in [a,b]$. To do this, we fix $t_0 \in (a,b)$ and we observe that $\varphi'(\theta(t_0)) \neq 0$. Since $\varphi' \circ \theta$ is a continuous function, we deduce that there exist $1 \leq k \leq n$ and $a < \alpha < t_0 < \beta < b$ such that

$$\varphi_k'(\theta(t)) \neq 0$$

for every $t \in [\alpha, \beta]$. We define

$$[\xi, \eta] := \theta([\alpha, \beta]).$$

Then the restriction of γ to $[\alpha, \beta]$ has the same trajectory as the restriction of φ to $[\xi, \eta]$. According to the previous lemma,

$$\theta : [\alpha, \beta] \to [\xi, \eta]$$

is a function of class C^1. Consequently,

$$\gamma'(t) = \varphi'(\theta(t)) \cdot \theta'(t)$$

for every $t \in [\alpha, \beta]$. Since, by hypothesis, $\gamma'(t)$ is a positive multiple of $\varphi'(\theta(t))$, we deduce that $\theta'(t) > 0$ for every $t \in [\alpha, \beta]$. In particular, θ is of class C^1 on a neighborhood of t_0 and $\theta'(t_0) > 0$.

In the case $t_0 = a$ we find $a < \beta < b$ such that

$$\varphi_k'(\theta(t)) \neq 0$$

for all $t \in [a, \beta]$. We now proceed as before but with $\alpha = a$ in order to finally deduce that θ is of class C^1 on $[a, \beta]$ and $\theta' > 0$ at every point of $[a, \beta]$. A similar argument covers the case $t_0 = b$. □

Definition 2.7.1. A closed path $\varphi : [c,d] \to \mathbb{R}^n$ is said to be *simple* if φ is piecewise C^1 and $\varphi|_{[c,d)}$ is injective.

Lemma 2.7.2. *Let $\varphi : [c,d] \to \mathbb{R}^n$ be a simple path. Then*

$$\varphi : (c,d) \to \varphi\big((c,d)\big)$$

is a homeomorphism.

Proof. Let $\{t_n\}$ be a sequence in (c,d) such that

$$\lim_{n\to\infty} \varphi(t_n) = \varphi(t_0)$$

for some $t_0 \in (c,d)$. In order to conclude that $\{t_n\}$ converges to t_0, it is enough to prove that each convergent subsequence of $\{t_n\}$ is convergent to t_0. Let us assume that $\{n_l\}$ is an increasing sequence of natural numbers and that

$$\lim_{l\to\infty} t_{n_l} = \xi .$$

Since $\xi \in [c,d]$ and φ is continuous, we obtain

$$\varphi(\xi) = \lim_{l\to\infty} \varphi(t_{n_l}) = \varphi(t_0).$$

From the injectivity of $\varphi|_{[c,d)}$ and $\varphi|_{(c,d]}$ we deduce $\xi = t_0$. Consequently, $\{t_n\}$ converges to t_0 and

$$\varphi^{-1} : \varphi\big((c,d)\big) \to (c,d)$$

is continuous. □

In saying that a path $\gamma : [a,b] \to \mathbb{R}^n$ is *piecewise smooth*, we mean that there is a partition

$$\{a = t_0 < t_1 < \cdots < t_l = b\},$$

such that there exists $\gamma'(t) \neq 0$ for every $t \neq t_j$ and γ admits nonzero one-sided derivatives at each point t_j.

Let us now assume that $\gamma : [a,b] \to \mathbb{R}^n$ and $\varphi : [c,d] \to \mathbb{R}^n$ are two simple paths, which are piecewise smooth and satisfy $\gamma([a,b]) = \varphi([c,d])$. We say that the two paths induce the same *orientation* if whenever $\gamma(t) = \varphi(s)$ for t and s such that γ is differentiable at t and φ is differentiable at s, then $\gamma'(t)$ is a positive multiple of $\varphi'(s)$.

Proposition 2.7.2. *Let $\gamma : [a,b] \to \mathbb{R}^n$ and $\varphi : [c,d] \to \mathbb{R}^n$ two simple paths that are piecewise smooth and satisfy $\gamma([a,b]) = \varphi([c,d]) \subset U$, where $U \subset \mathbb{R}^n$ is open. If the two paths induce the same orientation, then for every differential form ω of degree 1 and class C^1 on U, we have*

$$\int_\gamma \omega = \int_\varphi \omega .$$

Proof. Let us first assume that $\boldsymbol{\gamma}$ and $\boldsymbol{\varphi}$ are two piecewise smooth paths with the same initial point (and also the same final point, since they are closed paths). We introduce partitions of the intervals $[a,b]$ and $[c,d]$ as follows. In the optimal case that $\boldsymbol{\gamma}$ and $\boldsymbol{\varphi}$ are smooth on their respective domains, we consider arbitrary partitions $\{a=a_0<a_1<a_2=b\}$ and $\{c=c_0<c_1<c_2=d\}$ with the property that $\boldsymbol{\gamma}(a_1)=\boldsymbol{\varphi}(c_1)$. In the general case that either $\boldsymbol{\varphi}$ or $\boldsymbol{\gamma}$ is not smooth, let Q be a nonempty finite set containing all the points of the form $\boldsymbol{\gamma}(t)$, where $t\in(a,b)$ is such that $\boldsymbol{\gamma}$ is not differentiable at t, and also all the points of the form $\boldsymbol{\varphi}(s)$, where $s\in(c,d)$ for which $\boldsymbol{\varphi}$ is not differentiable at s. Then Q consists of $m-1$ elements for some $m\geq 2$. We can find partitions

$$\{a=a_0<a_1<\cdots<a_m=b\},\quad \{c=c_0<c_1<\cdots<c_m=d\}$$

of the intervals $[a,b]$ and $[c,d]$ with the property that

$$\boldsymbol{\gamma}\big(\{a_1,\ldots,a_{m-1}\}\big)=\boldsymbol{\varphi}\big(\{c_1,\ldots,c_{m-1}\}\big)=Q.$$

Observe that our partitions have at least three points and that

$$\boldsymbol{\gamma}|_{[a_{j-1},a_j]}\quad\text{and}\quad \boldsymbol{\varphi}|_{[c_{j-1},c_j]}$$

are injective and smooth paths. We now put

$$I_j:=(a_{j-1},a_j),\qquad \text{and hence}\quad \overline{I_j}=[a_{j-1},a_j].$$

By Lemma 2.7.2, $\boldsymbol{\gamma}(I_j)$ is an open and connected set in $\boldsymbol{\gamma}\big((a,b)\big)=\boldsymbol{\varphi}\big((c,d)\big)$, and hence, after applying again Lemma 2.7.2,

$$J_j:=\boldsymbol{\varphi}^{-1}\big(\boldsymbol{\gamma}(I_j)\big)$$

is an open (and proper) subinterval of (c,d). It is obvious that $\boldsymbol{\gamma}(I_j)=\boldsymbol{\varphi}(J_j)$. Moreover, $\boldsymbol{\gamma}(\overline{I_j})=\boldsymbol{\varphi}(\overline{J_j})$. In fact, if $t=\lim_{n\to\infty}t_n$ for $t_n\in I_j$, then

$$\boldsymbol{\gamma}(t)=\lim_{n\to\infty}\boldsymbol{\varphi}(s_n)$$

for some sequence $\{s_n\}$ in J_j. If $\{s_{n_k}\}$ is a subsequence convergent to $s\in\overline{J_j}$, then $\boldsymbol{\gamma}(t)=\boldsymbol{\varphi}(s)\in\boldsymbol{\varphi}(\overline{J_j})$. This proves that $\boldsymbol{\gamma}(\overline{I_j})\subset\boldsymbol{\varphi}(\overline{J_j})$. A similar argument gives the reverse inclusion. Since

$$\boldsymbol{\gamma}_{|\overline{I_j}}\quad\text{and}\quad \boldsymbol{\varphi}_{|\overline{J_j}}$$

are smooth and injective, we can apply Proposition 2.7.1 to conclude that these are equivalent paths. The sets $\{\overline{J}_1,\overline{J}_2,\ldots,\overline{J}_{m-1}\}$ are mutually disjoint (or have at most one point in common) and form a covering of $[c,d]$, so we can finally conclude that

$$\int_{\boldsymbol{\gamma}} \omega = \sum_{j=1}^{m} \int_{\boldsymbol{\gamma}_{|I_j}} \omega = \sum_{j=1}^{m} \int_{\boldsymbol{\varphi}_{|J_j}} \omega$$

$$= \int_{\boldsymbol{\varphi}} \omega .$$

We now analyze the case that φ and γ do not have the same initial (and final) point, that is, $\boldsymbol{\varphi}(c) \neq \boldsymbol{\gamma}(a)$. We take $a < t_0 < b$ such that $\boldsymbol{\gamma}(t_0) = \boldsymbol{\varphi}(c)$ and we define

$$\boldsymbol{\lambda} := \boldsymbol{\gamma}_{|[t_0,b]} \cup \boldsymbol{\gamma}_{|[a,t_0]},$$

which is a simple and piecewise smooth closed path. Moreover, the trace of λ coincides with

$$\boldsymbol{\gamma}([a,b]) = \boldsymbol{\varphi}([c,d]),$$

$\boldsymbol{\lambda}$ and $\boldsymbol{\varphi}$ have the same initial (and final) point, and the two paths $\boldsymbol{\lambda}$ and $\boldsymbol{\varphi}$ induce the same orientation. Hence

$$\int_{\boldsymbol{\lambda}} \omega = \int_{\boldsymbol{\varphi}} \omega .$$

Finally, we obtain

$$\int_{\boldsymbol{\lambda}} \omega = \int_{\boldsymbol{\gamma}_{|[t_0,b]}} \omega + \int_{\boldsymbol{\gamma}_{|[a,t_0]}} \omega$$

$$= \int_{\boldsymbol{\gamma}} \omega ,$$

and the proposition is proved. □

Remark 2.7.1. Physically, one might wish to consider two particles moving along the same closed trajectory with the same direction of movement but whose initial (and hence final) points differ. In this case, the two paths $\boldsymbol{\gamma}$ and $\boldsymbol{\varphi}$ are not equivalent. In fact, it follows from Definition 2.4.1 that two equivalent paths have the same initial (and final) point. For example, the paths $\gamma : [0,2\pi] \to \mathbb{R}^2$, $\gamma(t) = (\cos t, \sin t)$, and $\varphi : [\pi, 3\pi] \to \mathbb{R}^2$, $\varphi(s) = (\cos s, \sin s)$, are not equivalent. Another example of two inequivalent paths that satisfy the hypothesis of Proposition 2.7.2 are γ and $\beta : [\pi, 3\pi] \to \mathbb{R}^2$, $\beta(s) = (\cos(s + \frac{\pi}{2}), \sin(s + \frac{\pi}{2}))$.

2.8 Exercises

Exercise 2.8.1. Evaluate the length of the path

$$\boldsymbol{\gamma} : \left[0, \frac{\pi}{2}\right] \to \mathbb{R}^2, \quad \boldsymbol{\gamma}(t) = (\mathrm{e}^t \cos(t), \mathrm{e}^t \sin(t)).$$

Exercise 2.8.2. Obtain a parameterization, counterclockwise, of the ellipse

$$\left(\frac{x}{a}\right)^2 + \left(\frac{y}{b}\right)^2 = 1 \quad (a, b > 0).$$

Exercise 2.8.3. Integrate the vector field

$$\boldsymbol{F}(x,y) = (y^2, -2xy)$$

along the triangle with vertices $(0,0)$, $(1,0)$, $(0,1)$, oriented counterclockwise.

Exercise 2.8.4. (1) Find a path $\boldsymbol{\gamma}$ whose trajectory is the intersection of the cylinder $x^2 + y^2 = 1$ with the plane $x + y + z = 1$ and with the additional properties that the initial (and final) point is $(0, -1, 2)$ and the projection onto the xy-plane is oriented counterclockwise.
(2) Evaluate

$$\int_{\boldsymbol{\gamma}} xy \, \mathrm{d}x + yz \, \mathrm{d}y - x \, \mathrm{d}z.$$

Exercise 2.8.5. Let $\boldsymbol{\gamma}$ be a simple path whose trajectory is the intersection of the coordinate planes with the portion of the unit sphere in the first octant, oriented according to the sequence

$$(1,0,0), \ (0,1,0), \ (0,0,1), \ (1,0,0).$$

Evaluate

$$\int_{\boldsymbol{\gamma}} z \, \mathrm{d}x + x \, \mathrm{d}y + y \, \mathrm{d}z.$$

Exercise 2.8.6. Find a path whose trajectory is the intersection of the upper hemisphere of the sphere with radius $2a$,

$$x^2 + y^2 + z^2 = 4a^2,$$

with the cylinder

$$x^2 + (y-a)^2 = a^2.$$

Exercise 2.8.7. Determine whether the vector field

$$\boldsymbol{F} : \mathbb{R}^2 \to \mathbb{R}^2, \quad \boldsymbol{F}(x,y) = \left(x^3 y,\ x\right),$$

is conservative.

Exercise 2.8.8. Find a potential for the vector field

$$\boldsymbol{F} : \mathbb{R}^2 \to \mathbb{R}^2, \quad \boldsymbol{F}(x,y) = \left(2xy,\ x^2 - y^2\right).$$

Exercise 2.8.9. For the vector field

$$\boldsymbol{F} : \mathbb{R}^3 \to \mathbb{R}^3, \quad \boldsymbol{F}(x,y,z) = \left(2xy,\ x^2 + z^2,\ 2zy\right),$$

(1) show that $\boldsymbol{\nabla} \times \boldsymbol{F} = 0$;
(2) find a potential for $\boldsymbol{F}$.

Exercise 2.8.10. (1) Show that the vector field

$$\boldsymbol{F}(x,y) = \left(\mathrm{e}^x(\sin(x+y) + \cos(x+y)) + 1\ ,\ \mathrm{e}^x \cos(x+y)\right)$$

is conservative on $\mathbb{R}^2$ and find a potential.

(2) Evaluate the line integral

$$\int_{\boldsymbol{\gamma}} \boldsymbol{F},$$

where

$$\boldsymbol{\gamma} : [0,\pi] \to \mathbb{R}^2\ ;\ \boldsymbol{\gamma}(t) = \left(\sin(\pi \mathrm{e}^{\sin(t)}), \cos^5(t)\right).$$

Exercise 2.8.11. Evaluate

$$\int_{\boldsymbol{\gamma}} x\,\mathrm{d}x + y\,\mathrm{d}y + z\,\mathrm{d}z,$$

where $\boldsymbol{\gamma}(t) = \left(\cos^4(t), \sin^2(t) + \cos^3(t), t\right),\ 0 \le t \le \pi$.

Exercise 2.8.12. For the vector field

$$\boldsymbol{F} = (f_1, f_2) : \mathbb{R}^2 \setminus \{(0,0)\} \to \mathbb{R}^2$$

defined by

$$f_1(x,y) = \frac{-y}{x^2 + y^2}, \quad f_2(x,y) = \frac{x}{x^2 + y^2},$$

(1) Show that

$$\frac{\partial f_2}{\partial x}(x,y) = \frac{\partial f_1}{\partial y}(x,y)$$

for all $(x,y) \in \mathbb{R}^2 \setminus \{(0,0)\}$.

(2) Let $\boldsymbol{\gamma}$ be the unit circle oriented counterclockwise. Show that

$$\int_{\boldsymbol{\gamma}} \boldsymbol{F} = 2\pi.$$

Is this fact a contradiction to Poincaré's lemma?

(3) Argue whether this statement is true: for every closed and piecewise C^1 path $\boldsymbol{\alpha} : [a,b] \to \mathbb{R}^2 \setminus \{(0,0)\}$ such that $\alpha_1(t) \geq 0$ for all $t \in [a,b]$,

$$\int_{\boldsymbol{\alpha}} \boldsymbol{F} = 0.$$

(4) Evaluate $\int_{\boldsymbol{\gamma}} \boldsymbol{F}$, where

$$\boldsymbol{\gamma}(t) = \left(\cos(t),\ \sin^7(t)\right), \quad -\frac{\pi}{2} \leq t \leq \frac{\pi}{2}.$$

Exercise 2.8.13. Let $\boldsymbol{\gamma}(t) = (\cos(t),\ \sin(t))$ be given and let us consider the paths

$$\boldsymbol{\alpha} := \boldsymbol{\gamma}_{|[-\pi,\pi]} \text{ and } \boldsymbol{\beta} := \boldsymbol{\gamma}_{|[0,2\pi]}.$$

(1) Are $\boldsymbol{\alpha}$ and $\boldsymbol{\beta}$ equivalent paths?

(2) Justify why

$$\int_{\boldsymbol{\alpha}} \boldsymbol{F} = \int_{\boldsymbol{\beta}} \boldsymbol{F}$$

for any continuous vector field $\boldsymbol{F} : U \subset \mathbb{R}^2 \to \mathbb{R}^2$ defined on the unit circle.

Exercise 2.8.14. Let $\boldsymbol{\gamma}$ be the path whose trajectory is the union of the graph of $y = x^3$ from $(0,0)$ to $(1,1)$ and the segment from $(1,1)$ to $(0,0)$. Using Green's theorem, evaluate

$$\int_{\boldsymbol{\gamma}} (x^2 + y^2)\ \mathrm{d}x + (2xy + x^2)\ \mathrm{d}y.$$

Exercise 2.8.15. Use Green's theorem to evaluate the line integral

$$\int_{\boldsymbol{\gamma}} (x^2 + 3x^2y^2)\ \mathrm{d}x + (2x^3y + x^2)\ \mathrm{d}y,$$

where $\boldsymbol{\gamma}$ is the boundary of the region lying between the graphs of $y = 0$ and $y = 4 - x^2$, oriented counterclockwise.

Exercise 2.8.16. Let K be a region of type I or type II and let γ be a path whose trajectory is the boundary of K oriented counterclockwise. Then

$$\text{area}\,(K) = \int_\gamma \frac{1}{2}(-y\,\mathrm{d}x + x\,\mathrm{d}y) = \int_\gamma -y\,\mathrm{d}x = \int_\gamma x\,\mathrm{d}y.$$

Exercise 2.8.17. Evaluate the area bounded by the cycloid $\gamma : [0, 2\pi] \to \mathbb{R}^2$,

$$\gamma(t) = (at - a\sin(t), a - a\cos(t))$$

$(a > 0)$ and the x-axis.

Exercise 2.8.18. Evaluate the area bounded by the two coordinate axes and the path

$$\gamma : \left[0, \frac{\pi}{2}\right] \to \mathbb{R}^2, \quad \gamma(t) = (\sin^4(t), \cos^4(t)).$$

Exercise 2.8.19. Evaluate the area limited by the circles

$$C_1 := \left\{(x,y) \in \mathbb{R}^2 \ : \ x^2 + y^2 = a^2\right\}$$

and

$$C_2 := \left\{(x,y) \in \mathbb{R}^2 \ : \ x^2 + y^2 = 2ax\right\} \ (a > 0).$$

Chapter 3
Regular k-Surfaces

Roughly speaking, a regular surface in $\mathbb{R}^3$ is a two-dimensional set of points, in the sense that it can be locally described by two parameters (the local coordinates) and with the property that it is smooth enough (that is, there are no vertices, edges, or self-intersections) to guarantee the existence of a tangent plane to the surface at each point.

A. Galbis and M. Maestre, *Vector Analysis Versus Vector Calculus*, Universitext,
DOI 10.1007/978-1-4614-2200-6_3, © Springer Science+Business Media, LLC 2012

3.1 Coordinate Systems: Graphics

The objects we are interested in are usually referred to in texts on differential geometry as k-submanifolds of $\mathbb{R}^n$, but several authors call them k-manifolds [9, 10], even if the concept of k-manifold is a more general one. Following Do Carmo [6, p. 43], we prefer to call such a set a k-dimensional regular surface in $\mathbb{R}^n$.

We need to remind the student of some concepts from general topology. A bijective map $\boldsymbol{f} : A \subset \mathbb{R}^n \to B \subset \mathbb{R}^n$ is called a *homeomorphism* if both $\boldsymbol{f}$ and $\boldsymbol{f}^{-1} : B \to A$ are continuous mappings.

Given a subset A of $\mathbb{R}^n$, a set $D \subset A$ is said to be *open in A for the relative topology* if there exists an open subset U of $\mathbb{R}^n$ such that $D = A \cap U$. Analogously, a subset E of A is said to be *closed in A for the relative topology* if there exists a closed subset C of $\mathbb{R}^n$ such that $D = A \cap C$. An equivalent reformulation of these definitions is that there are open, respectively closed, sets in the topology of the metric space (A, d), where $d(\boldsymbol{x}, \boldsymbol{y}) = \|\boldsymbol{x} - \boldsymbol{y}\|$, for $\boldsymbol{x}, \boldsymbol{y} \in A$.

Definition 3.1.1. Let $M \subset \mathbb{R}^n$ be a nonempty set and $1 \leq k \leq n$. Then M is a *k-dimensional regular surface* (briefly, regular k-surface) of class C^p $(p \geq 1)$ if for each point $\boldsymbol{x}_0 \in M$ there exist an open set $A \subset \mathbb{R}^k$ and a mapping of class C^p,

$$\boldsymbol{\varphi} : A \subset \mathbb{R}^k \to M \subset \mathbb{R}^n,$$

such that
(S1) there exists an open set U in $\mathbb{R}^n$ such that $\boldsymbol{\varphi}(A) = M \cap U$, $\boldsymbol{x}_0 \in \boldsymbol{\varphi}(A)$, and

$$\boldsymbol{\varphi} : A \to \boldsymbol{\varphi}(A)$$

is a homeomorphism;
(S2) for each $\boldsymbol{t} \in A$, $\mathrm{d}\boldsymbol{\varphi}(\boldsymbol{t}) : \mathbb{R}^k \to \mathbb{R}^n$ is injective.

The pair $(A, \boldsymbol{\varphi})$ is called *a chart* or *local coordinate system* (Fig. 3.1). The mapping $\boldsymbol{\varphi}$ is a *parameterization*, and $\boldsymbol{\varphi}(A)$ is a *coordinate neighborhood* of $\boldsymbol{x}_0$. A family $\{(A_i, \boldsymbol{\varphi}_i)\}_{i \in I}$ of coordinate systems with the property that

$$M = \bigcup_{i \in I} \boldsymbol{\varphi}_i(A_i)$$

is called an *atlas* of M.

The condition (S1) is equivalent to the following:

(S1)$'$: $\boldsymbol{x}_0 \in \boldsymbol{\varphi}(A)$, $\boldsymbol{\varphi}$ is continuous, and for each open set $B \subset A$ there is an open set $V \subset \mathbb{R}^n$ such that

$$\boldsymbol{\varphi}(B) = V \cap M.$$

Indeed, let us assume that condition (S1) holds and fix an open set $B \subset A$. Since $\boldsymbol{\varphi} : A \to \boldsymbol{\varphi}(A)$ is a homeomorphism, it follows that $\boldsymbol{\varphi}(B)$ is open in $\boldsymbol{\varphi}(A)$ for the relative topology. Hence there is an open subset $W \subset \mathbb{R}^n$ such that

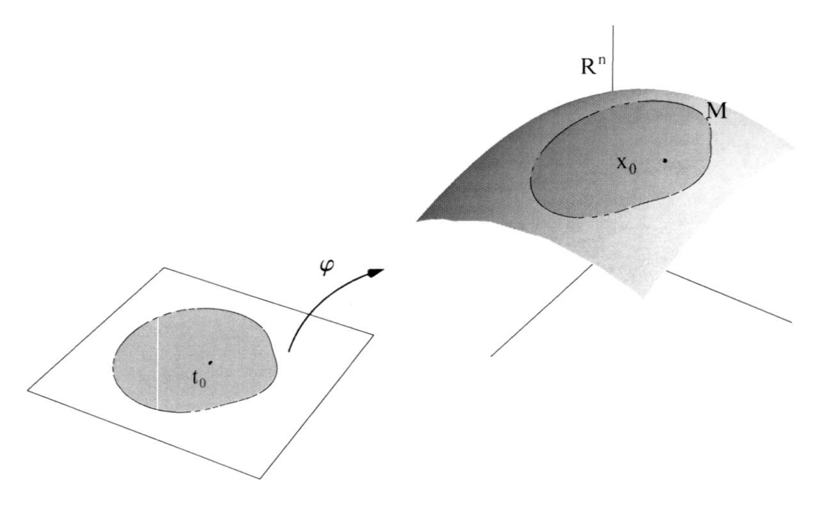

Fig. 3.1 A local coordinate system

$$\boldsymbol{\varphi}(B) = \boldsymbol{\varphi}(A) \cap W = M \cap V,$$

where $V = W \cap U$ is an open set in $\mathbb{R}^n$. Conversely, assume that condition (S1)$'$ holds. Given an open subset $B \subset A$, take V as in (S1)$'$. Then

$$(\boldsymbol{\varphi}^{-1})^{-1}(B) = \boldsymbol{\varphi}(B) = V \cap \boldsymbol{\varphi}(A),$$

which implies that

$$\boldsymbol{\varphi}^{-1} : \boldsymbol{\varphi}(A) \to A$$

is a continuous mapping. This together with condition (S1)$'$ implies condition (S1).

In general, condition (S1)$'$ is stronger than simply saying that $\boldsymbol{\varphi}^{-1} : \boldsymbol{\varphi}(A) \to A$ is a continuous mapping.

Condition (S2) is equivalent to the fact that the Jacobian matrix $\boldsymbol{\varphi}'(t)$ has rank k for every $t \in A$. If we set $\boldsymbol{\varphi} = (\varphi_1, \dots, \varphi_n)$, where

$$\varphi_j : A \subset \mathbb{R}^k \to \mathbb{R}, \; j = 1, \dots, n,$$

it turns out that the jth column of the Jacobian matrix at $t \in A$ coincides with the vector

$$\frac{\partial \boldsymbol{\varphi}}{\partial t_j}(\boldsymbol{t}) = \left(\frac{\partial \varphi_1}{\partial t_j}(\boldsymbol{t}), \dots, \frac{\partial \varphi_n}{\partial t_j}(\boldsymbol{t}) \right).$$

Hence, condition (S2) means that the k vectors in $\mathbb{R}^n$,

$$\left\{\frac{\partial \varphi}{\partial t_1}(\boldsymbol{t}), \frac{\partial \varphi}{\partial t_2}(\boldsymbol{t}), \ldots, \frac{\partial \varphi}{\partial t_k}(\boldsymbol{t})\right\},$$

are linearly independent for each $\boldsymbol{t} = (t_1, \ldots, t_k) \in A$.

In the case that $n = 3$ and $k = 2$, we usually denote by (s,t) the two variables of $\boldsymbol{\varphi}$, and then condition (S2) is equivalent to

$$\frac{\partial \varphi}{\partial s}(s,t) \times \frac{\partial \varphi}{\partial t}(s,t) \neq 0$$

for every $(s,t) \in A$.

The regular 2-surfaces in $\mathbb{R}^3$ are known simply as regular surfaces. The regular 1-surfaces in $\mathbb{R}^2$ or $\mathbb{R}^3$ are known as regular curves. Recall that an important concept in our study of *curves* (Chap. 2) was the notion of a path, that is, a vector-valued mapping defined on an interval of the real line. However, a regular surface in $\mathbb{R}^3$ is defined as a set of points rather than a mapping. This point of view leads to a better understanding of *orientation*, a concept to be studied in Chap. 5.

It is worth mentioning that Definition 3.1.1 is one of three equivalent statements given by Cartan [7, Proposition 4.7.1].

We need to recall one of the cornerstones of differential calculus in several variables, namely the inverse function theorem. A proof can be found in most texts dealing with analysis in several variables, for example [15, 9.24 Theorem, p. 221] or [9, p. 111].

Theorem 3.1.1 (Inverse function theorem). *Let $\boldsymbol{f}: G \subset \mathbb{R}^n \longrightarrow \mathbb{R}^n$ be a function of class C^p on the open set G such that the linear mapping $\mathrm{d}\boldsymbol{f}(\boldsymbol{a})$ is invertible for some $\boldsymbol{a} \in G$. Let $\boldsymbol{b} = \boldsymbol{f}(\boldsymbol{a})$. Then*

1. *There exist open sets U and V in $\mathbb{R}^n$ such that $\boldsymbol{a} \in U$, $\boldsymbol{b} \in V$, $\boldsymbol{f}(U) = V$ and the restriction of $\boldsymbol{f}$ to U is bijective on V.*
2. *If we denote the inverse of $\boldsymbol{f}: U \longrightarrow V$ by $\boldsymbol{g}: V \longrightarrow U$, then $\boldsymbol{g}$ is of class C^p in V.*

Example 3.1.1. The regular n-surfaces of class C^p in $\mathbb{R}^n$ are the open subsets of $\mathbb{R}^n$.

In fact, let us assume that $M \subset \mathbb{R}^n$ is a regular n-surface of class C^1. According to Definition 3.1.1, for each point $\boldsymbol{x}_0 \in M$ there exist an open set $A \subset \mathbb{R}^n$ and a mapping

$$\boldsymbol{\varphi}: A \subset \mathbb{R}^n \to \mathbb{R}^n$$

of class C^1 on A such that

$$\boldsymbol{x}_0 \in \boldsymbol{\varphi}(A) \subset M \tag{3.1}$$

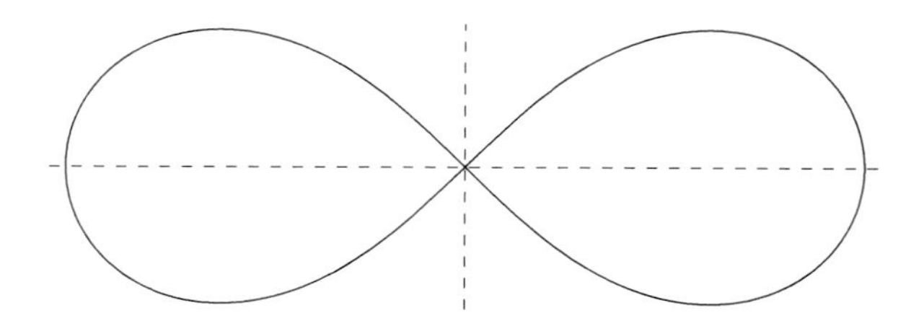

Fig. 3.2 Bernoulli lemniscate

and $d\boldsymbol{\varphi}(\boldsymbol{t}) : \mathbb{R}^n \to \mathbb{R}^n$ is injective for all $\boldsymbol{t} \in A$. It follows from the inverse function theorem that for each $\boldsymbol{t} \in A$, there exist two open sets $W_t \subset A$ and $Z_t \subset \mathbb{R}^n$ such that $\boldsymbol{t} \in W_t$ and $\boldsymbol{\varphi} : W_t \to Z_t$ is a bijection. Hence

$$\boldsymbol{\varphi}(A) = \bigcup_{t \in A} Z_t$$

is an open subset of $\mathbb{R}^n$. From condition (3.1), M is the union of a family of open sets and thus M is open. On the other hand, every open set M is a regular n-surface of class C^∞: we have only to consider $(A, \boldsymbol{\varphi})$, where $A := M$ and $\boldsymbol{\varphi} : A \to M$ is the identity mapping, $\boldsymbol{\varphi}(\boldsymbol{x}) = \boldsymbol{x}$, $\boldsymbol{x} \in A$.

Example 3.1.2. Let M be a regular k-surface of class C^p in $\mathbb{R}^n$ and let U be an open set in $\mathbb{R}^n$ such that $M \cap U \neq \varnothing$. Then $M \cap U$ is a regular k-surface of class C^p in $\mathbb{R}^n$.

Let $\boldsymbol{x}_0 \in M \cap U$ and let $(A, \boldsymbol{\varphi})$ be a coordinate system of M such that $\boldsymbol{x}_0 \in \boldsymbol{\varphi}(A)$. We define $B := \boldsymbol{\varphi}^{-1}(M \cap U)$, which is an open subset of A. If $\boldsymbol{\psi}$ is the restriction mapping of $\boldsymbol{\varphi}$ to the set B, then $(B, \boldsymbol{\psi})$ is a coordinate system of $M \cap U$ and $\boldsymbol{x}_0 \in \boldsymbol{\psi}(B)$.

Example 3.1.3. The *Bernoulli lemniscate* is not a regular curve (Fig. 3.2).

The equation of the lemniscate in polar coordinates is

$$\rho^2 = 2\cos(2\theta),$$

and it is given by

$$M := \{(x,y) \in \mathbb{R}^2 \ : \ (x^2 + y^2)^2 = 2(x^2 - y^2)\}.$$

If U is a ball centered at the origin of radius $0 < r < \sqrt{2}$, then $U \cap M$ is not homeomorphic to any open interval of the real line, since

$$(U \cap M) \setminus \{(0,0)\}$$

has four connected components. Consequently, there is no coordinate neighborhood of the origin satisfying the conditions of Definition 3.1.1. (The reader might like to compare Example 3.3.1.)

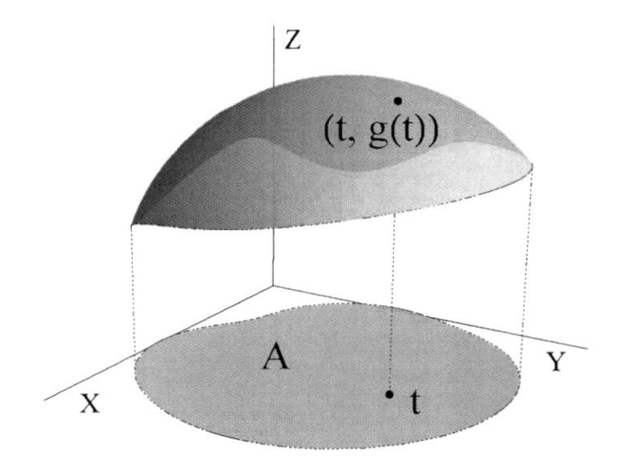

Fig. 3.3 The graph of a function

Example 3.1.4. Let $\boldsymbol{g} : A \subset \mathbb{R}^k \to \mathbb{R}^{n-k}$ be a mapping of class C^p on the open set A. Then the *graph* of $\boldsymbol{g}$,

$$G_g := \{(\boldsymbol{t}, \boldsymbol{g}(\boldsymbol{t})) \,:\, \boldsymbol{t} \in A\} \subset \mathbb{R}^k \times \mathbb{R}^{n-k} \simeq \mathbb{R}^n,$$

is a regular k-surface of class C^p with the coordinate system $(A, \boldsymbol{\varphi})$, $\boldsymbol{\varphi}(\boldsymbol{t}) = (\boldsymbol{t}, \boldsymbol{g}(\boldsymbol{t}))$.

In fact, we consider the mapping

$$\boldsymbol{\varphi} : A \subset \mathbb{R}^k \to \mathbb{R}^n$$

given by

$$\boldsymbol{\varphi}(\boldsymbol{t}) := (\boldsymbol{t}, \boldsymbol{g}(\boldsymbol{t})).$$

(1) Then $\boldsymbol{\varphi}$ is injective, and if $B \subset A$ is an open set, then (Fig. 3.3)

$$\boldsymbol{\varphi}(B) = \{(\boldsymbol{t}, \boldsymbol{g}(\boldsymbol{t})) \,:\, \boldsymbol{t} \in A\} \cap (B \times \mathbb{R}^{n-k}) = G_g \cap U,$$

where $U = B \times \mathbb{R}^{n-k}$ is an open subset of $\mathbb{R}^n$.

(2) The submatrix of

$$\boldsymbol{\varphi}'(\boldsymbol{t}) = \begin{pmatrix} 1 & 0 \;\; 0 \;\ldots & 0 \\ 0 & 1 \;\; 0 \;\ldots & 0 \\ \ldots & \ldots\;\ldots\;\ldots & \ldots \\ 0 & 0 \;\; 0 \;\ldots & 1 \\ \frac{\partial g_1}{\partial t_1}(\boldsymbol{t}) & \ldots\;\ldots\;\ldots & \frac{\partial g_1}{\partial t_k}(\boldsymbol{t}) \\ \ldots & \ldots\;\ldots\;\ldots & \ldots \\ \frac{\partial g_{n-k}}{\partial t_1}(\boldsymbol{t}) & \ldots\;\ldots\;\ldots & \frac{\partial g_{n-k}}{\partial t_k}(\boldsymbol{t}) \end{pmatrix}$$

formed by the first k rows is precisely the identity matrix in $\mathbb{R}^k$. Consequently, $\boldsymbol{\varphi}'(\boldsymbol{t})$ has rank k for each $\boldsymbol{t} \in A$.

Fig. 3.4 The unit sphere

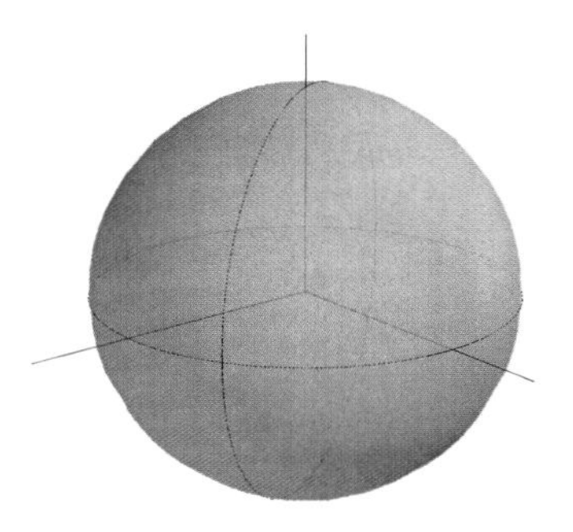

One of our aims is to show that regular k-surfaces of class C^p in $\mathbb{R}^n$ are *locally* nothing more than the graphs of C^p mappings. In other words, *locally*, Example 3.1.4 represents all possible regular k-surfaces in $\mathbb{R}^n$. The next example shows more clearly what is meant by being a graph *locally*.

Example 3.1.5. The unit sphere (Fig. 3.4) in $\mathbb{R}^3$,

$$S := \{(x,y,z) \in \mathbb{R}^3 \ : \ x^2+y^2+z^2 = 1\},$$

is a regular surface of class C^∞.

Consider $A := \{(s,t) \in \mathbb{R}^2 \ : \ s^2+t^2 < 1\}$ and let $g : A \to \mathbb{R}$ be the mapping of class C^∞ given by

$$g(s,t) = \sqrt{1-(s^2+t^2)}.$$

By Example 3.1.4, (A, φ_1),

$$\varphi_1(s,t) := (s,t,g(s,t)),$$

is a coordinate system of S. The set $\varphi_1(A)$ is the open hemisphere above the xy-plane. We plan to cover all the sphere with similar parameterizations. To this end, we take

$$\varphi_2(s,t) := (s,t,-g(s,t)).$$

Then $\varphi_1(A) \cup \varphi_2(A)$ covers all the sphere except the equator

$$\{(x,y,z) \in \mathbb{R}^3 \ : \ x^2+y^2 = 1 \ ; \ z = 0\}.$$

In order to cover the sphere, we have to consider also the following parameterizations (all of them defined on A) (See Fig. 3.5):

$$\varphi_3(s,t) = (s,g(s,t),t), \quad \varphi_4(s,t) = (s,-g(s,t),t),$$
$$\varphi_5(s,t) = (g(s,t),s,t), \quad \varphi_6(s,t) = (-g(s,t),s,t).$$

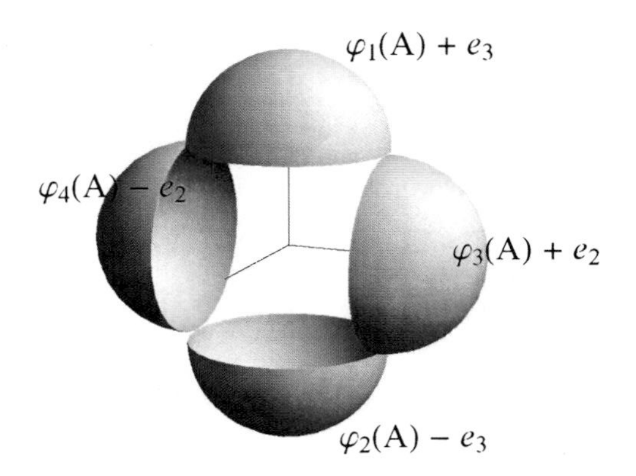

Fig. 3.5 Some charts of the sphere

We refer also to Example 3.3.2, where spherical coordinates are used to define a coordinate system that covers the complete sphere except a semicircle. As already mentioned, we will prove in Proposition 3.1.1 that every regular surface is, locally and up to a permutation of coordinates, the graph of a function of class C^1. However, it is generally not obvious how the surface may be locally represented as a graph. On the other hand, the surface often consists of the set of points where a certain function vanishes (see Sect. 3.2). In such cases, the local representation as a graph is a consequence of the implicit function theorem that we will state in next section.

Proposition 3.1.1. *Let M be a regular k-surface of class C^p. For each $\boldsymbol{x}_0 \in M$ there exist an open neighborhood U of $\boldsymbol{x}_0$, an open set $Z \subset \mathbb{R}^k$, and a mapping of class C^p, $\boldsymbol{g} : Z \to \mathbb{R}^{n-k}$, such that (up to a permutation of coordinates)*

$$M \cap U = G_g$$

and the pair $(Z, (\mathrm{Id}_Z, \boldsymbol{g}))$ is a coordinate system of M.

Proof. By hypothesis, there is a coordinate system $(A, \boldsymbol{\varphi})$ of M such that $\boldsymbol{x}_0 = \boldsymbol{\varphi}(\boldsymbol{t}_0)$ for some $\boldsymbol{t}_0 \in A$. We put $\boldsymbol{\varphi} = (\varphi_1, \varphi_2, \ldots, \varphi_n)$. Since $\boldsymbol{\varphi}'(\boldsymbol{t}_0)$ has rank k, we can select k indices $i_1, i_2, \ldots, i_k$ such that

$$\frac{\partial\left(\varphi_{i_1}, \varphi_{i_2}, \ldots, \varphi_{i_k}\right)}{\partial\left(t_1, t_2, \ldots, t_k\right)}(\boldsymbol{t}_0) \neq 0.$$

(We refer to the comments prior to Theorem 6.4.1 for an explanation of this notation.) For clarity of exposition we can assume, following a permutation of coordinates, that

$$i_1 = 1, i_2 = 2, \ldots, i_k = k.$$

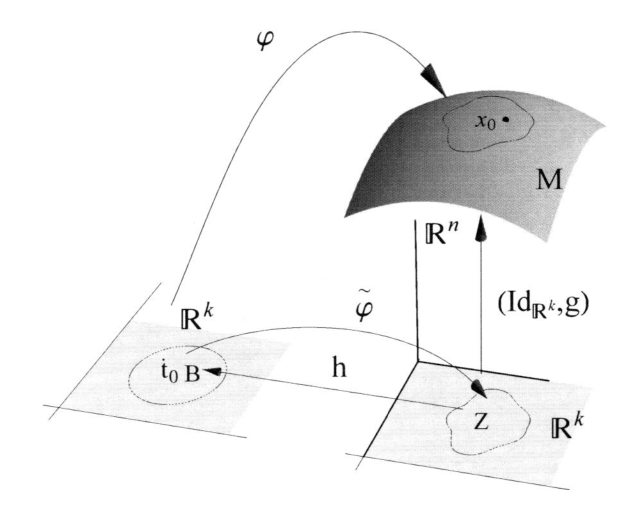

Fig. 3.6 Illustration of the proof of Proposition 3.1.1

We apply the inverse function theorem to the function

$$\tilde{\boldsymbol{\varphi}} := (\varphi_1, \ldots, \varphi_k)$$

to infer the existence of an open neighborhood B of $\boldsymbol{t}_0$ with the property that $Z := \tilde{\boldsymbol{\varphi}}(B)$ is an open subset of $\mathbb{R}^k$ and $\tilde{\boldsymbol{\varphi}} : B \to Z$ is a bijection with inverse

$$\boldsymbol{h} : Z \to B$$

of class C^p (Fig. 3.6). Finally, we consider

$$\boldsymbol{g} : Z \subset \mathbb{R}^k \to \mathbb{R}^{n-k},\ \boldsymbol{z} \mapsto (\varphi_{k+1}(\boldsymbol{h}(\boldsymbol{z})), \ldots \varphi_n(\boldsymbol{h}(\boldsymbol{z}))).$$

Then, for each $\boldsymbol{z} = (z_1, z_2, \ldots, z_k) \in Z$, we have

$$\begin{aligned}\boldsymbol{\varphi} \circ \boldsymbol{h}(\boldsymbol{z}) &= (\varphi_1 \circ \boldsymbol{h}(\boldsymbol{z}), \ldots, \varphi_k \circ \boldsymbol{h}(\boldsymbol{z}), \varphi_{k+1}(\boldsymbol{h}(\boldsymbol{z})), \ldots, \varphi_n(\boldsymbol{h}(\boldsymbol{z})))\\ &= (z_1, \ldots, z_k, \boldsymbol{g}(\boldsymbol{z})) = (\boldsymbol{z}, \boldsymbol{g}(\boldsymbol{z})).\end{aligned}$$

In particular, the graph of $\boldsymbol{g}$ coincides with the set $\boldsymbol{\varphi} \circ \boldsymbol{h}(Z) = \boldsymbol{\varphi}(B)$. Moreover, by (S1)$'$ there exists an open set U in $\mathbb{R}^n$ such that

$$M \cap U = \boldsymbol{\varphi}(B) = G_g.$$

□

We observe that the identity $\boldsymbol{\varphi} \circ \boldsymbol{h} = (\mathrm{Id}_Z, \boldsymbol{g})$ obtained in the previous proof means that

$$\boldsymbol{\varphi} = (\mathrm{Id}_Z, \boldsymbol{g}) \circ \boldsymbol{h}^{-1} = (\mathrm{Id}_Z, \boldsymbol{g}) \circ \tilde{\boldsymbol{\varphi}},$$

when $\boldsymbol{\varphi}$ is restricted to B.

Let $\boldsymbol{\pi} : \mathbb{R}^n \to \mathbb{R}^k$, $\boldsymbol{x} \mapsto (x_1, \dots, x_k)$, denote the projection onto the first k coordinates. From the identity (3.2) it follows that

$$\boldsymbol{\varphi}^{-1}(x) = \tilde{\boldsymbol{\varphi}}^{-1}(\mathrm{Id}_Z, \boldsymbol{g})^{-1}(\boldsymbol{x}) = \boldsymbol{h}(x_1, \dots, x_k) = \boldsymbol{h} \circ \boldsymbol{\pi}(\boldsymbol{x}) \tag{3.2}$$

for $\boldsymbol{x} \in M \cap U$. That is, $\boldsymbol{\varphi}^{-1} = \boldsymbol{h} \circ \boldsymbol{\pi}$, when restricted to $M \cap U$.

3.2 Level Surfaces

The *level sets* of a function $f : U \subset \mathbb{R}^n \to \mathbb{R}$ are defined, for every real number c, by

$$f^{-1}(c) = \{x \in U : f(x) = c\}.$$

The aim of this section is to analyze under what conditions the level sets of a function of two variables are regular curves in the plane and the level sets of a function of three variables are regular surfaces in $\mathbb{R}^3$ in the sense of Definition 3.1.1. We remark that several texts, for example Fleming [10], define a regular k-surface in $\mathbb{R}^n$ (or k-manifold in the notation of [10]) to be a set of points satisfying the hypothesis of Proposition 3.2.1.

To prove that proposition, we need the implicit function theorem, which again can be found in most textbooks on multivariable calculus, but the version we recall here is the one stated by Fleming [10, 4.6, pp. 148–149], since it perfectly suits our requirements. We employ the usual identification of $\mathbb{R}^n \times \mathbb{R}^p$ with $\mathbb{R}^{n+p}$, and if $\boldsymbol{L} : \mathbb{R}^{n+p} \longrightarrow \mathbb{R}^p$ is a linear mapping, we define linear mappings $L_1 : \mathbb{R}^n \longrightarrow \mathbb{R}^p$ by $\boldsymbol{L}_1(\boldsymbol{h}) := \boldsymbol{L}(\boldsymbol{h},0)$ and $\boldsymbol{L}_2 : \mathbb{R}^p \longrightarrow \mathbb{R}^p$ by $\boldsymbol{L}_2(\boldsymbol{k}) := \boldsymbol{L}(0,\boldsymbol{k})$. For $\boldsymbol{f} : G \subset \mathbb{R}^{n+p} \longrightarrow \mathbb{R}^p$ a function of class C^q on the open set G and each $(\boldsymbol{x},\boldsymbol{y})$ in G, we denote by $\tilde{J}\boldsymbol{f}(\boldsymbol{x},\boldsymbol{y}) := \frac{\partial(f_1,\dots,f_p)}{\partial(y_1,\dots y_p)}(\boldsymbol{x},\boldsymbol{y})$ the determinant of the $p \times p$ matrix obtained by taking the last p columns of the matrix of the linear mapping $\mathrm{d}\boldsymbol{f}(\boldsymbol{x},\boldsymbol{y})$.

Theorem 3.2.1 (Implicit function theorem). *Let $\boldsymbol{f} : G \subset \mathbb{R}^{n+p} \longrightarrow \mathbb{R}^p$ be a mapping of class C^q on the open set G. Let $(\boldsymbol{a},\boldsymbol{b})$ in G be such that $\boldsymbol{f}(\boldsymbol{a},\boldsymbol{b}) = \boldsymbol{0}$ and $\tilde{J}\boldsymbol{f}(\boldsymbol{a},\boldsymbol{b}) \neq 0$. Then there exist two open sets $U \subset \mathbb{R}^{n+p}$ and $A \subset \mathbb{R}^n$, with $(\boldsymbol{a},\boldsymbol{b}) \in U$ and $\boldsymbol{a} \in A$, that satisfy the following conditions:*

1. *$\tilde{J}\boldsymbol{f}(\boldsymbol{x},\boldsymbol{y}) \neq 0$ for every $(\boldsymbol{x},\boldsymbol{y}) \in U$.*
2. *For each $\boldsymbol{x} \in A$ there exists a unique $\boldsymbol{y} \in \mathbb{R}^p$ such that $(\boldsymbol{x},\boldsymbol{y}) \in U$ and $\boldsymbol{f}(\boldsymbol{x},\boldsymbol{y}) = \boldsymbol{0}$. In other words, there is a well-defined mapping $\boldsymbol{g} : A \longrightarrow \mathbb{R}^p$ such that*

$$\{(\boldsymbol{x},\boldsymbol{y}) \in U : \boldsymbol{f}(\boldsymbol{x},\boldsymbol{y}) = \boldsymbol{0}\} = \{(\boldsymbol{x},\boldsymbol{g}(\boldsymbol{x})) : \boldsymbol{x} \in A\}.$$

3. *The mapping $\boldsymbol{g}$ is of class C^q on A, and letting $\boldsymbol{L} = \mathrm{d}\boldsymbol{f}(\boldsymbol{x},\boldsymbol{g}(\boldsymbol{x}))$, we have*

$$\mathrm{d}\boldsymbol{g}(\boldsymbol{x}) = -\boldsymbol{L}_2^{-1} \circ \boldsymbol{L}_1$$

for all $\boldsymbol{x} \in A$.

Proposition 3.2.1. *Let $1 \leq k < n$ and $M \subset \mathbb{R}^n$, $M \neq \varnothing$, be given. Assume that for each $\boldsymbol{x}_0 \in M$ there exists a function of class C^p on the open set U,*

$$\boldsymbol{\Phi} : U \subset \mathbb{R}^n \to \mathbb{R}^{n-k},$$

such that $\boldsymbol{x}_0 \in U$ and the following conditions are satisfied:

(1) $M \cap U = \boldsymbol{\Phi}^{-1}(\boldsymbol{0})$,
(2) the rank of $\boldsymbol{\Phi}'(\boldsymbol{x})$ is $n-k$ for every $\boldsymbol{x} \in M \cap U$.

Then M is a regular k-surface of class C^p.

Proof. Let $\boldsymbol{x}_0 = (x_1^0, \ldots, x_n^0) \in M$ and $\boldsymbol{\Phi} = (\Phi_1, \ldots, \Phi_{n-k})$ be as in the hypothesis. After a permutation of the coordinates if necessary, we can assume that

$$\frac{\partial(\Phi_1, \ldots, \Phi_{n-k})}{\partial(x_{k+1}, \ldots, x_n)}(\boldsymbol{x}_0) \neq 0.$$

By the implicit function theorem, there exist an open set $A \subset \mathbb{R}^k$ with

$$\boldsymbol{t}_0 := (x_1^0, \ldots, x_k^0) \in A,$$

an open set $V \subset \mathbb{R}^n$, $V \subset U$, such that $\boldsymbol{x}_0 \in V$, and a mapping

$$\boldsymbol{g} : A \subset \mathbb{R}^k \to \mathbb{R}^{n-k}$$

of class C^p such that $\boldsymbol{g}(\boldsymbol{t}_0) = (x_{k+1}^0, \ldots, x_n^0)$ and

$$\{\boldsymbol{x} \in V \; : \; \boldsymbol{\Phi}(\boldsymbol{x}) = \boldsymbol{0}\} = \{(\boldsymbol{t}, \boldsymbol{g}(\boldsymbol{t})) \; : \; \boldsymbol{t} \in A\}.$$

That is, $M \cap V = G_g$. Now we define $\boldsymbol{\varphi} : A \to M$ by

$$\boldsymbol{\varphi}(\boldsymbol{t}) := (\boldsymbol{t}, \mathbf{g}(\boldsymbol{t})).$$

Then $M \cap V = \boldsymbol{\varphi}(A)$. Since $(A, \boldsymbol{\varphi})$ is a coordinate system (see Example 3.1.4), we conclude that M is a regular k-surface. □

The previous result means that under certain conditions, the intersection of an open set with a level curve is a regular 1-surface and the intersection of an open set of $\mathbb{R}^3$ with a level surface is a regular 2-surface. The next examples will clarify this statement.

Example 3.2.1. Let $f : U \subset \mathbb{R}^2 \to \mathbb{R}$ be a function of class C^p on the open set U and let c be a real number such that

$$\Gamma := \{(x, y) \in U \; : \; f(x, y) = c\}$$

is nonempty and $\boldsymbol{\nabla} f(x, y) \neq (0, 0)$ for all $(x, y) \in \Gamma$. Then Γ is a regular curve (regular surface of dimension 1) of class C^p. We refer to Γ as a *level curve* of the function f. To see this, it is enough to apply Proposition 3.2.1 with $\boldsymbol{\Phi} : U \to \mathbb{R}$,

$$\boldsymbol{\Phi}(x, y) = f(x, y) - c.$$

The level curve is the projection onto the xy-plane of the intersection of the graph of f with the horizontal plane $z = c$. The level curves of a function of two variables provide information about the graph of the function just as contours on a map indicate the topography of a landscape (Fig. 3.7).

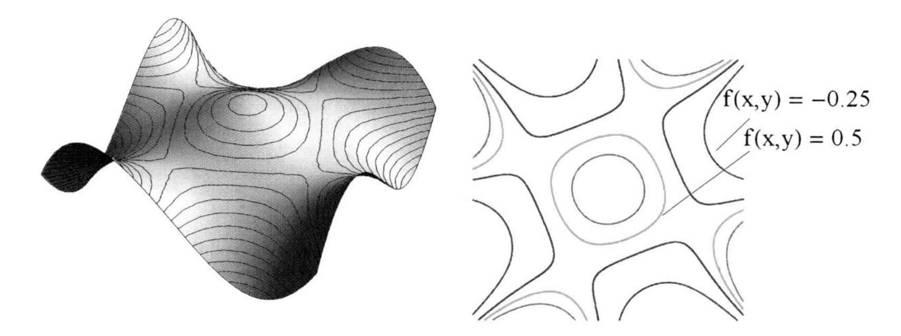

Fig. 3.7 Level curves

Example 3.2.2. The level surface

$$S := \{(x,y,z) \in U \ : f(x,y,z) = c\}$$

of the function $f : U \subset \mathbb{R}^3 \to \mathbb{R}$ of class C^p on the open set U is a regular surface of class C^p provided that $\boldsymbol{\nabla} f(x,y,z) \neq (0,0,0)$ for all $(x,y,z) \in S$. Again, it suffices to apply Proposition 3.2.1 to $\boldsymbol{\Phi}(x,y,z) = f(x,y,z) - c$.

Example 3.2.3. For every $a > 0$, the cone

$$x^2 + y^2 = az^2$$

with the vertex $(0,0,0)$ removed is a regular surface of class C^∞.

Let $U := \mathbb{R}^3 \setminus \{(0,0,0)\}$ and consider the function

$$\boldsymbol{\Phi} : U \to \mathbb{R},$$

of class C^∞, defined by

$$\boldsymbol{\Phi}(x,y,z) := x^2 + y^2 - az^2.$$

Then the cone without the vertex can be described as (Fig. 3.8)

$$S := \{(x,y,z) \in U \ : \ \boldsymbol{\Phi}(x,y,z) = 0\}.$$

Since

$$\boldsymbol{\nabla} \boldsymbol{\Phi}(x,y,z) = (2x, 2y, -2az) \neq (0,0,0),$$

whenever $(x,y,z) \in U$, it follows from Example 3.2.2 that S is a regular surface.

Example 3.2.4. The sphere

$$S^{n-1} := \left\{(x_1, \dots, x_n) \in \mathbb{R}^n : x_1^2 + \cdots + x_n^2 = 1\right\}$$

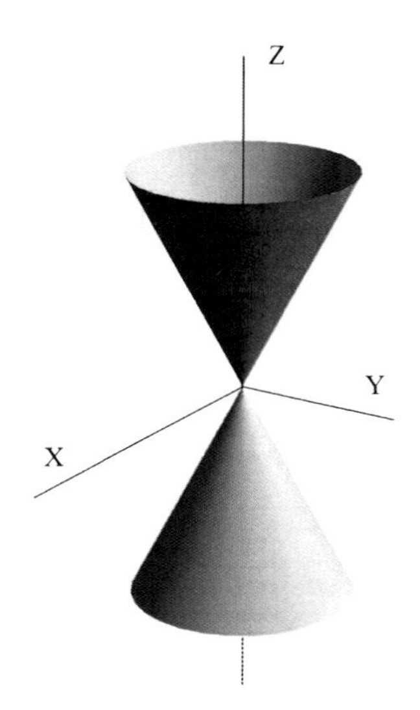

Fig. 3.8 The cone of Example 3.2.3

is a regular $(n-1)$-surface of class C^∞ in $\mathbb{R}^n$, $n \geq 2$.

Indeed, define $\boldsymbol{\Phi} : \mathbb{R}^n \to \mathbb{R}$ by

$$\boldsymbol{\Phi}(\boldsymbol{x}) := \left(\sum_{k=1}^{n} x_k^2 \right) - 1.$$

Then $S^{n-1} = \boldsymbol{\Phi}^{-1}(0)$ and $\boldsymbol{\Phi}'(\boldsymbol{x}) = (2x_1, \dots, 2x_n)$ has rank 1 for every $\boldsymbol{x} \in \mathbb{R}^n \setminus \{0\}$.

3.3 Change of Parameters

According to Definition 3.1.1, every point of a regular k-surface of class C^p in $\mathbb{R}^n$ belongs to a coordinate neighborhood, and each point of such a neighborhood is uniquely determined by k parameters (its local coordinates). However, a point can belong to two different coordinate neighborhoods, and we want to prove that certain properties of the surface, expressed in terms of the local coordinates, do not depend on the chosen parameterization. To this end, we show in this section that when a point $\boldsymbol{x}$ belongs to two coordinate neighborhoods with parameters $(t_1, t_2, \ldots, t_k)$, $(u_1, u_2, \ldots, u_k)$, it is possible to pass from one of these local coordinates to the other by means of a transformation of class C^p. The next lemma will be essential to this process.

Lemma 3.3.1. *Let M be a regular k-surface of class C^p in $\mathbb{R}^n$, $(A, \boldsymbol{\varphi})$ a coordinate system of M, and $\boldsymbol{\alpha} : V \subset \mathbb{R}^m \to M$ a mapping defined on the open set V. If $\boldsymbol{\alpha}(\boldsymbol{u}_0) \in \boldsymbol{\varphi}(A)$ and $\boldsymbol{\alpha}$ is differentiable at $\boldsymbol{u}_0 \in V$ (respectively $\boldsymbol{\alpha}$ is of class C^q on V), then*

$$\boldsymbol{\varphi}^{-1} \circ \boldsymbol{\alpha}$$

is differentiable at $\boldsymbol{u}_0$ (respectively it is of class C^r on its domain of definition, where $r = \min(p, q)$) (Fig. 3.9).

Proof. We provide the proof only for the case that $\boldsymbol{\alpha}$ is of class C^q on V. We take $\boldsymbol{t}_0 \in A$ such that

$$\boldsymbol{\alpha}(\boldsymbol{u}_0) = \boldsymbol{\varphi}(\boldsymbol{t}_0).$$

We now set

$$\boldsymbol{\varphi} = (\varphi_1, \ldots, \varphi_k, \varphi_{k+1}, \ldots, \varphi_n).$$

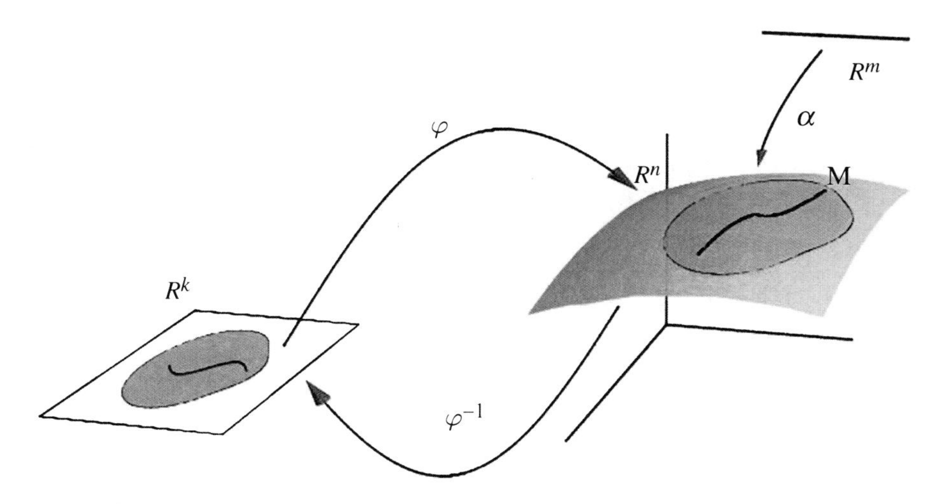

Fig. 3.9 Illustration of Lemma 3.3.1

Since $(A,\boldsymbol{\varphi})$ is a coordinate system of M, condition (S2) of Definition 3.1.1 implies that the matrix $\boldsymbol{\varphi}'(\boldsymbol{t}_0)$ has k linearly independent rows. For simplicity we assume that these are the first k rows. This means that

$$\frac{\partial(\varphi_1,\ldots,\varphi_k)}{\partial(t_1,\ldots,t_k)}(\boldsymbol{t}_0)\neq 0,$$

and consequently

$$\tilde{\boldsymbol{\varphi}}:=(\varphi_1,\ldots,\varphi_k)$$

has a local inverse of class C^p in a neighborhood of $\boldsymbol{t}_0$. According to (3.2), there exist an open neighborhood B of $\boldsymbol{t}_0$, $B\subset A$, an open set $Z\subset\mathbb{R}^k$, and a bijection $\boldsymbol{h}:Z\to B$ of class C^p such that $\boldsymbol{\varphi}^{-1}(\boldsymbol{x})=\boldsymbol{h}\circ\boldsymbol{\pi}(\boldsymbol{x})$ for every $\boldsymbol{x}\in\boldsymbol{\varphi}(B)$. We take $W:=\boldsymbol{\alpha}^{-1}(\boldsymbol{\varphi}(B))$, which is an open set in $\mathbb{R}^m$, since $W=\boldsymbol{\alpha}^{-1}(U)$, where U is an open set in $\mathbb{R}^n$ such that $\boldsymbol{\varphi}(B)=U\cap M$ (Definition 3.1.1) and $\boldsymbol{\alpha}$ is continuous. Then for every $\boldsymbol{u}\in W$,

$$\boldsymbol{\varphi}^{-1}\circ\boldsymbol{\alpha}(\boldsymbol{u})=\boldsymbol{h}\circ\boldsymbol{\pi}(\boldsymbol{\alpha}(\boldsymbol{u})),$$

which is obviously of class C^r. Since $\boldsymbol{u}_0\in W$, this finishes the proof. □

The first application of this result is the following important theorem.

Theorem 3.3.1. *Let $(A_1,\boldsymbol{\varphi}_1)$, $(A_2,\boldsymbol{\varphi}_2)$ be two coordinate systems of a regular k-surface M in $\mathbb{R}^n$ of class C^p such that*

$$D:=\boldsymbol{\varphi}_1(A_1)\cap\boldsymbol{\varphi}_2(A_2)\neq\varnothing.$$

Then

$$\boldsymbol{\varphi}_2^{-1}\circ\boldsymbol{\varphi}_1:\boldsymbol{\varphi}_1^{-1}(D)\subset\mathbb{R}^k\to\boldsymbol{\varphi}_2^{-1}(D)\subset\mathbb{R}^k$$

and

$$\boldsymbol{\varphi}_1^{-1}\circ\boldsymbol{\varphi}_2:\boldsymbol{\varphi}_2^{-1}(D)\subset\mathbb{R}^k\to\boldsymbol{\varphi}_1^{-1}(D)\subset\mathbb{R}^k$$

are functions of class C^p, each of which is an inverse of the other (Fig. 3.10).

Proof. It follows from Lemma 3.3.1, with $\boldsymbol{\varphi}=\boldsymbol{\varphi}_2$ and $\boldsymbol{\alpha}=\boldsymbol{\varphi}_1$, that $\boldsymbol{\varphi}_2^{-1}\circ\boldsymbol{\varphi}_1$ is a function of class C^p on its domain of definition, which is precisely $\boldsymbol{\varphi}_1^{-1}(D)$. □

An understanding of the importance of the previous theorem has been fundamental in the development of differential geometry. In discussing the notion of a regular surface in $\mathbb{R}^3$, Do Carmo [6, p. 33] says the following:

> The most serious problem with the above definition is its dependence on $\mathbb{R}^3$. Indeed the natural idea of an abstract surface is that of an object, that is, in a certain sense, two-dimensional, and to which we can apply, locally, the Differential Calculus in $\mathbb{R}^2$. Such an idea of an abstract surface (i.e. with no reference to an ambient space) has been foreseen since Gauss. It took, however, about a century for the definition to reach the definitive form that we present below. One of the reasons for this delay was that, even for surfaces in $\mathbb{R}^3$, the fundamental role of the change of parameters was not clearly understood.

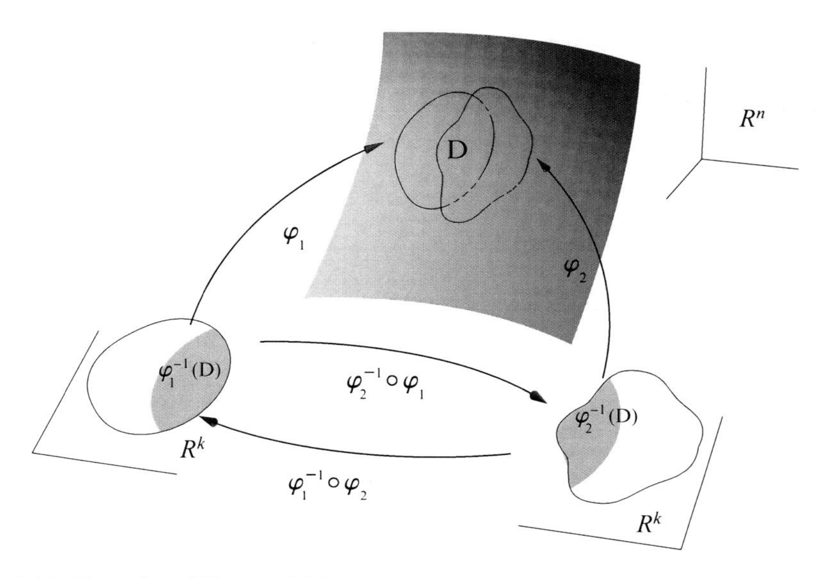

Fig. 3.10 Illustration of Theorem 3.3.1

Below we present the definition of an *abstract surface*, also known in geometry as a *differentiable manifold*, not necessarily contained in $\mathbb{R}^n$ or endowed with a topology. This general concept will not be used in the sequel.

Definition 3.3.1. A k-dimensional differentiable manifold of class C^p is a set M together with a family of injective mappings

$$f_\alpha : A_\alpha \subset \mathbb{R}^k \to M$$

such that:

(1) Each A_α is an open set and $\bigcup_\alpha f_\alpha(A_\alpha) = M$.

(2) For each pair α, β with

$$U := f_\alpha(A_\alpha) \cap f_\beta(A_\beta) \neq \varnothing,$$

the sets $f_\alpha^{-1}(U)$ and $f_\beta^{-1}(U)$ are open sets in $\mathbb{R}^k$ and

$$f_\alpha^{-1} \circ f_\beta$$

is a function of class C^p.

The inclusion of the *change of parameters* property in the definition above allows one to develop a differential calculus for abstract surfaces, and while that is not the direction that we focus on in this book, we do include the relevant definition of differentiable function.

Definition 3.3.2. Let M be a k-dimensional differentiable manifold of class C^p. A function $f : M \to \mathbb{R}$ is said to be differentiable, continuous, or C^r $(1 \leq r \leq \infty)$ if for each α the mapping

$$f \circ f_\alpha : A_\alpha \subset \mathbb{R}^k \to \mathbb{R}$$

is differentiable, continuous, or C^r respectively.

Let M be a regular k-surface in $\mathbb{R}^n$. The implication of our next result is that it is not necessary to prove that $\boldsymbol{\varphi} : A \to \boldsymbol{\varphi}(A)$ is a homeomorphism in order to show that a given mapping $\boldsymbol{\varphi}$ is a parameterization of M. More precisely, once we know that M is a regular k-surface, then condition (S1) in Definition 3.1.1 can be replaced by the injectivity of $\boldsymbol{\varphi}$. This makes it easier to conclude that $\boldsymbol{\varphi}$ is a parameterization of M, which is an important fact, given that in practical examples, proof that $\boldsymbol{\varphi} : A \to \boldsymbol{\varphi}(A)$ is a homeomorphism can be rather difficult or tedious.

Corollary 3.3.1. *Let M be a regular k-surface of class C^p in $\mathbb{R}^n$, $A \subset \mathbb{R}^k$ an open set, and*

$$\boldsymbol{\varphi} : A \subset \mathbb{R}^k \to M \subset \mathbb{R}^n$$

an injective mapping of class C^p such that $\boldsymbol{\varphi}'(\boldsymbol{t})$ has rank k for every $\boldsymbol{t} \in A$ (i.e., condition (S2) for a regular k-surface is satisfied). Then $(A, \boldsymbol{\varphi})$ is a coordinate system of M.

Proof. We fix $\boldsymbol{t}_0 \in A$ and put $\boldsymbol{x}_0 := \boldsymbol{\varphi}(\boldsymbol{t}_0)$. We may choose a coordinate system $(B, \boldsymbol{\psi})$ of M such that $\boldsymbol{x}_0 \in \boldsymbol{\psi}(B)$. By Theorem 3.3.1, the mapping

$$\boldsymbol{f} := \boldsymbol{\psi}^{-1} \circ \boldsymbol{\varphi} : W \subset \mathbb{R}^k \to \mathbb{R}^k$$

is defined on an open neighborhood W of $\boldsymbol{t}_0$ and is of class C^p. Moreover, since

$$\boldsymbol{\varphi}'(\boldsymbol{t}_0) = \boldsymbol{\psi}'(\boldsymbol{f}(\boldsymbol{t}_0)) \circ \boldsymbol{f}'(\boldsymbol{t}_0),$$

and $\boldsymbol{\varphi}'(\boldsymbol{t}_0)$ has rank k, the rank of $\boldsymbol{f}'(\boldsymbol{t}_0)$ is greater than or equal to k. That is, the rank of $\boldsymbol{f}'(\boldsymbol{t}_0)$ is also equal to k, or equivalently, the Jacobian of the mapping $\boldsymbol{f}$ at the point $\boldsymbol{t}_0$ is nonzero. By the inverse function theorem, there exists an open neighborhood W_0 of $\boldsymbol{t}_0$, $W_0 \subset W$, such that $\boldsymbol{f}(W_0)$ is an open set and

$$\boldsymbol{f} : W_0 \to \boldsymbol{f}(W_0)$$

is a homeomorphism. Consequently, after applying property (S1)$'$ of Definition 3.1.1 to the coordinate system $(B, \boldsymbol{\psi})$, it follows that $\boldsymbol{\varphi}_{|W_0} = \boldsymbol{\psi} \circ \boldsymbol{f}_{|W_0}$ satisfies property (S1)$'$ too, and hence $(A, \boldsymbol{\varphi})$ is a coordinate system. $\square$

The next example shows that we cannot remove the hypothesis that M is a regular k-surface from Corollary 3.3.1.

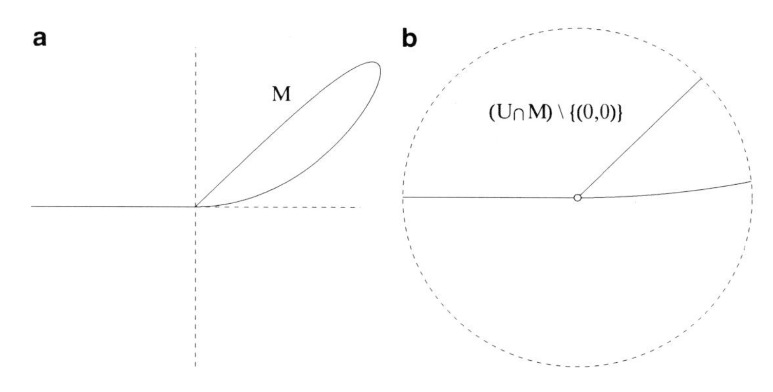

Fig. 3.11 (**a**) The curve of Example 3.3.1; (**b**) detail of that curve around the origin

Example 3.3.1. Let us consider the mapping

$$\boldsymbol{\varphi} : \left(-1, \frac{\pi}{2} \right) \to \mathbb{R}^2$$

defined by

$$\boldsymbol{\varphi}(t) = (t, 0)$$

for $-1 < t \leq 0$ and

$$\boldsymbol{\varphi}(t) = \left(t \cos t, t \frac{\sin(2t)}{2} \right)$$

for $0 \leq t < \frac{\pi}{2}$. Then $\boldsymbol{\varphi}$ in injective, of class C^1 on $(-1, \frac{\pi}{2})$, and in addition, for every $t \in (-1, \frac{\pi}{2})$,

$$\boldsymbol{\varphi}'(t) \neq 0.$$

However,

$$M := \boldsymbol{\varphi}\left(-1, \frac{\pi}{2} \right)$$

is not a regular curve in $\mathbb{R}^2$ (Fig. 3.11).

To see this, suppose there exists a coordinate system $(A, \boldsymbol{\gamma})$ with $(0,0) \in \boldsymbol{\gamma}(A)$. Let U be an open ball centered at $(0,0)$ such that $U \cap M \subset \boldsymbol{\gamma}(A)$. Since $\boldsymbol{\gamma} : A \to \boldsymbol{\gamma}(A)$ is a homeomorphism and $U \cap M$ is a connected set, it follows that $\boldsymbol{\gamma}^{-1}(U \cap M) = (a,b)$ and that $\boldsymbol{\gamma} : (a,b) \to U \cap M$ is a homeomorphism. We put $t_0 = \boldsymbol{\gamma}^{-1}(0,0) \in (a,b)$. Then also $\boldsymbol{\gamma} : (a,t_0) \cup (t_0,b) \to (U \cap M) \setminus \{(0,0)\}$ is a homeomorphism. This is a contradiction, since $(U \cap M) \setminus \{(0,0)\}$ has three connected components.

This example is not a contradiction to Corollary 3.3.1, because in order to apply that corollary to conclude that $((-1, \frac{\pi}{2}), \boldsymbol{\varphi})$ is a coordinate system, one must know a priori that M is a regular curve. Moreover, the previous example emphasizes the importance of condition (S1) in Definition 3.1.1.

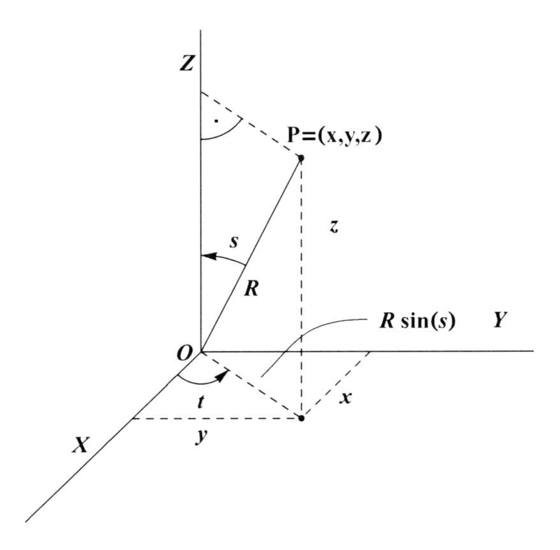

Fig. 3.12 Spherical coordinates

Next we parameterize the sphere of radius R centered at the origin in terms of the geographical coordinates *colatitude* (the complement of the latitude) s and *longitude* t. Recall that the Cartesian coordinates of a point on this sphere are given by

$$x = R\sin s\cos t, \quad y = R\sin s\sin t, \quad z = R\cos s.$$

Example 3.3.2 (Spherical coordinates). Let S be the sphere with radius R centered at the origin in $\mathbb{R}^3$ (Fig. 3.12),

$$A := \{(s,t) \in \mathbb{R}^2 \, : 0 < s < \pi, \, 0 < t < 2\pi\},$$

and let us consider

$$\varphi(s,t) := R\,(\sin s\cos t, \, \sin s\sin t, \, \cos s)\,.$$

Then (A, φ) is a coordinate system of the sphere S.

We show that $\varphi'(s,t)$ has rank 2 for every $(s,t) \in A$. To this end, we determine that

$$\frac{\partial(x,y)}{\partial(s,t)} = R^2 \cos s \sin s,$$

$$\frac{\partial(y,z)}{\partial(s,t)} = R^2 \sin^2 s\cos t,$$

$$\frac{\partial(x,z)}{\partial(s,t)} = R^2 \sin^2 s\sin t.$$

The three determinants do not vanish simultaneously, since for $0 < s < \pi$,

$$\cos^2 s \sin^2 s + \sin^4 s \cos^2 t + \sin^4 s \sin^2 t = \sin^2 s > 0.$$

It is rather intuitive that φ is injective and $\varphi(A)$ covers all the sphere S except the semicircle

$$\{(x,y,z) \in S \; : \; y = 0; \; x \geq 0\}.$$

3.4 Tangent and Normal Vectors

By a tangent vector to a surface we mean a tangent vector to a curve contained in that surface. It may seem obvious that the set of all tangent vectors to a regular k-surface at a given point is a k-dimensional vector subspace, but this statement is not so easy to prove. The main objectives of this section are to present a proof of this fact and to study the tangent plane at a given point of a regular surface in $\mathbb{R}^3$.

Definition 3.4.1. Let M be a regular k-surface of class C^p in $\mathbb{R}^n$. We say that $\boldsymbol{h} \in \mathbb{R}^n$ is a tangent vector (Fig. 3.13) to M at $\boldsymbol{x}_0 \in M$ if there is a function

$$\boldsymbol{\alpha} : (-\delta, \delta) \to M,$$

differentiable at $t = 0$, such that

$$\boldsymbol{\alpha}(0) = \boldsymbol{x}_0 \text{ and } \boldsymbol{\alpha}'(0) = \boldsymbol{h}.$$

The set of all tangent vectors to M at $\boldsymbol{x}_0$ is called the *tangent space* to the surface at $\boldsymbol{x}_0$ and is denoted by $T_{\boldsymbol{x}_0}M$. The set $\boldsymbol{x}_0 + T_{\boldsymbol{x}_0}M$ is called the *tangent plane*. This distinction is not always made clear, and the reader will observe that what is usually drawn in books as the tangent space $T_{\boldsymbol{x}_0}M$ is actually the translation of the tangent space to the point x_0, i.e., the tangent plane $\boldsymbol{x}_0 + T_{\boldsymbol{x}_0}M$.

Theorem 3.4.1. *$T_{\boldsymbol{x}_0}M$ is a k-dimensional vector subspace of $\mathbb{R}^n$. Moreover, if $(A, \boldsymbol{\varphi})$ is a coordinate system of M and $\boldsymbol{x}_0 = \boldsymbol{\varphi}(\boldsymbol{t}_0)$, then a basis of $T_{\boldsymbol{x}_0}M$ is given by*

$$\left\{ \frac{\partial \boldsymbol{\varphi}}{\partial t_1}(\boldsymbol{t}_0), \ldots, \frac{\partial \boldsymbol{\varphi}}{\partial t_k}(\boldsymbol{t}_0) \right\}.$$

Proof. We will show that $T_{\boldsymbol{x}_0}M = \mathrm{d}\boldsymbol{\varphi}(\boldsymbol{t}_0)(\mathbb{R}^k)$. To do this, we first take $\boldsymbol{v} \in \mathbb{R}^k$ and prove that $\mathrm{d}\boldsymbol{\varphi}(\boldsymbol{t}_0)(\boldsymbol{v}) \in T_{\boldsymbol{x}_0}M$. Choose $\delta > 0$ small enough that $\boldsymbol{t}_0 + s\boldsymbol{v} \in A$ whenever $|s| < \delta$ and let us consider

$$\boldsymbol{\alpha} : (-\delta, \delta) \to M$$

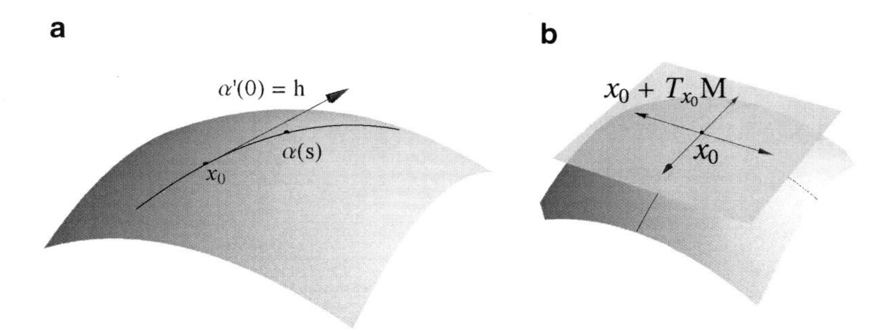

Fig. 3.13 (**a**) Tangent vector to a surface; (**b**) tangent plane

defined by

$$\boldsymbol{\alpha}(s) := \boldsymbol{\varphi}(\boldsymbol{t}_0 + s\boldsymbol{v}).$$

Then

$$\boldsymbol{\alpha}(0) = \boldsymbol{\varphi}(\boldsymbol{t}_0) = \boldsymbol{x}_0$$

and $\boldsymbol{\alpha}'(0)$ is the directional derivative of $\boldsymbol{\varphi}$ at the point t_0 in the direction $\boldsymbol{v}$. That is,

$$\boldsymbol{\alpha}'(0) = D_v\boldsymbol{\varphi}(\boldsymbol{t}_0) = \mathrm{d}\boldsymbol{\varphi}(\boldsymbol{t}_0)(\boldsymbol{v}),$$

which proves that

$$\mathrm{d}\boldsymbol{\varphi}(\boldsymbol{t}_0)(\boldsymbol{v}) \in T_{\boldsymbol{x}_0}M.$$

Let us now assume that $\boldsymbol{h} \in T_{\boldsymbol{x}_0}M$ and let us proceed to prove that there exists $\boldsymbol{v} \in \mathbb{R}^k$ such that

$$\boldsymbol{h} = \mathrm{d}\boldsymbol{\varphi}(\boldsymbol{t}_0)(\boldsymbol{v}).$$

By Definition 3.4.1 of the tangent vector, there is a differentiable function

$$\boldsymbol{\alpha} : (-\delta, \delta) \to M$$

with

$$\boldsymbol{\alpha}(0) = \boldsymbol{x}_0 \text{ and } \boldsymbol{\alpha}'(0) = \boldsymbol{h}.$$

Since $(A, \boldsymbol{\varphi})$ is a coordinate system, there is an open set U in $\mathbb{R}^n$ such that

$$\boldsymbol{\varphi}(A) = U \cap M.$$

Then

$$I := \boldsymbol{\alpha}^{-1}(\boldsymbol{\varphi}(A)) = \boldsymbol{\alpha}^{-1}(U \cap M) = \boldsymbol{\alpha}^{-1}(U)$$

is an open neighborhood of 0. By Lemma 3.3.1,

$$\boldsymbol{\beta} := \boldsymbol{\varphi}^{-1} \circ \boldsymbol{\alpha} : I \subset \mathbb{R} \to \mathbb{R}^k$$

is differentiable at $t = 0$. Moreover,

$$\boldsymbol{\beta}(0) = \boldsymbol{t}_0 \quad \text{and} \quad \boldsymbol{\alpha} = \boldsymbol{\varphi} \circ \boldsymbol{\beta}$$

on I and hence

$$\boldsymbol{h} = \boldsymbol{\alpha}'(0) = \mathrm{d}\boldsymbol{\varphi}(\boldsymbol{t}_0)(\boldsymbol{\beta}'(0)).$$

We finally obtain that

$$\boldsymbol{h} \in \mathrm{d}\boldsymbol{\varphi}(\boldsymbol{t}_0)(\mathbb{R}^k).$$

From the identity

$$T_{\boldsymbol{x}_0}M = \mathrm{d}\boldsymbol{\varphi}(\boldsymbol{t}_0)(\mathbb{R}^k)$$

we deduce that $T_{\boldsymbol{x}_0}M$ is a vector subspace. Moreover, since

$$\mathrm{d}\boldsymbol{\varphi}(\boldsymbol{t}_0) : \mathbb{R}^k \to \mathbb{R}^n$$

is injective, we conclude that a basis of $T_{\boldsymbol{x}_0}M$ is given by the vectors

$$\mathrm{d}\boldsymbol{\varphi}(\boldsymbol{t}_0)(\boldsymbol{e}_j) = \frac{\partial \boldsymbol{\varphi}}{\partial t_j}(\boldsymbol{t}_0),$$

where $1 \leq j \leq k$. □

This result explains why in Definition 3.1.1 we impose the requirement that the differential of $\boldsymbol{\varphi}$ be injective; i.e. the rank of φ' is k at every point. This condition guarantees that the tangent space at each point of the surface is a k-dimensional vector space.

In the particular case $n = 3$, $k = 2$, that is, if S is a regular surface in $\mathbb{R}^3$ and

$$\boldsymbol{\varphi} : A \subset \mathbb{R}^2 \to \mathbb{R}^3$$

defines a coordinate system $(A, \boldsymbol{\varphi})$ of S, then a basis of the tangent space to the surface at a point $(x_0, y_0, z_0) = \boldsymbol{\varphi}(s_0, t_0)$ of S is given by the vectors

$$\left\{ \frac{\partial \boldsymbol{\varphi}}{\partial s}(s_0, t_0), \frac{\partial \boldsymbol{\varphi}}{\partial t}(s_0, t_0) \right\}.$$

Consequently, the cross product

$$\frac{\partial \boldsymbol{\varphi}}{\partial s}(s_0, t_0) \times \frac{\partial \boldsymbol{\varphi}}{\partial t}(s_0, t_0)$$

is a normal vector to this tangent space (Fig. 3.14).

This fact will be crucial later in understanding why, from the point of view of *vector calculus*, orientation is thought of in terms of *normal vectors*, while in *vector analysis* we define orientation in terms of *tangent spaces*. See Chap. 5.

Example 3.4.1 (Tangent plane to the cone $x^2 + y^2 = 2z^2$ at the point $(1, 1, 1)$). We already proved in Example 3.2.3 that

$$S := \{(x, y, z) \neq (0, 0, 0) \ : \ x^2 + y^2 = 2z^2\}$$

is a regular surface of class C^∞. We now choose a coordinate system for this cone. We take

$$\boldsymbol{\varphi} : A \subset \mathbb{R}^2 \to \mathbb{R}^3 \ ; \ \boldsymbol{\varphi}(\rho, \theta) := \left(\rho \cos\theta, \rho \sin\theta, \frac{\rho}{\sqrt{2}} \right),$$

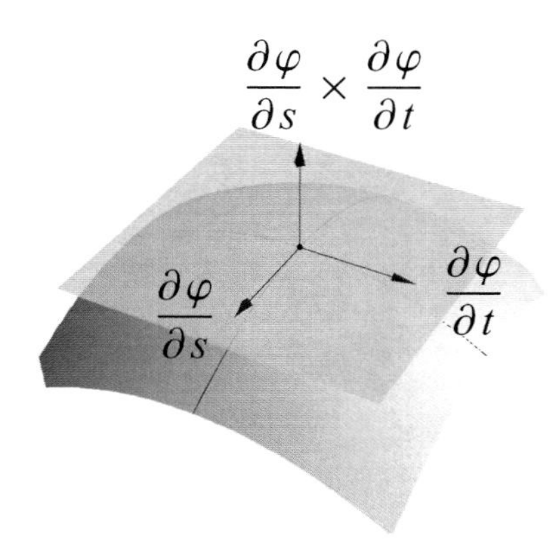

Fig. 3.14 Normal vector

where

$$A := \{(\rho,\theta)\ :\rho > 0,\ 0 < \theta < 2\pi\}.$$

Then

$$\frac{\partial\boldsymbol{\varphi}}{\partial\rho}(\rho,\theta) = \left(\cos\theta,\sin\theta,\frac{1}{\sqrt{2}}\right)$$

and

$$\frac{\partial\boldsymbol{\varphi}}{\partial\theta}(\rho,\theta) = (-\rho\sin\theta,\rho\cos\theta,0).$$

Observe that

$$\begin{aligned}\frac{\partial\boldsymbol{\varphi}}{\partial\rho}(\rho,\theta)\times\frac{\partial\boldsymbol{\varphi}}{\partial\theta}(\rho,\theta) &= \begin{vmatrix} e_1 & e_2 & e_3 \\ \cos\theta & \sin\theta & \frac{1}{\sqrt{2}} \\ -\rho\sin\theta & \rho\cos\theta & 0\end{vmatrix} \\ &= \left(-\frac{\rho}{\sqrt{2}}\cos\theta,-\frac{\rho}{\sqrt{2}}\sin\theta,\rho\right),\end{aligned}$$

which is clearly nonzero. Consequently, the vectors

$$\left\{\frac{\partial\boldsymbol{\varphi}}{\partial\rho}(\rho,\theta),\frac{\partial\boldsymbol{\varphi}}{\partial\theta}(\rho,\theta)\right\}$$

are linearly independent for every (ρ,θ).

It is not difficult to show that $\boldsymbol{\varphi}$ is injective and hence $(A,\boldsymbol{\varphi})$ is a coordinate system of the surface (see Corollary 3.3.1). Since $(1,1,1) = \boldsymbol{\varphi}(\sqrt{2},\frac{\pi}{4})$, we deduce that a normal vector to the tangent plane is

$$\frac{\partial\boldsymbol{\varphi}}{\partial\rho}\left(\sqrt{2},\frac{\pi}{4}\right)\times\frac{\partial\boldsymbol{\varphi}}{\partial\theta}\left(\sqrt{2},\frac{\pi}{4}\right) = \frac{1}{\sqrt{2}}(-1,-1,2).$$

Finally, the equation of the tangent plane is

$$\left\langle \frac{1}{\sqrt{2}}(-1,-1,2),(x-1,y-1,z-1)\right\rangle = 0,$$

or equivalently,

$$x+y-2z=0.$$

Definition 3.4.2. Let M be a regular k-surface in $\mathbb{R}^n$. The *normal space* to the surface at $\boldsymbol{x}_0 \in M$ is the orthogonal subspace to the tangent space $T_{x_0}M$, and we denote it by N_{x_0}. That is,

$$N_{\boldsymbol{x}_0} = \{\boldsymbol{h} \in \mathbb{R}^n \ : \ \langle \boldsymbol{h}, \boldsymbol{v}\rangle = 0 \ \text{ for all } \boldsymbol{v} \in T_{\boldsymbol{x}_0}M\}.$$

The set $\boldsymbol{x}_0 + N_{\boldsymbol{x}_0}$ is called the *normal plane* at x_0.

Proposition 3.4.1. *Let* $1 \le k < n$ *be given and let* $M = \boldsymbol{\Phi}^{-1}(\boldsymbol{0})$ *be the regular* k*-surface where*

$$\boldsymbol{\Phi} : U \subset \mathbb{R}^n \to \mathbb{R}^{n-k}, \quad \boldsymbol{\Phi} = (\Phi_1, \Phi_2, \ldots, \Phi_{n-k}),$$

is a function of class C^1 *on the open set* U *such that the rank of* $\boldsymbol{\Phi}'(\boldsymbol{x})$ *is* $n-k$ *for every* $\boldsymbol{x} \in M = \boldsymbol{\Phi}^{-1}(\boldsymbol{0})$*. Then for each* $\boldsymbol{x}_0 \in M$*, the vectors*

$$\{\boldsymbol{\nabla}\Phi_1(\boldsymbol{x}_0), \boldsymbol{\nabla}\Phi_2(\boldsymbol{x}_0), \ldots, \boldsymbol{\nabla}\Phi_{n-k}(\boldsymbol{x}_0)\}$$

form a basis of the normal space to M *at* $\boldsymbol{x}_0$*.*

Proof. It follows from Proposition 3.2.1 that M is a regular k-surface of class C^1 in $\mathbb{R}^n$. If $\boldsymbol{\alpha} : (-\delta, \delta) \to M$ is differentiable at $t = 0$ and $\boldsymbol{\alpha}(0) = \boldsymbol{x}_0$, then

$$\Phi_j \circ \boldsymbol{\alpha}(t) = 0$$

for all $t \in (-\delta, \delta)$. Hence

$$(\Phi_j \circ \boldsymbol{\alpha})'(0) = 0.$$

According to the chain rule,

$$\langle \boldsymbol{\nabla}\Phi_j(\boldsymbol{x}_0), \boldsymbol{\alpha}'(0)\rangle = 0,$$

for $j = 1, \ldots, n-k$. That is, every vector $\boldsymbol{\nabla}\Phi_j(\boldsymbol{x}_0)$ is in the normal space to the surface at $\boldsymbol{x}_0$. Since the rank of $\boldsymbol{\Phi}'(\boldsymbol{x}_0)$ is $n-k$, these $n-k$ vectors are linearly independent and the dimension of the normal space is precisely $n-k$, giving the conclusion. □

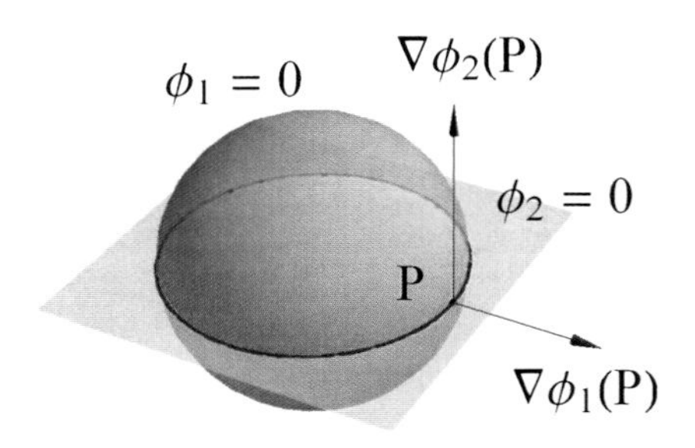

Fig. 3.15 A basis of the normal space to the circle

In Fig. 3.15, $\nabla\Phi_1(\boldsymbol{P})$ is a normal vector at $\boldsymbol{P}$ to the sphere with equation $\Phi_1(x,y,z)=0$, while $\nabla\Phi_2(\boldsymbol{P})$ is a normal vector to the plane $\Phi_2(x,y,z)=0$. A basis of the normal space at $\boldsymbol{P}$ to the circle obtained by intersecting the plane with the sphere is given by the two vectors

$$\{\nabla\Phi_1(\boldsymbol{P}), \nabla\Phi_2(\boldsymbol{P})\}.$$

Example 3.4.2 (Tangent hyperplane to a graph). Let $g : A \subset \mathbb{R}^{n-1} \to \mathbb{R}$ be a function of class C^1 on an open set A and

$$\boldsymbol{\varphi} : A \to \mathbb{R}^n \; ; \; \boldsymbol{\varphi}(\boldsymbol{t}) = (\boldsymbol{t}, g(\boldsymbol{t}))$$

a parameterization of the graph of g (see Example 3.1.4). Since the graph of g coincides with $\boldsymbol{\Phi}^{-1}(0)$, for

$$\boldsymbol{\Phi}(x_1,\dots,x_n) := g(x_1,\dots,x_{n-1}) - x_n$$

and $x = (x_1,...,x_n) \in A x \mathbb{R}$, it turns out that the normal space to G_g at the point $\boldsymbol{x}_0 = (\boldsymbol{t}_0, g(\boldsymbol{t}_0)) = (x_1^0,\dots,x_n^0)$ is the linear span of the vector

$$\nabla\boldsymbol{\Phi}(\boldsymbol{x}_0) = \left(\frac{\partial g}{\partial t_1}(\boldsymbol{t}_0),\dots,\frac{\partial g}{\partial t_{n-1}}(\boldsymbol{t}_0), -1\right).$$

Finally, the equation of the tangent hyperplane to the graph of g at the point $\boldsymbol{x}_0$ is

$$(x_1 - x_1^0)\frac{\partial g}{\partial t_1}(\boldsymbol{t}_0) + \cdots + (x_{n-1} - x_{n-1}^0)\frac{\partial g}{\partial t_{n-1}}(\boldsymbol{t}_0) = x_n - x_n^0.$$

Example 3.4.3 (Normal vector to a level surface). Let $f : U \subset \mathbb{R}^3 \to \mathbb{R}$ be a function of class C^1 on an open set U such that for all $(x,y,z) \in S$,

$$\nabla f(x,y,z) \neq (0,0,0),$$

where S represents the level surface

$$S := \{(x,y,z) \in U \; : f(x,y,z) = c\} \neq \varnothing.$$

From Example 3.2.2, S is a regular surface of class C^1. Moreover, the vector

$$\nabla f(x_0, y_0, y_0)$$

is normal to the surface at the point (x_0, y_0, z_0). In fact, it suffices to apply Proposition 3.4.1 with $\Phi := f - c$. The set $\{\nabla f(x_0, y_0, y_0)\}$ is a basis of the normal space.

The last result of this chapter shows the equivalence between the different definitions of regular k-surface (or k-submanifold of $\mathbb{R}^n$) appearing in the literature. Condition (2) coincides with condition (II) in the Proposition/Definition of Cartan [7, Proposition 4.7.1], while condition (3) coincides with the definition of differentiable manifold in $\mathbb{R}^n$ given by Fleming [10, 4.7 manifolds, p. 153]. We note that Edwards [9, II.5, p. 10] uses condition (2) as his definition, and Spivak [17, Chapter 5, p. 101] defines differentiable manifold as we have defined regular k-surface.

Theorem 3.4.2. *Let $M \subset \mathbb{R}^n$ be a nonempty set and $1 \leq k < n$. The following conditions are equivalent:*

1. *M is a k-dimensional regular surface in $\mathbb{R}^n$ of class C^p.*
2. *For every $\mathbf{x}_0 \in M$ there exist an open neighborhood U of $\mathbf{x}_0$, an open set $Z \subset \mathbb{R}^k$, and a mapping of class C^p, $\mathbf{g} : \mathbf{Z} \to \mathbb{R}^{\mathbf{n-k}}$, such that (up to permutation of the coordinates) $M \cap U$ coincides with $G_{\mathbf{g}}$, the graph of $\mathbf{g}$.*
3. *For every $\mathbf{x}_0 \in M$ there is a function of class C^p,*

$$\boldsymbol{\Phi} : U \subset \mathbb{R}^n \to \mathbb{R}^{n-k},$$

such that $\mathbf{x}_0 \in U$, $M \cap U = \boldsymbol{\Phi}^{-1}(\mathbf{0})$, and the rank of $\boldsymbol{\Phi}'(\mathbf{x})$ is $n-k$ for all $\mathbf{x} \in M \cap U$.

Proof. Some of the necessary implications have already been proved. For example, (1) implies (2) is proven in Proposition 3.1.1, while (2) implies (1) holds by Example 3.1.4. (3) implies (2) is contained in the proof of Proposition 3.2.1. We need only show that (2) implies (3). To do this, we represent the points $\mathbf{x} \in \mathbb{R}^n$ as

$$\mathbf{x} = (\mathbf{x}', \mathbf{x}''),$$

where $\mathbf{x}' \in \mathbb{R}^k$, $\mathbf{x}'' \in \mathbb{R}^{n-k}$. Now we define an open set

$$U := Z \times \mathbb{R}^{n-k}$$

in $\mathbb{R}^n$ and a function

$$\boldsymbol{\Phi} : U \subset \mathbb{R}^n \to \mathbb{R}^{n-k}, \quad \boldsymbol{\Phi}(\mathbf{x}) := \mathbf{x}'' - \mathbf{g}(\mathbf{x}').$$

Then $\boldsymbol{\Phi}$ is a function of class C^p on U and

$$\boldsymbol{\Phi}^{-1}(\boldsymbol{0}) = G_g = M \cap U.$$

Moreover,

$$\boldsymbol{\Phi}'(\boldsymbol{x}) = \begin{pmatrix} -\nabla g_1(\boldsymbol{x}') & 1 & \dots & 0 \\ \dots & \dots & \dots & \dots \\ -\nabla g_{n-k}(\boldsymbol{x}') & 0 & \dots & 1 \end{pmatrix},$$

and so the rank of $\boldsymbol{\Phi}'(\boldsymbol{x})$ is $n-k$ for every $\boldsymbol{x} \in U$. □

3.5 Exercises

Exercise 3.5.1. (1) Show that the cylinder

$$S := \{(x,y,z) \in \mathbb{R}^3 : x^2 + y^2 = a^2\}$$

is a regular surface.

(2) What about the portion of the cylinder

$$\{(x,y,z) \in S : 0 < z < 1\}?$$

Exercise 3.5.2. Show that the cone

$$S := \{(x,y,z) \in \mathbb{R}^3 : x^2 + y^2 = z^2,\ z \geq 0\}$$

is not a regular surface, but $S \setminus \{(0,0,0)\}$, the cone without the *vertex*, is.

Exercise 3.5.3. Determine whether the intersection of the cone

$$x^2 + y^2 = z^2$$

with the plane

$$x + y + z = 1$$

is a regular curve.

Exercise 3.5.4. Let

$$\boldsymbol{\gamma} : [a,b] \to U \subset \mathbb{R} \times (0,+\infty)$$

be a path such that $\boldsymbol{\gamma}([a,b])$ is a level curve . Show that the set obtained by rotating $\boldsymbol{\gamma}([a,b])$ around the x-axis is a level surface.

Exercise 3.5.5. Find a parameterization of the cylinder $x^2 + y^2 = a^2$ in a neighborhood of $(a,0,0)$.

Exercise 3.5.6. Let $\boldsymbol{\varphi} : \mathbb{R}^2 \to \mathbb{R}^3$ be the mapping

$$\boldsymbol{\varphi}(s,t) := (s+t,\ s-t,\ 4st).$$

Show that $(\mathbb{R}^2, \boldsymbol{\varphi})$ is a coordinate system of the regular surface with equation

$$z = x^2 - y^2.$$

Exercise 3.5.7. Explain how to obtain a coordinate system of the ellipsoid

$$\left(\frac{x}{a}\right)^2 + \left(\frac{y}{b}\right)^2 + \left(\frac{z}{c}\right)^2 = 1$$

from a coordinate system of the unit sphere.

Exercise 3.5.8. Let $0 < r < R$.

(1) Show that the torus obtained by rotation of the circle

$$x^2 + (y - R)^2 = r^2$$

about the x-axis is a regular surface.

(2) Find a coordinate system.

Exercise 3.5.9. Find the tangent plane to the paraboloid $z = x^2 + y^2$ at the point $\left(\frac{1}{2}, \frac{1}{2}, \frac{1}{2}\right)$.

Exercise 3.5.10. Find a tangent vector at $(0,0,0)$ to the curve determined by the equations

$$z = 3\,x^2 + 5\,y^2, \quad x^2 + y^2 = 4\,x.$$

Chapter 4
Flux of a Vector Field

In this chapter we concentrate on aspects of *vector calculus*. A common physical application of this theory is the fluid flow problem of calculating the amount of fluid passing through a permeable surface. The abstract generalization of this leads us to the *flux* of a vector field through a regular 2-surface in $\mathbb{R}^3$. More precisely, let the vector field F in $\mathbb{R}^3$ represent the velocity vector field of a fluid. We immerse a permeable surface S in that fluid, and we are interested in the amount of fluid flow across the surface S per unit time. This is the flux integral of the vector field F across the surface S.

A. Galbis and M. Maestre, *Vector Analysis Versus Vector Calculus*, Universitext, DOI 10.1007/978-1-4614-2200-6_4,

4.1 Area of a Parallelepiped

One of the goals in this chapter is to evaluate the k-dimensional area of a subset of a k-dimensional regular surface. The first situation to consider is that of a subset A of a k-dimensional subspace $V \subset \mathbb{R}^n$. In this case we proceed as follows. First we choose a linear mapping

$$\boldsymbol{\Psi} : V \to \mathbb{R}^k$$

preserving the inner product: it suffices to select an orthonormal basis of V and consider a linear mapping transforming such a basis into the canonical basis of $\mathbb{R}^k$. Then $\boldsymbol{\Psi}$ is an isometry, that is, it preserves distances, and it seems reasonable to define the k-dimensional area of A as the k-dimensional Lebesgue measure of $\boldsymbol{\Psi}(A)$ (Fig. 4.1).

Next we proceed to develop this idea in the particular case that $A = P$ is the parallelepiped spanned by k vectors in $\mathbb{R}^n$. We proceed essentially as in Edwards [9, p. 324], which we recommend for further details and related topics, in particular for a general Pythagorean theorem [9, Theorem 3.11, p. 329].

Calculation of the area of a subset of a k-dimensional regular surface will be dealt with in Sect. 4.2.

Definition 4.1.1. Let $\{\boldsymbol{a}_1, \ldots, \boldsymbol{a}_k\}$ be k vectors in $\mathbb{R}^n$ $(1 \leq k \leq n)$. By the *k-dimensional parallelepiped* spanned by $a_1, \ldots, a_k$, we mean the compact set

$$P := \left\{ \sum_{i=1}^{k} t_i \boldsymbol{a}_i \ : \ t_i \in [0,1] \right\}.$$

A parallelepiped spanned by two vectors in $\mathbb{R}^2$ or $\mathbb{R}^3$ is called a parallelogram.

Let $V \subset \mathbb{R}^n$ be a k-dimensional subspace of $\mathbb{R}^n$ containing the vectors $\{\boldsymbol{a}_1, \ldots, \boldsymbol{a}_k\}$ and let

$$\boldsymbol{\Psi} : V \to \mathbb{R}^k$$

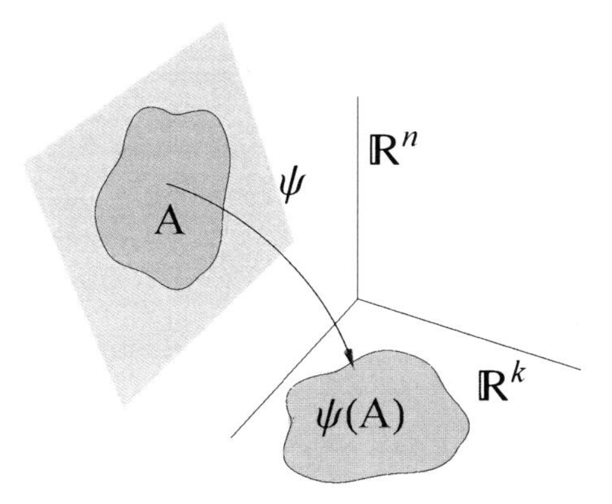

Fig. 4.1 Isometry used to define the k-dimensional area of A

be a linear mapping preserving the inner product, i.e.,

$$\langle \boldsymbol{\Psi}(u), \boldsymbol{\Psi}(v) \rangle = \langle u, v \rangle$$

for every $u, v \in V$.

Definition 4.1.2. The k-dimensional area (or simply k-area) of the parallelepiped P is

$$k\text{-area}(P) := \int_{\mathbb{R}^k} \chi_{\boldsymbol{\Psi}(P)}(\boldsymbol{t})\, \mathrm{d}\boldsymbol{t} = \int_{\boldsymbol{\Psi}(P)} \mathrm{d}\boldsymbol{t}.$$

If the vectors $\{\boldsymbol{a}_1, \ldots, \boldsymbol{a}_k\}$ are linearly dependent, then $\boldsymbol{\Psi}(P)$ is a compact set contained in a hyperplane of $\mathbb{R}^k$; hence its k-dimensional Lebesgue measure is zero. The interesting case occurs when the vectors $\{\boldsymbol{a}_1, \ldots, \boldsymbol{a}_k\}$ are linearly independent (and consequently, V coincides with the linear span of these vectors). The following theorem shows how to compute the area in this case, and actually shows that the concept of k-area does not depend on the mapping $\boldsymbol{\Psi}$ preserving the dot product.

Theorem 4.1.1. *Let P be the parallelepiped spanned by the linearly independent vectors $\{\boldsymbol{a}_1, \ldots, \boldsymbol{a}_k\}$. Then*

$$k\text{-area}(P) = \sqrt{\det(\langle \boldsymbol{a}_i, \boldsymbol{a}_j \rangle)} = \sqrt{\det(B^{\mathrm{T}} \circ B)},$$

where B is the $n \times k$ matrix whose column vectors are $[\boldsymbol{a}_1, \ldots, \boldsymbol{a}_k]$.

Proof. Let V be the k-dimensional vector subspace spanned by $\{\boldsymbol{a}_1, \ldots, \boldsymbol{a}_k\}$ and let $\boldsymbol{\Psi} : V \to \mathbb{R}^k$ be a linear isomorphism preserving the inner product. Consider the linear mapping

$$\boldsymbol{L} : \mathbb{R}^k \to \mathbb{R}^k$$

defined by

$$\boldsymbol{L}(\boldsymbol{e}_i) := \boldsymbol{\Psi}(\boldsymbol{a}_i), \quad 1 \leq i \leq k.$$

Then

$$\boldsymbol{\Psi}(P) = \left\{ \sum_{i=1}^{k} t_i \boldsymbol{\Psi}(\boldsymbol{a}_i) \ : \ t_i \in [0,1] \right\} = \boldsymbol{L}([0,1]^k)$$

because $\boldsymbol{L}(t_1, \ldots, t_k) = \sum_{i=1}^{k} t_i \boldsymbol{\Psi}(\boldsymbol{a}_i)$. Since $\boldsymbol{L}$ is a linear isomorphism, we can apply the change of variables theorem (Theorems 6.5.1 and 6.5.2) to conclude that

$$k\text{-area}(P) = \int_{\boldsymbol{\Psi}(P)} \mathrm{d}\boldsymbol{t} = \int_{\boldsymbol{L}([0,1]^k)} \mathrm{d}\boldsymbol{t} = \int_{[0,1]^k} |\det \boldsymbol{L}|\, \mathrm{d}\boldsymbol{t} = |\det \boldsymbol{L}|.$$

Finally, let A be the matrix of $\boldsymbol{L}$ with respect to the canonical basis. Then $A = [\boldsymbol{\Psi}(\boldsymbol{a}_1), \ldots, \boldsymbol{\Psi}(\boldsymbol{a}_k)]$, and denoting by A^{T} the transposed matrix, we obtain

$$\begin{aligned} |\det \boldsymbol{L}| = |\det A| &= \sqrt{\det (A^{\mathrm{T}} \circ A)} \\ &= \sqrt{\det (\langle \boldsymbol{\Psi}(\boldsymbol{a}_i), \boldsymbol{\Psi}(\boldsymbol{a}_j) \rangle)} \\ &= \sqrt{\det (\langle \boldsymbol{a}_i, \boldsymbol{a}_j \rangle)}, \end{aligned}$$

as claimed. □

Theorem 4.1.1 implies that the definition of k-dimensional area of a parallelepiped (Definition 4.1.2) is independent of the choice of the linear mapping $\boldsymbol{\Psi}$.

Example 4.1.1 (Parallelogram spanned by two vectors in $\mathbb{R}^3$). Let

$$\boldsymbol{a} = (a_1, a_2, a_3) \text{ and } \boldsymbol{b} = (b_1, b_2, b_3)$$

be two vectors in $\mathbb{R}^3$. Recall (see Definition 1.1.4) that the cross product of the two vectors is defined by means of the formal expression

$$\boldsymbol{a} \times \boldsymbol{b} := \begin{vmatrix} \boldsymbol{e}_1 & \boldsymbol{e}_2 & \boldsymbol{e}_3 \\ a_1 & a_2 & a_3 \\ b_1 & b_2 & b_3 \end{vmatrix}.$$

That is,

$$\boldsymbol{a} \times \boldsymbol{b} := \left(\begin{vmatrix} a_2 & a_3 \\ b_2 & b_3 \end{vmatrix}, - \begin{vmatrix} a_1 & a_3 \\ b_1 & b_3 \end{vmatrix}, \begin{vmatrix} a_1 & a_2 \\ b_1 & b_2 \end{vmatrix} \right).$$

As already mentioned after Definition 1.1.4, one can see that

$$\| \boldsymbol{a} \times \boldsymbol{b} \|^2 = \| \boldsymbol{a} \|^2 \cdot \| \boldsymbol{b} \|^2 - |\langle \boldsymbol{a}, \boldsymbol{b} \rangle|^2,$$

and thus deduce from Theorem 4.1.1 that the 2-dimensional area of the parallelogram spanned by $\boldsymbol{a}$ and $\boldsymbol{b}$ is

$$\text{area} = \sqrt{\begin{vmatrix} \langle \boldsymbol{a}, \boldsymbol{a} \rangle & \langle \boldsymbol{a}, \boldsymbol{b} \rangle \\ \langle \boldsymbol{b}, \boldsymbol{a} \rangle & \langle \boldsymbol{b}, \boldsymbol{b} \rangle \end{vmatrix}} = \| \boldsymbol{a} \times \boldsymbol{b} \| .$$

4.2 Area of a Regular Surface

When investigating the concept of area for k-dimensional regular surfaces one might expect to take an approach analogous to that of path length, with suitable inscribed parallelepipeds taking the role of inscribed polygonal lines. However, such an approach has some very counterintuitive consequences. A very instructive example can be found on page 119 (and on the cover) of the book of Spivak [17], where it is shown that in contrast to the usual procedure in topography, it is not a good idea to triangulate a terrain in order to approximate its area. We must therefore employ a slightly different method. The basic idea consists in generalizing the formula given in Theorem 2.1.3 after replacing the length of the vector tangent to the curve by the area of the k-dimensional parallelepiped spanned by an appropriate basis of the tangent space to the surface.

To obtain a better intuition of this viewpoint we first analyze the particular case of a regular 2-surface in $\mathbb{R}^3$ and restrict our attention to the situation in which the surface is defined only by a coordinate system (A,φ), that is, $S=\varphi(A)$. Moreover, we assume that

$$K := [a,b]\times[c,d]$$

is contained in A.

Let

$$a = s_0 < s_1 < \cdots < t_m = b$$

be a partition of the interval $[a,b]$ into subintervals of length h and

$$c = t_0 < t_1 < \cdots < t_l = d$$

a partition of the interval $[c,d]$ into subintervals of length k. We put $I_{i,j} := [s_i,s_{i+1}]\times[t_j,t_{j+1}]$, and so

$$\varphi(K) = \bigcup_{i=0}^{m-1}\bigcup_{j=0}^{l-1}\varphi(I_{i,j}).$$

The portion of the surface $\varphi(I_{i,j})$ can be approximated by the parallelepiped spanned by the vectors

$$\left\{h\cdot\frac{\partial\varphi}{\partial s}(s_i,t_j)\ ,\ k\cdot\frac{\partial\varphi}{\partial t}(s_i,t_j)\right\}$$

(suitably translated so that the point $\varphi(s_i,t_j)$ is one of its vertices). As proven in Sect. 4.1, the area of that parallelepiped is given by

$$hk\cdot\left\|\frac{\partial\varphi}{\partial s}(s_i,t_j)\times\frac{\partial\varphi}{\partial t}(s_i,t_j)\right\|.$$

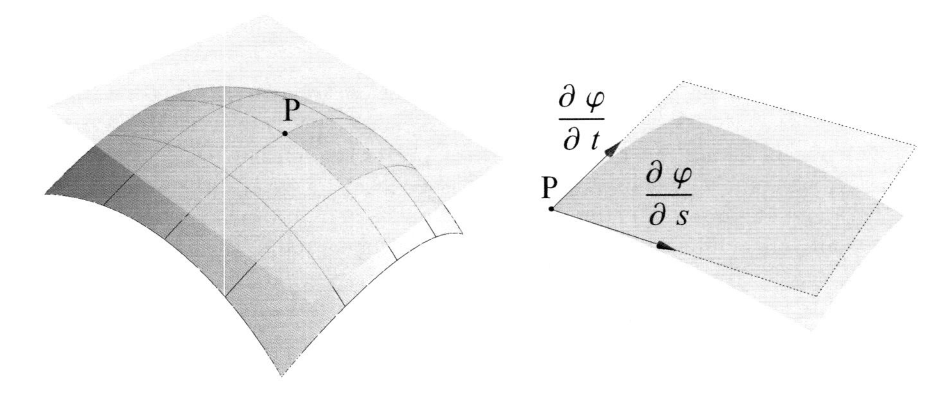

Fig. 4.2 Approximation of a small piece of a surface with a parallelogram in the tangent plane

Therefore, a good approximation to the area of $\varphi(K)$ should be (Fig. 4.2)

$$\sum_{i=0}^{m-1}\sum_{j=0}^{l-1} hk \cdot \left\| \frac{\partial \varphi}{\partial s}(s_i,t_j) \times \frac{\partial \varphi}{\partial t}(s_i,t_j) \right\|. \tag{4.1}$$

We evaluate the limit of the above expression as h and k tend to zero.

Lemma 4.2.1. *Let $K := [a,b] \times [c,d]$,*

$$a = s_0 < s_1 < \cdots < t_m = b$$

a partition of $[a,b]$ into subintervals of length h, and

$$c = t_0 < t_1 < \cdots < t_l = d$$

a partition of $[c,d]$ into subintervals of length k. If $F : K \to \mathbb{R}$ is a continuous function, then

$$\lim_{(h,k)\to(0,0)} \sum_{i=0}^{m-1}\sum_{j=0}^{l-1} hk \cdot F(s_i,t_j) = \iint_K F(s,t)\, \mathrm{d}(s,t).$$

Proof. Define $I_{i.j} := [s_i, s_{i+1}] \times [t_j, t_{j+1}]$. Then

$$\iint_K F(s,t)\, \mathrm{d}(s,t) = \sum_{i=0}^{m-1}\sum_{j=0}^{l-1} \iint_{I_{i,j}} F(s,t)\, \mathrm{d}(s,t)$$

and

$$hk \cdot F(s_i,t_j) = \iint_{I_{i,j}} F(s_i,t_j)\, \mathrm{d}(s,t).$$

Hence

$$\left|\sum_{i=0}^{m-1}\sum_{j=0}^{l-1} hk \cdot F(s_i,t_j) - \iint_K F(s,t)\, \mathrm{d}(s,t)\right|$$

is less than or equal to

$$\sum_{i=0}^{m-1}\sum_{j=0}^{l-1} \iint_{I_{i,j}} |F(s_i,t_j) - F(s,t)|\, \mathrm{d}(s,t). \tag{4.2}$$

By the Heine–Cantor theorem, F is uniformly continuous on K, which means that for every $\varepsilon > 0$ there exist $h_0 > 0$ and $k_0 > 0$ such that

$$|F(\tilde{s},\tilde{t}) - F(s,t)| \leq \frac{\varepsilon}{(b-a)(d-c)},$$

whenever

$$|\tilde{s} - s| < h_0 \quad \text{and} \quad |\tilde{t} - t| < k_0.$$

Therefore, if $0 < h < h_0$ and $0 < k < k_0$, then

$$|F(s_i,t_j) - F(s,t)| \leq \frac{\varepsilon}{(b-a)(d-c)}$$

for every $(s,t) \in I_{i,j}$, and the expression (4.2) is less than or equal to

$$\sum_{i=0}^{m-1}\sum_{j=0}^{l-1} \frac{\varepsilon}{(b-a)(d-c)} \iint_{I_{i,j}} \mathrm{d}(s,t).$$

From

$$\sum_{i=0}^{m-1}\sum_{j=0}^{l-1} \iint_{I_{i,j}} \mathrm{d}(s,t) = \sum_{i=0}^{m-1}\sum_{j=0}^{l-1} (t_{j+1} - t_j)(s_{i+1} - s_i) = (b-a)(d-c),$$

we conclude that

$$\left|\sum_{i=0}^{m-1}\sum_{j=0}^{l-1} hk \cdot F(s_i,t_j) - \iint_K F(s,t)\, \mathrm{d}(s,t)\right| \leq \varepsilon.$$

□

By Lemma 4.2.1, the limit, as h and k go to zero, of the expression (4.1) is the integral

$$\iint_K \left\| \frac{\partial \varphi}{\partial s}(s,t) \times \frac{\partial \varphi}{\partial t}(s,t) \right\| \mathrm{d}(s,t).$$

We are ready to introduce the concept of area of a regular surface. Again we refer to [9] for further details.

Definition 4.2.1. Let (A,φ) be a coordinate system of a regular 2-surface S in $\mathbb{R}^3$ and let K be a compact subset of A. The area of (K,φ) is defined by

$$\operatorname{area}(K,\varphi) := \iint_K \left\| \frac{\partial \varphi}{\partial s}(s,t) \times \frac{\partial \varphi}{\partial t}(s,t) \right\| \mathrm{d}(s,t).$$

Next, suppose (A,φ) is a coordinate system of a k-dimensional regular surface. It is natural to proceed as in Definition 4.2.1 but integrating instead the area of the k-dimensional parallelepiped spanned by the vectors

$$\left\{ \frac{\partial \varphi}{\partial t_1}(t), \ldots, \frac{\partial \varphi}{\partial t_k}(t) \right\}.$$

Since $\{\frac{\partial \varphi}{\partial t_1}(t), \ldots, \frac{\partial \varphi}{\partial t_k}(t)\}$ are the column vectors of the Jacobian matrix $\varphi'(t)$, it follows from Theorem 4.1.1 that the area of this k-parallelepiped is

$$\sqrt{\det(\varphi'^{\mathrm{T}}(t) \circ \varphi'(t))}.$$

This leads to the following.

Definition 4.2.2. Let (A,φ) be a coordinate system of a k-dimensional regular surface M and let $K \subset A$ be a Lebesgue measurable subset. By the *area* (*actually k-area*) of (K,φ) we mean

$$\operatorname{area}(K,\varphi) := \int_K \sqrt{\det(\varphi'^{\mathrm{T}}(t) \circ \varphi'(t))}\mathrm{d}t,$$

where $\varphi'^{\mathrm{T}}(t)$ denotes the transposed matrix of $\varphi'(t)$.

In cases in which the positive and continuous function $f(t) := \sqrt{\det(\varphi'^{\mathrm{T}}(t) \circ \varphi'(t))}$ is not integrable on K, we take $\operatorname{area}(K,\varphi) = +\infty$.

The quantity $\operatorname{area}(K,\varphi)$ is well defined, because if (U,ψ) is another coordinate system of M and $B \subset U$ is a measurable subset such that $\varphi(K) = \psi(B)$, then

$$\operatorname{area}(K,\varphi) \;=\; \operatorname{area}(B,\psi).$$

A proof of this fact will be provided in Chap. 7. The main reason for this postponement is that a direct proof using the change of variables theorem is rather tedious, while a more satisfying proof can be given once elementary properties of differential forms (Chap. 6) and the theory of integration of forms (Chap. 7) have been developed.

Example 4.2.1. Let $\varphi : \mathbb{R}^k \to \mathbb{R}^n$ be a linear mapping preserving the dot product. Then $V = \varphi(\mathbb{R}^k)$ is a k-dimensional vector subspace of $\mathbb{R}^n$ and $(\mathbb{R}^k,\varphi)$ is a coordinate system of V. If K is a measurable subset of $\mathbb{R}^k$, then

$$\text{area}(K,\boldsymbol{\varphi}) = \int_{\mathbb{R}^k} \chi_K(\boldsymbol{t})\,\mathrm{d}\boldsymbol{t}.$$

In fact, since $\boldsymbol{\varphi}$ is a linear mapping, we have that $\boldsymbol{\varphi}'(\boldsymbol{t})$ is the matrix of $\boldsymbol{\varphi}$ with respect to the canonical basis for every $\boldsymbol{t} \in \mathbb{R}^k$. From the fact that $\boldsymbol{\varphi}$ preserves dot products we deduce that

$$\begin{aligned}\boldsymbol{\varphi}'^{\mathrm{T}}(\boldsymbol{t}) \circ \boldsymbol{\varphi}'(\boldsymbol{t}) &= \left(\langle \boldsymbol{\varphi}(\boldsymbol{e}_i), \boldsymbol{\varphi}(\boldsymbol{e}_j)\rangle\right)_{i,j} \\ &= \left(\langle \boldsymbol{e}_i, \boldsymbol{e}_j\rangle\right)_{i,j}\end{aligned}$$

is the identity matrix. Therefore

$$\begin{aligned}\text{area}(K,\boldsymbol{\varphi}) &= \int_K \sqrt{\det(\boldsymbol{\varphi}'^{\mathrm{T}}(\boldsymbol{t}) \circ \boldsymbol{\varphi}'(\boldsymbol{t}))}\mathrm{d}\boldsymbol{t} \\ &= \int_{\mathbb{R}^k} \chi_K(\boldsymbol{t})\,\mathrm{d}\boldsymbol{t}.\end{aligned}$$

Next we discuss how to evaluate the area of a k-dimensional regular surface that is not covered by a unique coordinate neighborhood. The reader may wish to recall the notion of atlas given after Definition 3.1.1.

Lemma 4.2.2. *Let M be a regular k-surface in $\mathbb{R}^n$ with a finite atlas*

$$\{(A_j, \boldsymbol{\varphi}_j)\}_{j=1}^m.$$

Then there is a partition of M into subsets $\{B_j\}$ with the property that $B_j = \boldsymbol{\varphi}_j(K_j)$, where K_j is a measurable subset of A_j.

Proof. According to Definition 3.1.1, for every $j = 1,\dots,m$ there is an open set U_j in $\mathbb{R}^n$ such that $\boldsymbol{\varphi}_j(A_j) = U_j \cap M$. Define $B_1 := \boldsymbol{\varphi}_1(A_1)$, and for each $j = 2,\dots m$,

$$\begin{aligned}B_j :&= \boldsymbol{\varphi}_j(A_j) \setminus \left(\bigcup_{1\le i<j} \boldsymbol{\varphi}_i(A_i)\right) \\ &= M \cap U_j \cap \left(\mathbb{R}^n \setminus \bigcup_{1\le i<j} U_i\right).\end{aligned}$$

Some of the sets B_j may be empty, but the nonempty sets of the family $\{B_j : j = 1,\dots,m\}$ constitute a partition of M, and since $B_j \subset \boldsymbol{\varphi}_j(A_j)$, we can put $B_j = \boldsymbol{\varphi}_j(K_j)$, where $K_1 = A_1$, and for every $j \neq 1$ such that $B_j \neq \varnothing$,

$$K_j = \boldsymbol{\varphi}_j^{-1}\left(U_j \cap \left(\mathbb{R}^n \setminus \bigcup_{1\le i<j} U_i\right)\right).$$

Finally, observe that K_j is a measurable subset of A_j, since it is the intersection of an open set and a closed set. □

Definition 4.2.3. Let M be a regular k-surface in $\mathbb{R}^n$ that admits a finite partition into subsets B_j, $1 \leq j \leq m$, such that $B_j = \boldsymbol{\varphi}_j(K_j)$, where $(A_j, \boldsymbol{\varphi}_j)$ is a coordinate system of M and K_j is a measurable subset of A_j. The *k-dimensional area* of M is defined by

$$\text{area}(M) := \sum_{j=1}^{m} \text{area}(K_j, \boldsymbol{\varphi}_j).$$

Here we follow the convention $a + \infty = \infty + \infty = \infty$.

This definition of area (Definition 4.2.3) is independent of both the concrete atlas and the selected partition. Again, we postpone a proof of this fact to Chap. 7.

Let us say some words of justification about the name "k-area" that we have associated to a regular k-surface M in $\mathbb{R}^n$. Another natural name could be k-volume of M in $\mathbb{R}^n$. But what we have in mind is the generalization of the concept of area of a surface in $\mathbb{R}^3$, and we have already used the word k-surface in $\mathbb{R}^n$ to generalize that classical concept of surface. Hence, we feel that the word k-area is the best to generalize the concept of usual area. Moreover, speaking in an informal way, the word "volume" conveys intuitively the idea of the "measure" of sets with nonempty interior in $\mathbb{R}^3$. But in our case, except in the trivial case $k = n$, our k-surfaces have empty interior in $\mathbb{R}^n$. The reader must be also aware that the concept of 1-area of a 1-surface in $\mathbb{R}^n$ is the generalization of the length of a set in $\mathbb{R}$.

Example 4.2.2 (Area of the sphere with radius r in $\mathbb{R}^3$). Let S be the sphere with radius r,

$$U := \{(s,t) \ : 0 < s < \pi \ , \ 0 < t < 2\pi\}$$

and let us consider

$$\boldsymbol{\varphi}(s,t) := (r \sin s \cos t, \ r \sin s \sin t, \ r \cos s)\,.$$

As we proved in Example 3.3.2, $(U, \boldsymbol{\varphi})$ is a coordinate system of S. Also, S is the union of $\boldsymbol{\varphi}(U)$, the two poles of the sphere and a meridian. Since the meridian can be obtained as $\boldsymbol{\Psi}(B)$ where $(A, \boldsymbol{\Psi})$ is a coordinate system of S and B is a null subset of A (see Exercise 4.4.2 for fuller details), the previous definitions give

$$\text{area sphere} = \text{area}(U, \boldsymbol{\varphi}).$$

From

$$\left\| \frac{\partial \boldsymbol{\varphi}}{\partial s} \times \frac{\partial \boldsymbol{\varphi}}{\partial t}(s,t) \right\| = r^2 \sin s,$$

we obtain

$$\text{area sphere} = \int_0^{\pi} \left(\int_0^{2\pi} r^2 \sin s \ \mathrm{d}t \right) \mathrm{d}s = 4\pi r^2.$$

Example 4.2.3. Let $g : A \subset \mathbb{R}^2 \to \mathbb{R}$ be a mapping of class C^1 on the open set A. Then $S := G_g$ is a regular surface in $\mathbb{R}^3$. We plan to express the area of S in terms of the function g. We already know that

$$\varphi : A \subset \mathbb{R}^2 \to \mathbb{R}^3, \quad (s,t) \mapsto (s,t,g(s,t)),$$

is a parameterization of S and that

$$\begin{aligned}\frac{\partial \varphi}{\partial s}(s,t) \times \frac{\partial \varphi}{\partial t}(s,t) &= \begin{pmatrix} e_1 & e_2 & e_3 \\ 1 & 0 & \frac{\partial g}{\partial s} \\ 0 & 1 & \frac{\partial g}{\partial t} \end{pmatrix} \\ &= \left(-\frac{\partial g}{\partial s}(s,t), -\frac{\partial g}{\partial t}(s,t), 1\right).\end{aligned}$$

Hence

$$\left\|\frac{\partial \varphi}{\partial s}(s,t) \times \frac{\partial \varphi}{\partial t}(s,t)\right\| = \sqrt{1+\left(\frac{\partial g}{\partial s}(s,t)\right)^2+\left(\frac{\partial g}{\partial t}(s,t)\right)^2}.$$

We conclude that

$$\operatorname{area}(S) = \iint_A \sqrt{1+\left(\frac{\partial g}{\partial s}(s,t)\right)^2+\left(\frac{\partial g}{\partial t}(s,t)\right)^2}\, \mathrm{d}(s,t).$$

4.3 Flux of a Vector Field

Suppose that a surface S, contained in a plane of $\mathbb{R}^3$, is submerged in a fluid having constant velocity field $\boldsymbol{F}_0$. We fix a unit normal vector $\boldsymbol{N}$ to the plane and are interested in the amount of fluid crossing the surface per unit time in the direction given by the vector $\boldsymbol{N}$.

We begin with an analysis of the simplest case, which occurs when the fluid is moving in the direction given by $\boldsymbol{N}$, that is, $\boldsymbol{F}_0 = \| \boldsymbol{F}_0 \| \cdot \boldsymbol{N}$. We imagine the fluid as a solid block that in time t moves a distance $\| \boldsymbol{F}_0 \| \cdot t$ in the direction given by $\boldsymbol{N}$. Hence, the amount of fluid crossing S in time t coincides with the volume of the *moved block*, precisely $\| \boldsymbol{F}_0 \| \cdot t \cdot \text{area}(S)$. Thus the amount of fluid crossing S per unit time is

$$\| \boldsymbol{F}_0 \| \cdot \text{area}(S).$$

Similarly, if $\boldsymbol{F}_0 = - \| \boldsymbol{F}_0 \| \cdot \boldsymbol{N}$, we get $- \| \boldsymbol{F}_0 \| \cdot \text{area}(S)$, where the minus sign indicates that the fluid is moving in the opposite direction to that given by $\boldsymbol{N}$.

For arbitrary $\boldsymbol{F}_0$ we decompose it as

$$\boldsymbol{F}_0 = \langle \boldsymbol{F}_0 \,,\, \boldsymbol{N} \rangle \cdot \boldsymbol{N} + \boldsymbol{G}_0,$$

where $\boldsymbol{G}_0$ is a vector parallel to the plane containing the surface S. It is rather clear that the movement in a direction parallel to the surface does not contribute to the total amount of fluid crossing the surface, which is (Fig. 4.3)

$$\langle \boldsymbol{F}_0 \,,\, \boldsymbol{N} \rangle \cdot \text{area}(S).$$

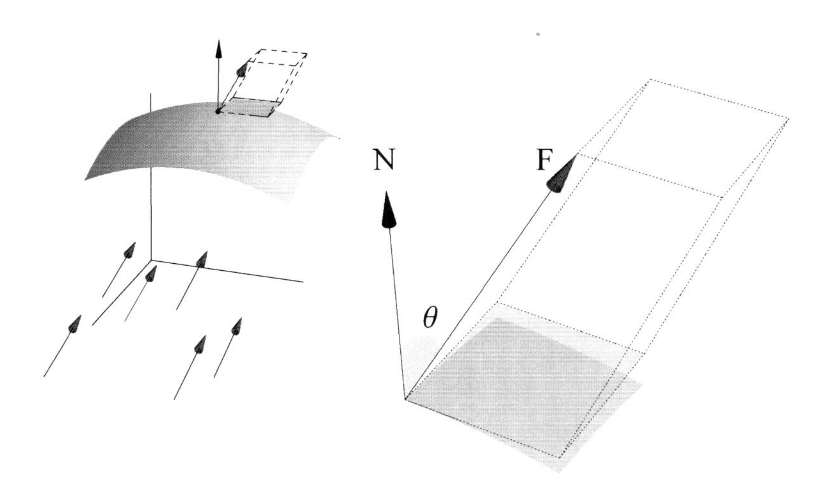

Fig. 4.3 Amount of fluid crossing the surface per unit time

Suppose now that $\boldsymbol{F}$ is an arbitrary continuous vector field representing the velocity field of a fluid, and let us consider a regular surface S submerged in that fluid. To begin with, we restrict our attention to the case that the surface S is covered by a unique coordinate system $(A, \boldsymbol{\varphi})$, that is, $S = \boldsymbol{\varphi}(A)$. We also assume that $K := [a,b] \times [c,d]$ is contained in A. Consider the unit normal vector to S at the point $\boldsymbol{\varphi}(s,t)$:

$$\boldsymbol{N} := \boldsymbol{N}(s,t) := \frac{\frac{\partial \boldsymbol{\varphi}}{\partial s}(s,t) \times \frac{\partial \boldsymbol{\varphi}}{\partial t}(s,t)}{\left\| \frac{\partial \boldsymbol{\varphi}}{\partial s}(s,t) \times \frac{\partial \boldsymbol{\varphi}}{\partial t}(s,t) \right\|}.$$

Let

$$a = s_0 < s_1 < \cdots < s_m = b$$

be a partition of $[a,b]$ into subintervals of length h and let

$$c = t_0 < t_1 < \cdots < t_l = d$$

be a partition of $[c,d]$ into subintervals of length k. Put $I_{i,j} := [s_i, s_{i+1}] \times [t_j, t_{j+1}]$, so that

$$\boldsymbol{\varphi}(K) = \bigcup_{i=0}^{m-1} \bigcup_{j=0}^{l-1} \boldsymbol{\varphi}(I_{i,j}).$$

The portion of the surface $\boldsymbol{\varphi}(I_{i,j})$ can be approximated by the parallelepiped spanned by the vectors

$$\left\{ h \cdot \frac{\partial \boldsymbol{\varphi}}{\partial s}(s_i, t_j) \ , \ k \cdot \frac{\partial \boldsymbol{\varphi}}{\partial t}(s_i, t_j) \right\}$$

(translated so that the point $\boldsymbol{\varphi}(s_i, t_j)$ is one of its vertices). As we proved in Example 4.1.1, the area of that parallelepiped is

$$hk \cdot \left\| \frac{\partial \boldsymbol{\varphi}}{\partial s}(s_i, t_j) \times \frac{\partial \boldsymbol{\varphi}}{\partial t}(s_i, t_j) \right\|.$$

Moreover, if h and k are small enough, we can assume that $\boldsymbol{F}$ is constant on $\boldsymbol{\varphi}(I_{i,j})$ and the amount of fluid crossing $\boldsymbol{\varphi}(I_{i,j})$ in the direction of vector $\boldsymbol{N}(s_i, t_j)$ is approximately

$$\left\langle \boldsymbol{F}(\boldsymbol{\varphi}(s_i, t_j)) \ , \ \boldsymbol{N}(s_i, t_j) \right\rangle \cdot hk \cdot \left\| \frac{\partial \boldsymbol{\varphi}}{\partial s}(s_i, t_j) \times \frac{\partial \boldsymbol{\varphi}}{\partial t}(s_i, t_j) \right\|.$$

From Lemma 4.2.1, the limit as h and k go to zero of

$$\sum_{i=0}^{m-1} \sum_{j=0}^{l-1} \left\langle \boldsymbol{F}(\boldsymbol{\varphi}(s_i, t_j)) \ , \ \boldsymbol{N}(s_i, t_j) \right\rangle \cdot hk \cdot \left\| \frac{\partial \boldsymbol{\varphi}}{\partial s}(s_i, t_j) \times \frac{\partial \boldsymbol{\varphi}}{\partial t}(s_i, t_j) \right\|$$

is given by the integral

$$\iint_K \langle \boldsymbol{F}(\boldsymbol{\varphi}(s,t)) \, , \, \boldsymbol{N} \rangle \cdot \left\| \frac{\partial \boldsymbol{\varphi}}{\partial s}(s,t) \times \frac{\partial \boldsymbol{\varphi}}{\partial t}(s,t) \right\| \, \mathrm{d}(s,t).$$

This discussion justifies the next definition.

Definition 4.3.1. Let $(A, \boldsymbol{\varphi})$ be a coordinate system of a regular surface $S := \boldsymbol{\varphi}(A)$ in $\mathbb{R}^3$, $\boldsymbol{F}$ a continuous vector field defined on S, and

$$\boldsymbol{N}(s,t) := \frac{\frac{\partial \boldsymbol{\varphi}}{\partial s}(s,t) \times \frac{\partial \boldsymbol{\varphi}}{\partial t}(s,t)}{\left\| \frac{\partial \boldsymbol{\varphi}}{\partial s}(s,t) \times \frac{\partial \boldsymbol{\varphi}}{\partial t}(s,t) \right\|}.$$

The flux of the vector field $\boldsymbol{F}$ across the surface S in the direction of the unit normal vector $\boldsymbol{N}$ is defined by

$$\text{Flux} := \iint_A \langle \boldsymbol{F}(\boldsymbol{\varphi}(s,t)) \, , \, \boldsymbol{N}(s,t) \rangle \cdot \left\| \frac{\partial \boldsymbol{\varphi}}{\partial s}(s,t) \times \frac{\partial \boldsymbol{\varphi}}{\partial t}(s,t) \right\| \, \mathrm{d}(s,t),$$

when this integral exists (in the sense of Lebesgue).

We observe that if $\boldsymbol{F} = (f_1, f_2, f_3)$ and $\boldsymbol{\varphi}(s,t) = (x,y,z)$, then the flux can be written thus:

$$\begin{aligned} \text{Flux} &= \iint_A \left\langle \boldsymbol{F}(\boldsymbol{\varphi}(s,t)) \, , \frac{\partial \boldsymbol{\varphi}}{\partial s}(s,t) \times \frac{\partial \boldsymbol{\varphi}}{\partial t}(s,t) \right\rangle \, \mathrm{d}(s,t) \\ &= \iint_A \begin{vmatrix} f_1 & f_2 & f_3 \\ \frac{\partial x}{\partial s} & \frac{\partial y}{\partial s} & \frac{\partial z}{\partial s} \\ \frac{\partial x}{\partial t} & \frac{\partial y}{\partial t} & \frac{\partial z}{\partial t} \end{vmatrix} \, \mathrm{d}(s,t). \end{aligned}$$

When M is a compact set contained in the regular surface S, the set $K := \boldsymbol{\varphi}^{-1}(M)$ is compact (by property (S1) in Definition 3.1.1) and then we define

$$\text{Flux}_M := \iint_{\boldsymbol{\varphi}^{-1}(M)} \left\langle \boldsymbol{F}(\boldsymbol{\varphi}(s,t)) \, , \frac{\partial \boldsymbol{\varphi}}{\partial s}(s,t) \times \frac{\partial \boldsymbol{\varphi}}{\partial t}(s,t) \right\rangle \, \mathrm{d}(s,t),$$

which always exists. This situation is frequently the case in applications.

In Chap. 7 we will show that Definition 4.3.1 is independent of the chosen parameterization but it does depend on the field of unit normal vectors $\boldsymbol{N}$. We will also discuss the case in which the surface cannot be covered by a unique coordinate neighborhood.

4.4 Exercises

Exercise 4.4.1. Find the area of the triangle in $\mathbb{R}^3$ with vertices

$$\boldsymbol{P}_1 = 2\,\boldsymbol{e}_3\ ,\ \boldsymbol{P}_2 = \boldsymbol{e}_1 - \boldsymbol{e}_2 + 2\,\boldsymbol{e}_3\ ,\ \boldsymbol{P}_3 = \boldsymbol{e}_1 + 3\,\boldsymbol{e}_3.$$

For the next exercise, one needs the following fact concerning the Lebesgue integral. *If N is a null subset of $\mathbb{R}^n$, then any function $h : N \to \mathbb{R}$ is Lebesgue integrable and its Lebesgue integral on that set is* 0*, i.e.,*

$$\int_N h = 0.$$

This statement may sound surprising, but it is an immediate consequence of the fact that if f and $g : \mathbb{R}^n \to R$ are two functions that coincide almost everywhere (that is, there exists N_1 null such that $f(x) = g(x)$ for all $x \in \mathbb{R}^n \setminus N_1$), then f is Lebesgue integrable if and only if g is Lebesgue integrable, and in that case,

$$\int_{\mathbb{R}^n} f = \int_{\mathbb{R}^n} g.$$

Indeed, the function $h\chi_N$ vanishes outside the null set N and hence coincides almost everywhere with the constant function 0 (which is Lebesgue integrable). Thus $h\chi_N$ is Lebesgue integrable on $\mathbb{R}^n$ and its integral is equal to 0:

$$\int_N h = \int_{\mathbb{R}^n} h\chi_N = \int_{\mathbb{R}^n} 0 = 0.$$

Exercise 4.4.2. Let M be a regular surface of class C^1 in $\mathbb{R}^3$ and suppose that $(W, \boldsymbol{\varphi})$ is a coordinate system of M,

$$\boldsymbol{\varphi} : W \subset \mathbb{R}^2 \to M \subset \mathbb{R}^3,$$

with the property that $M \setminus \boldsymbol{\varphi}(W)$ can be decomposed as the union of a finite family of subsets of M each of which is the image of a null set through some parameterization of M. Then

$$\text{area}\, M = \text{area}\, (W, \boldsymbol{\varphi}).$$

Exercise 4.4.3. Let $\boldsymbol{\gamma} = (\gamma_1, \gamma_2) : (a,b) \to \mathbb{R}^2$ be a parameterization of a regular curve of class C^1, with $\gamma_2 \geq 0$. Evaluate the area of the surface of revolution formed by rotating $\boldsymbol{\gamma}(]a,b[)$ around the x-axis.

Exercise 4.4.4. Find the area of the part of the plane

$$\frac{x}{a} + \frac{y}{b} + \frac{z}{c} = 1$$

$(a, b, c$ are positive constants) that lies in the first octant.

Exercise 4.4.5. Evaluate the area of the part of the cone $x^2+y^2=z^2$ that lies above the plane $z=0$ and inside the sphere $x^2+y^2+z^2=4ax$ with $a>0$.

Exercise 4.4.6. Find the area of the part of the cylinder $x^2+y^2=1$ bounded by the planes $x+y+z=0$ and $x+y+z=1$.

Exercise 4.4.7. Determine the flux of the vector field

$$\boldsymbol{F}(x,y,z)=(x,y,2z)$$

across the part of the sphere

$$x^2+y^2+z^2=1$$

that lies in the first octant.

Exercise 4.4.8. Let M be the part of the paraboloid

$$x^2+y^2+z=4R^2$$

that lies inside the cylinder

$$(y-R)^2+x^2=R^2.$$

(1) Find a coordinate system of M such that the associated vector field $\boldsymbol{N}$ of unit normal vectors points upward.
(2) Evaluate the flux of the vector field

$$\boldsymbol{F}(x,y,z)=(x,0,0)$$

across M in the direction of $\boldsymbol{N}$.

Chapter 5
Orientation of a Surface

We know from Chap. 4 that in order to evaluate the flux of a vector field across a regular surface S, we need to choose a unit normal vector at each point of S in such a way that the resulting vector field is continuous. For instance, if we submerge a permeable sphere into a fluid and we select the field of unit normal *outward* vectors on the sphere, then the flux of the velocity field of the fluid across the sphere gives the amount of fluid leaving the sphere per unit time. However, if we select the field of unit normal *inward* vectors on the sphere, then the flux of the velocity field of the fluid across the sphere gives the amount of fluid entering the sphere per unit time (which is the negative of the flux obtained in the first case). So, it is a natural question to ask which (if not all) regular surfaces admit a continuous field of unit normal vectors. The regular surfaces admitting such a continuous vector field are called orientable surfaces. Most common surfaces, such as spheres, paraboloids, and planes, are orientable. However, there do exist surfaces that are not orientable.

A. Galbis and M. Maestre, *Vector Analysis Versus Vector Calculus*, Universitext,
DOI 10.1007/978-1-4614-2200-6_5,

5.1 Orientation of Vector Spaces

Before we present a formal definition of orientation for a real vector space, we discuss the particular cases of a line and a plane in $\mathbb{R}^3$.

To orient a one-dimensional vector space means to choose a direction in that space, and it is obvious that there are two possibilities.

If our objective is to obtain a version of Green's formula from which we can calculate the work done by a force field on a particle as it moves along a circle contained in a plane of $\mathbb{R}^3$, then a quick look at Theorem 2.6.2 will convince us of the need to interpret what is meant by moving along the circle *counterclockwise*. This is not so obvious; indeed, it rather depends on our point of view. After all, a sheet of paper has a (very small) thickness, and all of us would agree whether a circle is traversed clockwise or counterclockwise (this is because the circle appears on one side of the paper only), however the situation is completely different if we replace a sheet of paper by an abstract plane in $\mathbb{R}^3$, which has no thickness and so the same circle can be viewed from both sides of the plane, leading to different interpretations of whether a motion is clockwise or counterclockwise. To fix an orientation of the plane means to choose one interpretation over the other. In a sense, this is rather like placing a mark to indicate which is the appropriate side of the plane. Mathematically, this is done by fixing a basis of the plane (a two-dimensional vector space) and identifying this basis, via some linear isomorphism, with the canonical basis of $\mathbb{R}^2$, or equivalently, ordering the basis.

Definition 5.1.1. Let V be a n-dimensional real vector space. The ordered bases $\{\boldsymbol{v}_j\}_{j=1}^n$ and $\{\boldsymbol{w}_j\}_{j=1}^n$ of V are said to *have the same orientation,* denoted by $\{\boldsymbol{v}_j\}_{j=1}^n \sim \{\boldsymbol{w}_j\}_{j=1}^n$, if the unique linear transformation $\boldsymbol{f} : V \to V$ that takes $\boldsymbol{v}_i$ to $\boldsymbol{w}_i$ has a positive determinant.

That is, the change-of-basis matrix has a positive determinant.

The property of having the same orientation is an equivalence relation, and there are two equivalence classes. These classes are the two possible orientations of V. If $\mathscr{O}$ is one of the orientations, we will write $-\mathscr{O}$ for the other.

Definition 5.1.2. An *oriented vector space* is a pair $(V, \mathscr{O})$, where V is a real vector space and $\mathscr{O}$ is one of the two possible orientations. A basis $\mathscr{B} = \{\boldsymbol{v}_j\}_{j=1}^n$ of V is said to be positively oriented if $\mathscr{B} \in \mathscr{O}$. In that case, we usually write $\mathscr{O} = [\boldsymbol{v}_1, \dots, \boldsymbol{v}_n]$.

It is natural to call the chosen orientation $\mathscr{O}$ the *positive orientation* and the other orientation $-\mathscr{O}$ the *negative orientation.* Let us emphasize that the notion of positively oriented basis depends on the choice of positive orientation $\mathscr{O}$ and is not an intrinsic property of the basis.

Example 5.1.1. The bases $\{\boldsymbol{v}_1, \boldsymbol{v}_2, \boldsymbol{v}_3\}$ and $\{\boldsymbol{v}_3, \boldsymbol{v}_1, \boldsymbol{v}_2\}$ of $\mathbb{R}^3$ define the same orientation. However, $\{\boldsymbol{v}_1, \boldsymbol{v}_2, \boldsymbol{v}_3\}$ and $\{\boldsymbol{v}_2, \boldsymbol{v}_1, \boldsymbol{v}_3\}$ define different orientations.

In the first case, the change-of-basis matrix is

$$\begin{pmatrix} 0 & 1 & 0 \\ 0 & 0 & 1 \\ 1 & 0 & 0 \end{pmatrix},$$

which has determinant equal to 1. In the second case, the change-of-basis matrix is

$$\begin{pmatrix} 0 & 1 & 0 \\ 1 & 0 & 0 \\ 0 & 0 & 1 \end{pmatrix},$$

which has determinant -1.

Unless we indicate otherwise, $\mathbb{R}^n$ *is always taken to have the orientation defined by the canonical basis,* and we will call this orientation the *positive orientation* of $\mathbb{R}^n$. It follows from Definition 5.1.2 that a basis $\{v_1, v_2, \ldots, v_n\}$ of $\mathbb{R}^n$ is positively oriented if and only if the matrix with vector columns $[v_1, v_2, \ldots, v_n]$ has positive determinant.

Proposition 5.1.1. *Let*

$$\varphi : [0,1] \to \mathbb{R}^n \times \mathbb{R}^n \times \cdots \times \mathbb{R}^n, \quad \varphi = (\varphi_1, ..., \varphi_n)$$

be a continuous mapping such that $\{\varphi_j(t)\}_{j=1}^n$ *is a basis of* $\mathbb{R}^n$, $\varphi(0) = \{e_j\}_{j=1}^n$ *(canonical basis), and* $\varphi(1) = \{v_j\}_{j=1}^n$. *Then the basis* $\{v_j\}_{j=1}^n$ *is positively oriented.*

Proof. Let $f(t)$ denote the determinant of the matrix with vector columns $[\varphi_1(t), \varphi_2(t), \ldots, \varphi_n(t)]$. Then f is a continuous function on the closed interval $[0,1]$ that does not vanish; hence $f(0)$ and $f(1)$ have the same sign. Since $f(0) = 1$, we have

$$\det [v_1, v_2, \ldots, v_n] = f(1) > 0,$$

which means that $\{v_j\}_{j=1}^n$ has the same orientation as the canonical basis of $\mathbb{R}^n$ (Definition 5.1.1). □

One interpretation of Proposition 5.1.1 is that taking the parameter t to represent time, if we begin with the canonical basis and as time goes on we continuously move the basis vectors in such a way that at all times we still have a basis, then all of these bases are positively oriented. Let us consider the canonical basis of $\mathbb{R}^3$ and imagine that the index finger of your right hand points in the positive direction of the x-axis while the middle finger points toward the positive direction of the y-axis. Then the thumb points toward the positive direction of the z-axis. Since a basis (v_1, v_2, v_3) of $\mathbb{R}^3$ is positively oriented when $(v_1, v_2, v_3) \in [e_1, e_2, e_3]$, Proposition 5.1.1 provides the following practical rule to determine the orientation of an orthogonal basis of $\mathbb{R}^3$:

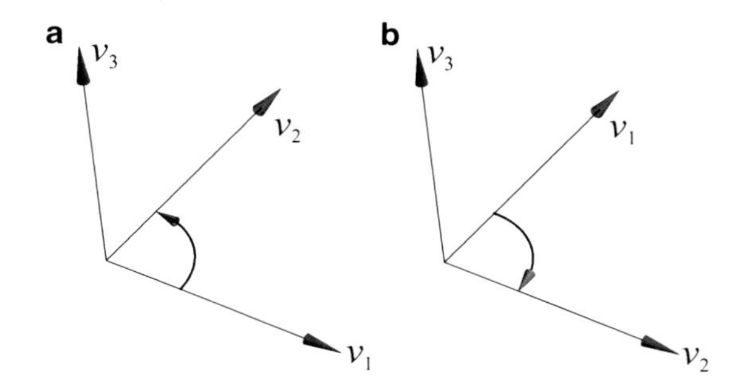

Fig. 5.1 (**a**) Basis positively oriented; (**b**) basis negatively oriented

In $\mathbb{R}^3$, an orthogonal basis $(\boldsymbol{v}_1, \boldsymbol{v}_2, \boldsymbol{v}_3)$ is positively oriented if the three vectors $\{\boldsymbol{v}_1, \boldsymbol{v}_2, \boldsymbol{v}_3\}$, in this order, correspond with the directions of the index finger, middle finger, and thumb of the right hand.

This rule can also be expressed as follows:

If we stand at the origin with our head pointing in the direction of $\boldsymbol{v}_3$ and face in the direction of $\boldsymbol{v}_1$, then the vector $\boldsymbol{v}_2$ lies to our left (Fig. 5.1).

Next, we consider the orientation of a given two-dimensional vector subspace V in $\mathbb{R}^3$. Since V is generated by two linearly independent vectors, the following result will be useful.

Proposition 5.1.2. *Let $\{\boldsymbol{v}_1, \boldsymbol{v}_2\}$ be two linearly independent vectors in $\mathbb{R}^3$. Then the ordered basis $\{\boldsymbol{v}_1 \times \boldsymbol{v}_2, \boldsymbol{v}_1, \boldsymbol{v}_2\}$ of $\mathbb{R}^3$ is positively oriented.*

Proof. With respect to the canonical basis, the matrix of the linear transformation $f : \mathbb{R}^3 \to \mathbb{R}^3$ defined by $\boldsymbol{f}(\boldsymbol{e}_1) = \boldsymbol{v}_1 \times \boldsymbol{v}_2$, $\boldsymbol{f}(\boldsymbol{e}_2) = \boldsymbol{v}_1$, and $\boldsymbol{f}(\boldsymbol{e}_3) = \boldsymbol{v}_2$ has the vector columns $[\boldsymbol{v}_1 \times \boldsymbol{v}_2, \boldsymbol{v}_1, \boldsymbol{v}_2]$. Since the determinant of a matrix does not change under matrix transposition, we get

$$\begin{aligned} \det \boldsymbol{f} &= \det\,[\boldsymbol{v}_1 \times \boldsymbol{v}_2, \boldsymbol{v}_1, \boldsymbol{v}_2] \\ &= \det \begin{bmatrix} \boldsymbol{v}_1 \times \boldsymbol{v}_2 \\ \boldsymbol{v}_1 \\ \boldsymbol{v}_2 \end{bmatrix}. \end{aligned}$$

Hence $\det \boldsymbol{f}$ coincides with the triple scalar product (Definition 1.1.5) of vectors $\{\boldsymbol{v}_1 \times \boldsymbol{v}_2, \boldsymbol{v}_1, \boldsymbol{v}_2\}$. Finally,

$$\begin{aligned} \det \boldsymbol{f} &= \langle \boldsymbol{v}_1 \times \boldsymbol{v}_2, \boldsymbol{v}_1 \times \boldsymbol{v}_2 \rangle \\ &= \| \boldsymbol{v}_1 \times \boldsymbol{v}_2 \|^2 > 0. \end{aligned}$$

□

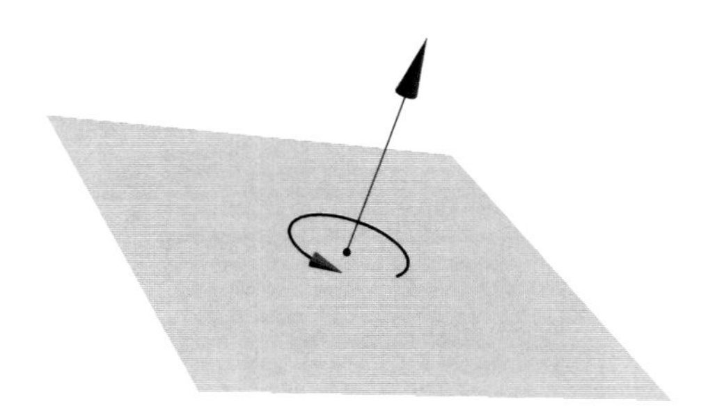

Fig. 5.2 Orientation of a two-dimensional vector subspace

Proposition 5.1.3. *Let V be a two-dimensional vector subspace of $\mathbb{R}^3$ and let $\mathscr{B}_1 := \{\boldsymbol{v}_1, \boldsymbol{v}_2\}$ and $\mathscr{B}_2 := \{\boldsymbol{w}_1, \boldsymbol{w}_2\}$ be two ordered bases of V. The following are equivalent:*

(1) The two bases $\mathscr{B}_1$ and $\mathscr{B}_2$ define the same orientation on V.
(2) There is $\lambda > 0$ such that $\boldsymbol{v}_1 \times \boldsymbol{v}_2 = \lambda(\boldsymbol{w}_1 \times \boldsymbol{w}_2)$.

Proof. Both $\boldsymbol{v}_1 \times \boldsymbol{v}_2$ and $\boldsymbol{w}_1 \times \boldsymbol{w}_2$ are orthogonal vectors to V, and so (since V has a one-dimensional orthogonal complement) there exists $\lambda \neq 0$ such that $\boldsymbol{v}_1 \times \boldsymbol{v}_2 = \lambda(\boldsymbol{w}_1 \times \boldsymbol{w}_2)$. We denote by

$$B = \begin{bmatrix} b_{11} & b_{12} \\ b_{21} & b_{22} \end{bmatrix}$$

the change-of-basis matrix from $\{\boldsymbol{v}_1, \boldsymbol{v}_2\}$ to $\{\boldsymbol{w}_1, \boldsymbol{w}_2\}$. If A is the change-of-basis matrix from $\{\boldsymbol{v}_1 \times \boldsymbol{v}_2, \boldsymbol{v}_1, \boldsymbol{v}_2\}$ to $\{\boldsymbol{w}_1 \times \boldsymbol{w}_2, \boldsymbol{w}_1, \boldsymbol{w}_2\}$, then

$$A = \begin{bmatrix} \lambda & 0 & 0 \\ 0 & b_{11} & b_{12} \\ 0 & b_{21} & b_{22} \end{bmatrix},$$

and therefore

$$\det A = \lambda \det B.$$

Since the two bases $\{\boldsymbol{v}_1 \times \boldsymbol{v}_2, \boldsymbol{v}_1, \boldsymbol{v}_2\}$ and $\{\boldsymbol{w}_1 \times \boldsymbol{w}_2, \boldsymbol{w}_1, \boldsymbol{w}_2\}$ define the same (positive) orientation in $\mathbb{R}^3$, we have that $\det A > 0$ and $\det B$ and λ have the same sign. We deduce from Definition 5.1.1 that the bases $\mathscr{B}_1$ and $\mathscr{B}_2$ define the same orientation on V if and only if $\lambda > 0$. □

We note that an ordered basis $\mathscr{B} := \{\boldsymbol{v}_1, \boldsymbol{v}_2\}$ of a two-dimensional vector subspace of $\mathbb{R}^3$ defines a unit normal vector to that subspace:

$$\boldsymbol{N} := \frac{\boldsymbol{v}_1 \times \boldsymbol{v}_2}{\| \boldsymbol{v}_1 \times \boldsymbol{v}_2 \|}.$$

According to Proposition 5.1.3, two ordered bases of V define the same orientation if and only if they define the same unit normal vector $\boldsymbol{N}$. In other words, to specify the orientation of a two-dimensional vector subspace V of $\mathbb{R}^3$, it suffices to select a normal vector to that subspace (Fig. 5.2).

5.2 Orientation of Surfaces

If we choose an orientation of the tangent space T_pM to a regular surface $M \subset \mathbb{R}^3$ (i.e., a regular 2-surface) at a point $\boldsymbol{p} \in M$, we are effectively orientating the surface in a neighborhood $U_{\boldsymbol{p}}$ of $\boldsymbol{p}$. If it is possible to choose an orientation of each tangent space so that on the intersection of two arbitrary neighborhoods $U_{\boldsymbol{p}}$ and $U_{\boldsymbol{q}}$ the orientations *induced* by T_pM and T_qM coincide, then the surface M is said to be orientable. In this section we formalize this intuitive idea and we show that the orientable regular surfaces are those admitting a continuous vector field consisting of unit and normal vectors.

If $\boldsymbol{x}_0 \in M$ and $(A, \boldsymbol{\varphi})$ is a coordinate system of a regular k-surface M in $\mathbb{R}^n$ such that

$$\boldsymbol{x}_0 = \boldsymbol{\varphi}(\boldsymbol{t}_0),$$

then on $T_{\boldsymbol{x}_0}M$ we consider the orientation defined by the ordered basis

$$\left\{ \frac{\partial \boldsymbol{\varphi}}{\partial t_1}(\boldsymbol{t}_0), \ldots, \frac{\partial \boldsymbol{\varphi}}{\partial t_k}(\boldsymbol{t}_0) \right\}.$$

We need to find conditions that ensure that the orientation of $T_{\boldsymbol{x}_0}M$ is independent of the coordinate system chosen.

Lemma 5.2.1. *Let V and W be n-dimensional real vector spaces,*

$$\boldsymbol{L} : V \to W$$

a linear isomorphism, and $\{\boldsymbol{u}_j\}_{j=1}^n$, $\{\boldsymbol{v}_j\}_{j=1}^n$ two bases of V. Then

$$\{\boldsymbol{u}_j\}_{j=1}^n \sim \{\boldsymbol{v}_j\}_{j=1}^n \quad \text{if and only if} \quad \{\boldsymbol{L}\boldsymbol{u}_j\}_{j=1}^n \sim \{\boldsymbol{L}\boldsymbol{v}_j\}_{j=1}^n.$$

Proof. Denote by $\boldsymbol{f} : V \to V$ and $\boldsymbol{g} : W \to W$ the linear isomorphisms defined by $\boldsymbol{f}(\boldsymbol{u}_i) = \boldsymbol{v}_i$, $\boldsymbol{g}(\boldsymbol{L}\boldsymbol{u}_i) = \boldsymbol{L}\boldsymbol{v}_i$. Then $(\boldsymbol{L}^{-1} \circ \boldsymbol{g} \circ \boldsymbol{L})(\boldsymbol{u}_i) = \boldsymbol{f}(\boldsymbol{u}_i)$ for every $1 \le i \le n$, and hence $\boldsymbol{L}^{-1} \circ \boldsymbol{g} \circ \boldsymbol{L} = \boldsymbol{f}$, from which it follows that

$$\begin{aligned} \det \boldsymbol{f} = \det\,(\boldsymbol{L}^{-1} \circ \boldsymbol{g} \circ \boldsymbol{L}) &= \det(\boldsymbol{L}^{-1}) \det(\boldsymbol{g}) \det(\boldsymbol{L}) \\ &= (\det(\boldsymbol{L}))^{-1} \det(\boldsymbol{g}) \det(\boldsymbol{L}) \\ &= \det \boldsymbol{g}. \end{aligned}$$

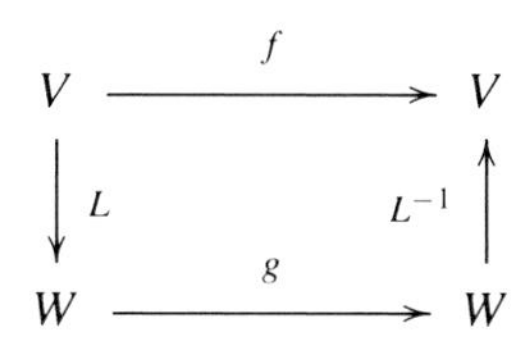

□

Proposition 5.2.1. *Let M be a regular k-surface of class C^p in $\mathbb{R}^n$ and let $(A,\boldsymbol{\varphi})$ and $(B,\boldsymbol{\psi})$ be two coordinate systems such that $D := \boldsymbol{\varphi}(A) \cap \boldsymbol{\psi}(B) \neq \varnothing$. Let $\boldsymbol{x}_0 \in D$, $\boldsymbol{t}_0 \in A$, and $\boldsymbol{u}_0 \in B$ be given with $\boldsymbol{x}_0 = \boldsymbol{\varphi}(\boldsymbol{t}_0) = \boldsymbol{\psi}(\boldsymbol{u}_0)$. Then the following are equivalent:*

(1) The bases

$$\mathscr{B}_1 := \left\{ \frac{\partial \boldsymbol{\varphi}}{\partial t_1}(\boldsymbol{t}_0), \ldots, \frac{\partial \boldsymbol{\varphi}}{\partial t_k}(\boldsymbol{t}_0) \right\} \text{ and } \mathscr{B}_2 := \left\{ \frac{\partial \boldsymbol{\psi}}{\partial u_1}(\boldsymbol{u}_0), \ldots, \frac{\partial \boldsymbol{\psi}}{\partial u_k}(\boldsymbol{u}_0) \right\}$$

of $T_{\boldsymbol{x}_0}M$ have the same orientation.
(2) $J(\boldsymbol{\varphi}^{-1} \circ \boldsymbol{\psi})(\boldsymbol{u}_0) > 0$.

Proof. Put $\boldsymbol{L} := \mathrm{d}\boldsymbol{\varphi}(\boldsymbol{t}_0)$, $\boldsymbol{T} := \mathrm{d}\boldsymbol{\psi}(\boldsymbol{u}_0)$, and $\boldsymbol{R} := \mathrm{d}(\boldsymbol{\varphi}^{-1} \circ \boldsymbol{\psi})(\boldsymbol{u}_0)$. Since

$$\boldsymbol{\psi}(\boldsymbol{u}) = \left(\boldsymbol{\varphi} \circ \boldsymbol{\varphi}^{-1} \circ \boldsymbol{\psi}\right)(\boldsymbol{u})$$

in an open neighborhood of $\boldsymbol{u}_0$ and $\left(\boldsymbol{\varphi}^{-1} \circ \boldsymbol{\psi}\right)(\boldsymbol{u}_0) = \boldsymbol{t}_0$, we can apply the chain rule to get $\boldsymbol{T} = \boldsymbol{L} \circ \boldsymbol{R}$. Also

$$\mathscr{B}_1 = \{\boldsymbol{L}\boldsymbol{e}_1, \ldots, \boldsymbol{L}\boldsymbol{e}_k\},$$

while

$$\mathscr{B}_2 = \{\boldsymbol{T}\boldsymbol{e}_1, \ldots, \boldsymbol{T}\boldsymbol{e}_k\} = \{\boldsymbol{L}(\boldsymbol{R}\boldsymbol{e}_1), \ldots, \boldsymbol{L}(\boldsymbol{R}\boldsymbol{e}_k)\}.$$

By Lemma 5.2.1, $\mathscr{B}_1 \sim \mathscr{B}_2$ if and only if $\{\boldsymbol{e}_1, \ldots, \boldsymbol{e}_k\}$ and $\{\boldsymbol{R}\boldsymbol{e}_1, \ldots, \boldsymbol{R}\boldsymbol{e}_k\}$ are two bases of $\mathbb{R}^k$ with the same orientation. This in turn is equivalent to

$$J(\boldsymbol{\varphi}^{-1} \circ \boldsymbol{\psi})(\boldsymbol{u}_0) = \det \boldsymbol{R} > 0.$$

□

Definition 5.2.1. Let M be a regular k-surface of class C^p in $\mathbb{R}^n$. Then M is said to be *orientable* if there exists an atlas $\{(A_i, \boldsymbol{\varphi}_i) : i \in I\}$ such that for each $\boldsymbol{x} \in M$ there is an orientation $\theta_{\boldsymbol{x}}$ of $T_{\boldsymbol{x}}M$ with the property that whenever $(A,\boldsymbol{\varphi})$ is a coordinate system in the given atlas and $\boldsymbol{x} = \boldsymbol{\varphi}(\boldsymbol{t})$, one has $\theta_{\boldsymbol{x}} = \left[\frac{\partial \boldsymbol{\varphi}}{\partial t_1}(\boldsymbol{t}), \ldots, \frac{\partial \boldsymbol{\varphi}}{\partial t_k}(\boldsymbol{t})\right]$ (Fig. 5.3).

Definition 5.2.2. Under the conditions of the previous definition, we say that the atlas defines an orientation on M. Moreover, an arbitrary coordinate system $(B,\boldsymbol{\psi})$ of M is said to be compatible with that orientation if for every $\boldsymbol{t} \in B$, the basis

$$\left\{ \frac{\partial \boldsymbol{\psi}}{\partial t_1}(\boldsymbol{t}), \ldots, \frac{\partial \boldsymbol{\psi}}{\partial t_k}(\boldsymbol{t}) \right\}$$

of the tangent space $T_{\boldsymbol{\psi}(\boldsymbol{t})}M$ belongs to the orientation $\theta_{\boldsymbol{\psi}(\boldsymbol{t})}$ (Fig. 5.4).

From Proposition 5.2.1 we get the following characterization of orientable surfaces.

Fig. 5.3 We choose an orientation at each tangent space

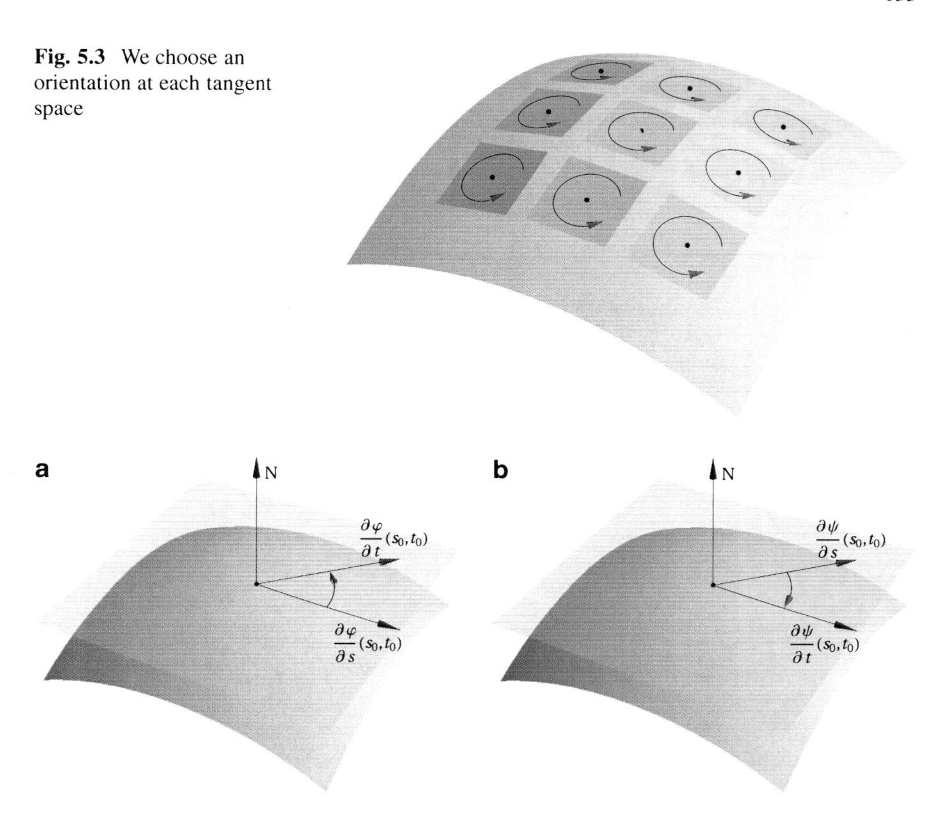

Fig. 5.4 (**a**) Compatible coordinate system; (**b**) incompatible coordinate system

Theorem 5.2.1. *Let M be a regular k-surface of class C^p in $\mathbb{R}^n$. Then M is orientable if and only if there exists an atlas $\{(A_i, \varphi_i) : i \in I\}$ such that for each pair of coordinate systems (A_1, φ_1) and (A_2, φ_2) with $\varphi_1(A_1) \cap \varphi_2(A_2) \neq \varnothing$ we have $J(\varphi_1^{-1} \circ \varphi_2) > 0$ on its domain of definition.*

Remark 5.2.1. In some texts (for instance Edwards [9]) the orientable surfaces *are defined* as those regular k-surfaces which have an atlas $\{(A_i, \varphi_i) : i \in I\}$ such that for each pair of coordinate systems (A_1, φ_1) and (A_2, φ_2) with $\varphi_1(A_1) \cap \varphi_2(A_2) \neq \varnothing$ we have $J(\varphi_1^{-1} \circ \varphi_2) > 0$ on its domain of definition.

The orientable regular 2-surfaces in $\mathbb{R}^3$ are precisely those that admit a continuous vector field of unit normal vectors. To prove this, let us first show that there is a simple way to reverse the orientation defined by a coordinate system.[1]

---

[1] *Reverse orientation* and *opposite orientation* are terms used synonymously to indicate the orientation with negated sign.

Lemma 5.2.2. *Let $(A,\boldsymbol{\varphi})$ be a coordinate system of a regular k-surface M in $\mathbb{R}^n$ and define $\boldsymbol{L}:\mathbb{R}^k\to\mathbb{R}^k$ by $\boldsymbol{L}(\boldsymbol{t})=(t_1,\ldots,t_{k-2},t_k,t_{k-1})$. If $B=\boldsymbol{L}^{-1}(A)$ and $\boldsymbol{\psi}=\boldsymbol{\varphi}\circ\boldsymbol{L}$, then $(B,\boldsymbol{\psi})$ is a coordinate system of M, $\boldsymbol{\varphi}(A)=\boldsymbol{\psi}(B)$, and $(B,\boldsymbol{\psi})$ reverses the orientation induced by $(A,\boldsymbol{\varphi})$.*

Proof. It is clear that $(B,\boldsymbol{\psi})$ is a coordinate system of M. Set $S:=\boldsymbol{\varphi}(A)=\boldsymbol{\psi}(B)$. For every

$$\boldsymbol{x}=\boldsymbol{\varphi}(\boldsymbol{t})\in S,$$

the orientation of $T_{\boldsymbol{x}}M$ given by the coordinate system $(A,\boldsymbol{\varphi})$ is

$$\theta_{\boldsymbol{x}}=\left[\frac{\partial\boldsymbol{\varphi}}{\partial t_1}(\boldsymbol{t}),\ldots\frac{\partial\boldsymbol{\varphi}}{\partial t_k}(\boldsymbol{t})\right].$$

We put $\boldsymbol{s}=\boldsymbol{L}^{-1}(\boldsymbol{t})$. Then the orientation defined by $(B,\boldsymbol{\psi})$ on the tangent space at the same point $\boldsymbol{x}=\boldsymbol{\psi}(\boldsymbol{s})$ is

$$\begin{aligned}\nu_{\boldsymbol{x}}&=\left[\frac{\partial\boldsymbol{\psi}}{\partial s_1}(\boldsymbol{s}),\ldots,\frac{\partial\boldsymbol{\psi}}{\partial s_{k-2}}(\boldsymbol{s}),\frac{\partial\boldsymbol{\psi}}{\partial s_{k-1}}(\boldsymbol{s}),\frac{\partial\boldsymbol{\psi}}{\partial s_k}(\boldsymbol{s})\right]\\&=\left[\frac{\partial\boldsymbol{\varphi}}{\partial t_1}(\boldsymbol{t}),\ldots,\frac{\partial\boldsymbol{\varphi}}{\partial t_{k-2}}(\boldsymbol{t}),\frac{\partial\boldsymbol{\varphi}}{\partial t_k}(\boldsymbol{t}),\frac{\partial\boldsymbol{\varphi}}{\partial t_{k-1}}(\boldsymbol{t})\right]\\&=-\theta_{\boldsymbol{x}}.\end{aligned}$$

That is, $\theta_{\boldsymbol{x}}$ and $\nu_{\boldsymbol{x}}$ are different orientations of the tangent space. □

The following lemma is not used in this chapter but will be required in Chap. 8.

Lemma 5.2.3. *Let M be an orientable regular k-surface of class C^p and let $(A,\boldsymbol{\varphi})$ be a coordinate system with $A\subset\mathbb{R}^k$ connected and open. Consider $\boldsymbol{L}:\mathbb{R}^k\to\mathbb{R}^k$ defined by $\boldsymbol{L}(\boldsymbol{t})=(t_1,\ldots,t_{k-2},t_k,t_{k-1})$. Then either the coordinate system $(A,\boldsymbol{\varphi})$ or the coordinate system $(\boldsymbol{L}^{-1}(A),\boldsymbol{\varphi}\circ\boldsymbol{L})$ is compatible with the orientation.*

Proof. Let $\{\theta_{\boldsymbol{x}}:\boldsymbol{x}\in M\}$ be the orientation of the regular k-surface. Fix $\boldsymbol{x}_0\in\boldsymbol{\varphi}(A)$ and choose $\boldsymbol{t}_0\in A$ with $\boldsymbol{\varphi}(\boldsymbol{t}_0)=\boldsymbol{x}_0$. Either

$$(i)\quad\left[\frac{\partial\boldsymbol{\varphi}}{\partial t_1}(\boldsymbol{t}_0),\ldots,\frac{\partial\boldsymbol{\varphi}}{\partial t_k}(\boldsymbol{t}_0)\right]=\theta_{\boldsymbol{x}_0}$$

or

$$(ii)\quad\left[\frac{\partial\boldsymbol{\varphi}}{\partial t_1}(\boldsymbol{t}_0),\ldots,\frac{\partial\boldsymbol{\varphi}}{\partial t_k}(\boldsymbol{t}_0)\right]=-\theta_{\boldsymbol{x}_0}.$$

In the second case we would have (by Lemma 5.2.2)

$$\left[\frac{\partial(\boldsymbol{\varphi}\circ\boldsymbol{L})}{\partial t_1}(\boldsymbol{L}^{-1}(\boldsymbol{t}_0)),\ldots,\frac{\partial(\boldsymbol{\varphi}\circ\boldsymbol{L})}{\partial t_k}(\boldsymbol{L}^{-1}(\boldsymbol{t}_0))\right]=\theta_{\boldsymbol{x}_0},$$

and so by switching coordinate system (if necessary) we can assume that we are dealing with case (i). We need to show therefore that

$$\left[\frac{\partial \boldsymbol{\varphi}}{\partial t_1}(\boldsymbol{t}),\ldots,\frac{\partial \boldsymbol{\varphi}}{\partial t_k}(\boldsymbol{t})\right] = \theta_{\boldsymbol{\varphi}(\boldsymbol{t})}$$

for all $\boldsymbol{t} \in A$. Given $t \in A$, we know that there is a polygonal arc

$$[\boldsymbol{u}_0,\boldsymbol{u}_1] \cup [\boldsymbol{u}_1,\boldsymbol{u}_2] \cup \cdots \cup [\boldsymbol{u}_{s-1},\boldsymbol{u}_s] \subset A$$

with $\boldsymbol{u}_0 = \boldsymbol{t}_0$ and $\boldsymbol{u}_s = t$. By induction, it suffices to prove

$$\left[\frac{\partial \boldsymbol{\varphi}}{\partial t_1}(\boldsymbol{u}_1),\ldots,\frac{\partial \boldsymbol{\varphi}}{\partial t_k}(\boldsymbol{u}_1)\right] = \theta_{\boldsymbol{\varphi}(\boldsymbol{u}_1)}.$$

For this, let

$$B := \left\{\boldsymbol{u} \in [\boldsymbol{u}_0,\boldsymbol{u}_1] \,:\, \left[\frac{\partial \boldsymbol{\varphi}}{\partial t_1}(\boldsymbol{u}),\ldots,\frac{\partial \boldsymbol{\varphi}}{\partial t_k}(\boldsymbol{u})\right] = \theta_{\boldsymbol{\varphi}(\boldsymbol{u})}\right\}.$$

From $\boldsymbol{u}_0 \in B$ we have $B \neq \varnothing$. The set B is open relative to $[\boldsymbol{u}_0,\boldsymbol{u}_1]$. Indeed, given $\boldsymbol{x} = \boldsymbol{\varphi}(\boldsymbol{u})$, $\boldsymbol{u} \in B$, there is a coordinate system $(G,\boldsymbol{\psi})$, compatible with the orientation, with $\boldsymbol{\varphi}(\boldsymbol{u}) \in \boldsymbol{\psi}(G)$. By Proposition 5.2.1 we have

$$J(\boldsymbol{\psi}^{-1} \circ \boldsymbol{\varphi})(\boldsymbol{u}) > 0.$$

Select $r > 0$ such that the open ball in $\mathbb{R}^k$ centered at $\boldsymbol{u}$ and with radius r satisfies $B(\boldsymbol{u},r) \subset A$ and $\boldsymbol{\varphi}(B(\boldsymbol{u},r)) \subset \boldsymbol{\psi}(G)$. Then $J(\boldsymbol{\psi}^{-1} \circ \boldsymbol{\varphi})(\boldsymbol{t}) \neq 0$ for all $\boldsymbol{t} \in B(\boldsymbol{u},r)$ and $J(\boldsymbol{\psi}^{-1} \circ \boldsymbol{\varphi})(\boldsymbol{u}) > 0$; hence, due to the continuity, $J(\boldsymbol{\psi}^{-1} \circ \boldsymbol{\varphi})(\boldsymbol{t}) > 0$ for every $\boldsymbol{t} \in B(\boldsymbol{u},r)$ and we obtain, by Proposition 5.2.1, that

$$B(\boldsymbol{u},r) \cap [\boldsymbol{u}_0,\boldsymbol{u}_1] \subset B.$$

The same argument shows that the complement of B in $[\boldsymbol{u}_0,\boldsymbol{u}_1]$, namely

$$\left\{\boldsymbol{u} \in [\boldsymbol{u}_0,\boldsymbol{u}_1] \,:\, \left[\frac{\partial \boldsymbol{\varphi}}{\partial t_1}(\boldsymbol{u}),\ldots,\frac{\partial \boldsymbol{\varphi}}{\partial t_k}(\boldsymbol{u})\right] = -\theta_{\boldsymbol{\varphi}(\boldsymbol{u})}\right\},$$

is also open relative to $[u_0,u_1]$. Hence B is closed.

We have proved that B is a nonempty open and closed subset of the connected set $[\boldsymbol{u}_0,\boldsymbol{u}_1]$, from which we conclude that $B = [\boldsymbol{u}_0,\boldsymbol{u}_1]$, and thus the orientation defined by the coordinate system $(A,\boldsymbol{\varphi})$ in $\boldsymbol{\varphi}(\boldsymbol{u}_1)$ is compatible with the orientation of M. □

Lemma 5.2.4. *Let $\boldsymbol{f},\boldsymbol{g} : A \subset \mathbb{R}^p \to \mathbb{R}^n \setminus \{0\}$ be continuous on the open set A and let $h : A \subset \mathbb{R}^p \to \mathbb{R}$ be such that $\boldsymbol{f}(\boldsymbol{x}) = h(\boldsymbol{x})\boldsymbol{g}(\boldsymbol{x})$ for every x in A. Then h is also continuous on A.*

Proof. For any $\boldsymbol{x}_0 \in A$ there exists $j \in \{1,\dots,n\}$ such that $g_j(\boldsymbol{x}_0) \neq 0$, and by continuity, there then exists $r > 0$ with $g_j(\boldsymbol{x}) \neq 0$ for all $\boldsymbol{x} \in B(\boldsymbol{x}_0, r)$. It follows that

$$h(\boldsymbol{x}) = \frac{f_j(\boldsymbol{x})}{g_j(\boldsymbol{x})} \text{ for all } \boldsymbol{x} \in B(\boldsymbol{x}_0, r),$$

which shows that h is continuous on the ball $B(\boldsymbol{x}_0, r)$, since it is the quotient of two continuous functions. □

Theorem 5.2.2. *Let M be a regular 2-surface of class C^1 in $\mathbb{R}^3$. Then M is orientable if and only if there is a continuous function $\boldsymbol{F} : M \to \mathbb{R}^3$ such that $\boldsymbol{F}(\boldsymbol{x})$ is orthogonal to $T_{\boldsymbol{x}}M$ for each $\boldsymbol{x} \in M$ and $\|\boldsymbol{F}(\boldsymbol{x})\| = 1$ (in particular, $\{\boldsymbol{F}(\boldsymbol{x})\}$ is a basis of $N_{\boldsymbol{x}}M$ for every $\boldsymbol{x} \in M$).*

Proof. Suppose that M is orientable and let $\{(A_j, \boldsymbol{\varphi}_j) : j \in J\}$ be the atlas whose existence is guaranteed by Theorem 5.2.1. For each $j \in J$ we put $M_j := \boldsymbol{\varphi}_j(A_j)$ and define $\boldsymbol{F}_j : M_j \to \mathbb{R}^3$ as follows. We first consider

$$\boldsymbol{G}_j(s,t) := \frac{\frac{\partial \boldsymbol{\varphi}_j}{\partial s}(s,t) \times \frac{\partial \boldsymbol{\varphi}_j}{\partial t}(s,t)}{\left\| \frac{\partial \boldsymbol{\varphi}_j}{\partial s}(s,t) \times \frac{\partial \boldsymbol{\varphi}_j}{\partial t}(s,t) \right\|},$$

which is a continuous function on A_j because $\boldsymbol{\varphi}_j$ is of class C^1 on A_j. Then we take

$$\boldsymbol{F}_j(\boldsymbol{x}) := \boldsymbol{G}_j(\boldsymbol{\varphi}_j^{-1}(\boldsymbol{x})) \text{ for } \boldsymbol{x} \in \boldsymbol{\varphi}_j(A_j).$$

For $\boldsymbol{x} = \boldsymbol{\varphi}_j(s,t)$, it is obvious that $\boldsymbol{F}_j(\boldsymbol{x})$ is a unit vector that is orthogonal to the tangent space $T_{\boldsymbol{x}}M$. Also, since $\boldsymbol{F}_j = \boldsymbol{G}_j \circ \boldsymbol{\varphi}_j^{-1}$, we conclude that $\boldsymbol{F}_j : M_j \to \mathbb{R}^3$ is a continuous function. If $j_1, j_2 \in J$, then $\boldsymbol{F}_{j_1}$ coincides with $\boldsymbol{F}_{j_2}$ on $M_{j_1} \cap M_{j_2}$ by Proposition 5.1.3. Since the sets $\{M_j : j \in J\}$ are open in M and they cover all of M, it turns out that there is a unique continuous function

$$\boldsymbol{F} : M \to \mathbb{R}^3$$

such that $\boldsymbol{F}(\boldsymbol{x}) = \boldsymbol{F}_j(\boldsymbol{x})$ whenever $\boldsymbol{x} \in M_j$.

Suppose now that there is a continuous function $\boldsymbol{F} : M \to \mathbb{R}^3$ such that $\boldsymbol{F}(\boldsymbol{x})$ is orthogonal to $T_{\boldsymbol{x}}M$ and $\|\boldsymbol{F}(\boldsymbol{x})\| = 1$ for each $\boldsymbol{x} \in M$. For every $\boldsymbol{x} \in M$ we select a basis $\{\boldsymbol{v}_1, \boldsymbol{v}_2\}$ of the tangent space $T_{\boldsymbol{x}}M$ with

$$\boldsymbol{F}(\boldsymbol{x}) = \frac{\boldsymbol{v}_1 \times \boldsymbol{v}_2}{\| \boldsymbol{v}_1 \times \boldsymbol{v}_2 \|} \tag{5.1}$$

and we consider the orientation $\theta_{\boldsymbol{x}} := [\boldsymbol{v}_1, \boldsymbol{v}_2]$ of $T_{\boldsymbol{x}}M$. By Proposition 5.1.3, the chosen orientation $\theta_{\boldsymbol{x}}$ does not depend on the chosen basis of the tangent space as long as condition (5.1) holds. Now we fix $\boldsymbol{x}_0 \in M$ and let $(A, \boldsymbol{\varphi})$ be a coordinate system of M with $\boldsymbol{x}_0 \in \boldsymbol{\varphi}(A)$. Take $(s_0,t_0) \in A$ with $\boldsymbol{x}_0 = \boldsymbol{\varphi}(s_0,t_0)$ and $r > 0$ such that $B := B\big((s_0,t_0),r\big) \subset A$. Finally, we consider the coordinate system $(B,\boldsymbol{\varphi})$ and put

$$\boldsymbol{N}(s,t) := \frac{\frac{\partial \boldsymbol{\varphi}}{\partial s}(s,t) \times \frac{\partial \boldsymbol{\varphi}}{\partial t}(s,t)}{\left\| \frac{\partial \boldsymbol{\varphi}}{\partial s}(s,t) \times \frac{\partial \boldsymbol{\varphi}}{\partial t}(s,t) \right\|}.$$

Since $\boldsymbol{N}(s,t)$ and $\boldsymbol{F}(\boldsymbol{\varphi}(s,t))$ are both unit vectors orthogonal to the tangent space to M at $\boldsymbol{\varphi}(s,t)$, it follows that for every $(s,t) \in B$,

$$\boldsymbol{N}(s,t) = \lambda(s,t)\boldsymbol{F}(\boldsymbol{\varphi}(s,t)),$$

where

$$\lambda(s,t) = \pm 1.$$

Lemma 5.2.4 guarantees that the scalar function λ is continuous on B. Since the ball B is connected, there are only two possibilities. Either

$$\lambda(s,t) = 1 \text{ for all } (s,t) \in B \tag{5.2}$$

or

$$\lambda(s,t) = -1 \text{ for all } (s,t) \in B. \tag{5.3}$$

In the first case, (5.2),

$$\boldsymbol{N}(s,t) = \boldsymbol{F}(\boldsymbol{\varphi}(s,t))$$

for every $(s,t) \in B$, which means that

$$\theta_{\boldsymbol{x}} = \left[\frac{\partial \boldsymbol{\varphi}}{\partial s}(s,t), \frac{\partial \boldsymbol{\varphi}}{\partial t}(s,t)\right]$$

for all $\boldsymbol{x} = \boldsymbol{\varphi}(s,t)$. However, when case (5.3) prevails, the basis

$$\left\{\frac{\partial \boldsymbol{\varphi}}{\partial s}(s,t), \frac{\partial \boldsymbol{\varphi}}{\partial t}(s,t)\right\}$$

does not define the orientation $\theta_{\boldsymbol{x}}$, and we define

$$C := \{(t,s) \, : (s,t) \in B\}$$

and

$$\boldsymbol{\psi}(t,s) := \boldsymbol{\varphi}(s,t).$$

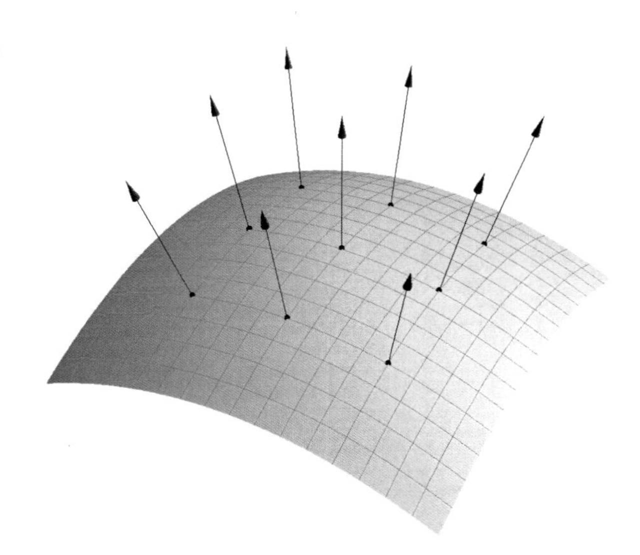

Fig. 5.5 Unit normal vectors to a surface

Then by Lemma 5.2.2, $(C, \boldsymbol{\psi})$ is a coordinate system of M, $\boldsymbol{x}_0 = \boldsymbol{\psi}(t_0, s_0)$, and for each

$$\boldsymbol{x} = \boldsymbol{\psi}(t,s) = \boldsymbol{\varphi}(s,t)$$

the bases

$$\left\{ \frac{\partial \boldsymbol{\varphi}}{\partial s}(s,t), \frac{\partial \boldsymbol{\varphi}}{\partial t}(s,t) \right\} \quad \text{and} \quad \left\{ \frac{\partial \boldsymbol{\psi}}{\partial t}(t,s), \frac{\partial \boldsymbol{\psi}}{\partial s}(t,s) \right\}$$

define different orientations in $T_{\boldsymbol{x}}M$. That is,

$$\theta_{\boldsymbol{x}} = \left[\frac{\partial \boldsymbol{\psi}}{\partial t}(t,s), \frac{\partial \boldsymbol{\psi}}{\partial s}(t,s) \right].$$

According to Definition 5.2.1, M is orientable. □

Remark 5.2.2. Let M be a regular surface in $\mathbb{R}^3$ that is orientable and connected. Then an orientation of M is determined by choice of a normal unit vector at any single point $\boldsymbol{x}_0 \in M$ (Fig. 5.5).

Theorem 5.2.2 motivates the following definition.

Definition 5.2.3. Let M be a regular surface in $\mathbb{R}^3$ and let $\boldsymbol{F} : M \to \mathbb{R}^3$ be a continuous function such that for every $\boldsymbol{x} \in M$, $\boldsymbol{F}(\boldsymbol{x})$ is orthogonal to $T_{\boldsymbol{x}}M$ and $\|\boldsymbol{F}(\boldsymbol{x})\| = 1$. By the *orientation induced by* $\boldsymbol{F}$ *in* M we mean the orientation with the property that for all $\boldsymbol{x} \in M$, the orientation of the tangent space $T_{\boldsymbol{x}}M$ is

$$\theta_{\boldsymbol{x}} = [\boldsymbol{v}_1(\boldsymbol{x}), \boldsymbol{v}_2(\boldsymbol{x})],$$

where $\{\boldsymbol{v}_1(\boldsymbol{x}), \boldsymbol{v}_2(\boldsymbol{x})\}$ is a basis of $T_{\boldsymbol{x}}M$ such that

$$\{\boldsymbol{F}(\boldsymbol{x}), \boldsymbol{v}_1(\boldsymbol{x}), \boldsymbol{v}_2(\boldsymbol{x})\}$$

is a positively oriented basis of $\mathbb{R}^3$.

This last condition is equivalent to the fact that $\boldsymbol{F}(\boldsymbol{x})$ is a positive multiple of the cross product $\boldsymbol{v}_1(\boldsymbol{x}) \times \boldsymbol{v}_2(\boldsymbol{x})$.

Example 5.2.1. The sphere $M := \{(x,y,z) \in \mathbb{R}^3 \ : x^2+y^2+z^2 = R^2\}$ is orientable.

Proof. It suffices to consider the continuous vector field $\boldsymbol{F} : M \to \mathbb{R}^3$ defined by $\boldsymbol{F}(x,y,z) := \frac{1}{R}(x,y,z)$. It is obvious that $\boldsymbol{F}(x,y,z)$ is normal to the sphere at the point (x,y,z). We could also apply Proposition 3.4.1, after considering the function $\boldsymbol{\Phi}(x,y,z) = x^2+y^2+z^2-R^2$ and observing that

$$\boldsymbol{F}(x,y,z) = \frac{\nabla \boldsymbol{\Phi}(x,y,z)}{\| \nabla \boldsymbol{\Phi}(x,y,z) \|}.$$

Moreover, if we consider the orientation of the sphere induced by the vector field $\boldsymbol{F}$ (see Definition 5.2.3), then the coordinate system $(A, \boldsymbol{\varphi})$ of M (see Example 3.3.2) defined as

$$A := \{(s,t) \ : 0 < s < \pi \ , \ 0 < t < 2\pi\}$$

and

$$\boldsymbol{\varphi}(s,t) := (R \sin s \cos t, \ R \sin s \sin t, \ R \cos s)$$

is compatible with the orientation of M. For this, it suffices to show that

$$\left(\frac{\partial \boldsymbol{\varphi}}{\partial s} \times \frac{\partial \boldsymbol{\varphi}}{\partial t}\right)(s,t)$$

is a positive multiple of $\boldsymbol{F}(\boldsymbol{\varphi}(s,t))$, and this follows from

$$\begin{aligned}
\left(\frac{\partial \boldsymbol{\varphi}}{\partial s} \times \frac{\partial \boldsymbol{\varphi}}{\partial t}\right)(s,t) &= \begin{vmatrix} \boldsymbol{e}_1 & \boldsymbol{e}_2 & \boldsymbol{e}_3 \\ R\cos s \cos t & R \cos s \sin t & -R \sin s \\ -R \sin s \sin t & R \sin s \cos t & 0 \end{vmatrix} \\
&= (R^2 \sin^2 s \cos t, \ R^2 \sin^2 s \sin t, \ R^2 \sin s \cos s) \\
&= R^2 \sin s \, \boldsymbol{F}\Big(\boldsymbol{\varphi}(s,t)\Big).
\end{aligned}$$

□

We will not prove the following result. A proof can be found in Samelson [16].

Theorem 5.2.3. *Every compact regular 2-surface in $\mathbb{R}^3$ is orientable.*

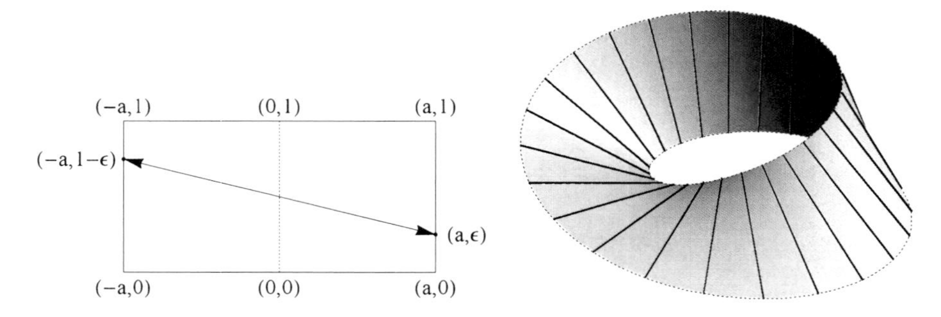

Fig. 5.6 Forming a Möbius strip

Fig. 5.7 The Möbius strip is not orientable

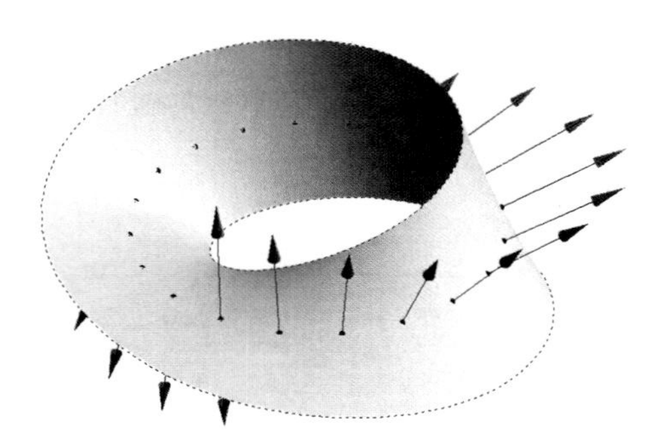

Example 5.2.2 is contained in Do Carmo [5, p. 106]. There, one can find a detailed description of an atlas of the surface and a rigorous proof that it is not orientable.

Example 5.2.2 (A nonorientable regular surface: the Möbius strip). We consider a circle centered at the origin in the xy-plane and a line segment (without endpoints) in the yz-plane whose center is located at a point of the circle (Fig. 5.6). We move the center of the segment along the circle, and as we do so, we rotate the segment about the tangent line to the circle in such a manner that when the center of the segment has rotated on the circle by an angle θ, then the segment has rotated by an angle $\frac{\theta}{2}$. The surface thus obtained is called the *Möbius strip*. It can also be obtained from a rectangle $[-a,a] \times (0,1)$ in the xy-plane by *joining* the two vertical sides so that the point (a,ε) coincides with $(-a,1-\varepsilon)$. (That is, we join the two ends of the rectangle but incorporate a "half twist" in doing so.)

If the surface is orientable, then there is a continuous field $\boldsymbol{F}$ of unit normal vectors to the surface (see Theorem 5.2.2). An analysis of these normal vectors along the circle used to generate the Möbius strip allows us to conclude that after a complete revolution of the strip, the normal vector $\boldsymbol{F}$ returns to its original position as $-\boldsymbol{F}$, which is a contradiction (Fig. 5.7).

To finish, we mention that a coordinate system of the Möbius strip is (A, φ), where $A = (0, 2\pi) \times (-1, 1)$ and

$$\varphi(s,t) = \left(\left(2 - t\sin\left(\frac{s}{2}\right) \right) \sin(s), \left(2 - t\sin\left(\frac{s}{2}\right) \right) \cos(s), t\cos\left(\frac{s}{2}\right) \right).$$

This coordinate system is sufficient to draw the surface (with the help, perhaps, of some graphical software), since $\varphi(A)$ covers all the surface except an open segment. In [5] it is proved that the Möbius strip admits an atlas with only two coordinate systems (A, φ) and (B, ψ) with the property that $\varphi(A) \cap \psi(B)$ has two connected components.

5.3 Exercises

Exercise 5.3.1. If V is a two-dimensional vector space and $\mathscr{B}_1 = \{\mathbf{v}_1, \mathbf{v}_2\}$ is a basis of V, does the basis $\mathscr{B}_2 = \{\mathbf{v}_1 + \mathbf{v}_2, \mathbf{v}_1 - \mathbf{v}_2\}$ have the same orientation as $\mathscr{B}_1$?

Exercise 5.3.2. The sets

$$\{\mathbf{e}_1 - \mathbf{e}_2, \mathbf{e}_1 - \mathbf{e}_3\} \text{ and } \{\mathbf{e}_1 - \mathbf{e}_3, 2\mathbf{e}_1 - \mathbf{e}_2 - \mathbf{e}_3\}$$

span the same vector subspace V of $\mathbb{R}^3$. Determine whether the two bases define the same orientation in V.

Exercise 5.3.3. Determine whether the following coordinate systems of the cylindrical surface

$$S := \{(x,y,z) \in \mathbb{R}^3 \ ; \ x^2 + y^2 = 1, \ 0 < z < 1\}$$

define the same orientation in the tangent space to S at each common point

$$(x_0, y_0, z_0) = \boldsymbol{\varphi}(s_0, t_0) = \boldsymbol{\psi}(u_0, v_0) :$$

1. $\boldsymbol{\varphi} : (0, 2\pi) \times (0,1) \to S, \quad \boldsymbol{\varphi}(s,t) = (\cos s, \sin s, t)$,
2. $\boldsymbol{\psi} : (0,1) \times (0,1) \to S, \quad \boldsymbol{\psi}(u,v) = \left(u, \sqrt{1-u^2}, v\right)$.

Exercise 5.3.4. Find a parameterization of the paraboloid

$$z = 1 - x^2 - y^2 \ \ (0 < z < 1)$$

compatible with the orientation induced by a field of normal vectors *pointing inward.*

Exercise 5.3.5. Prove that any level surface M in $\mathbb{R}^3$ is orientable.

Chapter 6
Differential Forms

6.1 Differential Forms of Degree k

The expression

$$\iint_A \begin{vmatrix} f_1 & f_2 & f_3 \\ \frac{\partial x}{\partial s} & \frac{\partial y}{\partial s} & \frac{\partial z}{\partial s} \\ \frac{\partial x}{\partial t} & \frac{\partial y}{\partial t} & \frac{\partial z}{\partial t} \end{vmatrix} \, \mathrm{d}(s,t)$$

obtained in Chap. 4 for calculating the flux of a vector field across a surface defined by a unique coordinate system suggests that it may be convenient to identify the vector $\boldsymbol{F}(\boldsymbol{x}) = (f_1(\boldsymbol{x}), f_2(\boldsymbol{x}), f_3(\boldsymbol{x}))$ with the alternating bilinear form on $\mathbb{R}^3$ defined by

$$(u,v) \mapsto \begin{vmatrix} f_1(\boldsymbol{x}) & f_2(\boldsymbol{x}) & f_3(\boldsymbol{x}) \\ u_1 & u_2 & u_3 \\ v_1 & v_2 & v_3 \end{vmatrix},$$

where $\boldsymbol{x} = (x,y,z)$. It is natural then to identify a vector field on an open set $U \subset \mathbb{R}^3$ with the mapping that to each point $\boldsymbol{x}$ of U associates the bilinear form defined above.

More generally, we will study mappings $\boldsymbol{\omega}$ defined on an open set $U \subset \mathbb{R}^n$ that associate an alternating k-linear form on $(\mathbb{R}^n)^k$ to each $\boldsymbol{x} \in U$.

Definition 6.1.1. We denote by $\Lambda^k(\mathbb{R}^n)$ the vector space of all *alternating k-linear forms,* that is, the mappings $\varphi : (\mathbb{R}^n)^k \to \mathbb{R}$ that satisfy the following conditions:

(a) $\varphi(\boldsymbol{v}_1,\ldots,\boldsymbol{v}_j+\boldsymbol{w}_j,\ldots,\boldsymbol{v}_k) = \varphi(\boldsymbol{v}_1,\ldots,\boldsymbol{v}_j,\ldots,\boldsymbol{v}_k) + \varphi(\boldsymbol{v}_1,\ldots,\boldsymbol{w}_j,\ldots,\boldsymbol{v}_k)$,
(b) $\varphi(\boldsymbol{v}_1,\ldots,\alpha_j\boldsymbol{v}_j,\ldots,\boldsymbol{v}_k) = \alpha_j\varphi(\boldsymbol{v}_1,\ldots,\boldsymbol{v}_j,\ldots,\boldsymbol{v}_k)$,
(c) $\varphi(\boldsymbol{v}_1,\ldots,\boldsymbol{v}_i,\ldots,\boldsymbol{v}_j,\ldots,\boldsymbol{v}_k) = -\varphi(\boldsymbol{v}_1,\ldots,\boldsymbol{v}_j,\ldots,\boldsymbol{v}_i,\ldots,\boldsymbol{v}_k)$,

for every $\boldsymbol{v}_1,\ldots,\boldsymbol{v}_k,\boldsymbol{w}_j \in \mathbb{R}^n$, $\alpha_j \in \mathbb{R}$, and $j \in \{1,\ldots,k\}$.

A. Galbis and M. Maestre, *Vector Analysis Versus Vector Calculus*, Universitext,
DOI 10.1007/978-1-4614-2200-6_6, © Springer Science+Business Media, LLC 2012

Definition 6.1.2. Let $\varphi_j := \mathrm{d}x_{i_j} : \mathbb{R}^n \to \mathbb{R}$, $1 \le j \le k$, be given and define the k-linear mapping

$$\varphi_1 \wedge \varphi_2 \wedge \cdots \wedge \varphi_k : (\mathbb{R}^n)^k \to \mathbb{R}$$

by

$$(\varphi_1 \wedge \varphi_2 \wedge \cdots \wedge \varphi_k)(\boldsymbol{v}_1, \boldsymbol{v}_2, \ldots, \boldsymbol{v}_k) := \det\left(\varphi_i(\boldsymbol{v}_j)\right)$$

for every $(\boldsymbol{v}_1, \ldots, \boldsymbol{v}_k) \in (\mathbb{R}^n)^k$. This operation is called the *exterior product* of $\{\varphi_1, \ldots, \varphi_k\}$

From the properties of determinants we get

$$\varphi_1 \wedge \varphi_2 \wedge \cdots \wedge \varphi_k \in \Lambda^k(\mathbb{R}^n).$$

For each k-tuple $\mathbf{I} = (i_1, \ldots, i_k)$, $1 \le i_j \le n$, we put

$$\mathrm{d}x_{\mathbf{I}} := \mathrm{d}x_{i_1} \wedge \cdots \wedge \mathrm{d}x_{i_k} \in \Lambda^k(\mathbb{R}^n), \quad \boldsymbol{e}_{\mathbf{I}} := (\boldsymbol{e}_{i_1}, \ldots, \boldsymbol{e}_{i_k}) \in (\mathbb{R}^n)^k.$$

It is natural to ask why we did not allow arbitrary linear forms $\varphi_1, \ldots, \varphi_k$ on $\mathbb{R}^n$ in Definition 6.1.2. The answer is in Corollary 6.2.1.

For the following proposition, recall that if π is a permutation of $\mathbf{I} = (i_1, \ldots, i_k)$ and we express π as the product of σ *transpositions*, i.e., of permutations

$$(1, 2, \ldots, i, \ldots, j, \ldots, n) \mapsto (1, 2, \ldots, j, \ldots, i, \ldots, n),$$

then the *signature* of π is the number σ.

Proposition 6.1.1. *(1) If $f, g \in \Lambda^k(\mathbb{R}^n)$ and $f(e_{\mathbf{I}}) = g(e_{\mathbf{I}})$ for any strictly increasing k-tuple $\mathbf{I}$, then $f = g$.*

(2) If any index in $\mathbf{I}$ is repeated, then $\mathrm{d}x_{\mathbf{I}} = 0$. In particular, $\mathrm{d}x_{\mathbf{I}} = 0$ whenever $\mathbf{I} = (i_1, \ldots, i_k)$ and $k > n$.

(3) Let π be a permutation of $\mathbf{I} = (i_1, \ldots, i_k)$. Then

$$\mathrm{d}x_{i_1} \wedge \cdots \wedge \mathrm{d}x_{i_k} = (-1)^{\sigma} \mathrm{d}x_{\pi(i_1)} \wedge \cdots \wedge \mathrm{d}x_{\pi(i_k)},$$

where σ is the signature of the permutation.

(4) Let $\mathbf{I} = (i_1, \ldots, i_k)$, $\mathbf{J} = (j_1, \ldots, j_k)$ be given and suppose $1 \le i_1 < i_2 < \cdots < i_k \le n$ and $1 \le j_1 < j_2 < \cdots < j_k \le n$. Then

$$\mathrm{d}x_{\mathbf{I}}(e_{\mathbf{J}}) = \begin{cases} 1 & \text{if } i_1 = j_1, \ldots, i_k = j_k, \\ 0 & \text{otherwise.} \end{cases}$$

That is, using Kronecker δ notation,

$$\mathrm{d}x_{\mathbf{I}}(e_{\mathbf{J}}) = \delta_{\mathbf{IJ}}.$$

Proof. (1) After applying property (c) of Definition 6.1.1 as many times as necessary, we deduce that $f(\boldsymbol{e}_{i_1},\ldots,\boldsymbol{e}_{i_k}) = g(\boldsymbol{e}_{i_1},\ldots,\boldsymbol{e}_{i_k})$ for all $i_1,\ldots,i_k$. It then follows from properties (a) and (b) that $f = g$.

(2) From Definition 6.1.2, $\mathrm{d}x_{\mathbf{I}}$ is the determinant of a matrix with two identical rows and hence it vanishes.

(3) This is a consequence of the fact that the sign of the determinant of a matrix changes if two rows are interchanged.

(4) If $j_1 < i_1$, then $\mathrm{d}x_{i_l}(\boldsymbol{e}_{j_1}) = 0$ for all l and $\mathrm{d}x_{\mathbf{I}}(e_{\mathbf{J}})$ is the determinant of a matrix with a zero column. Hence $\mathrm{d}x_{\mathbf{I}}(e_{\mathbf{J}}) = 0$. If $i_1 < j_1$, then $\mathrm{d}x_{i_1}(\boldsymbol{e}_{j_l}) = 0$ for all l and $\mathrm{d}x_{\mathbf{I}}(e_{\mathbf{J}})$ is the determinant of a matrix with a zero row. Hence $\mathrm{d}x_{\mathbf{I}}(e_{\mathbf{J}}) = 0$. Finally, in the case $i_1 = j_1$ we have that the first row in the matrix $\big(\mathrm{d}x_{i_l}(\boldsymbol{e}_{j_m})\big)$ is $(1,0,\ldots,0)$. It follows from the properties of determinants that $\mathrm{d}x_{\mathbf{I}}(e_{\mathbf{J}}) = \mathrm{d}x_{\mathbf{I}'}(e_{\mathbf{J}'})$, where $\mathbf{I}' = (i_2,\ldots,i_k)$ and $\mathbf{J}' = (j_2,\ldots,j_k)$. Now an induction argument provides the conclusion.

Let $\mathscr{J}$ denote the set of strictly increasing k-tuples.

Proposition 6.1.2. *For any $n \in \mathbb{N}$ and $1 \le k \le n$, the set*

$$\{\mathrm{d}x_{\mathbf{I}} : \mathbf{I} \in \mathscr{J}\}$$

is a basis of $\Lambda^k(\mathbb{R}^n)$.

Proof. Let us assume

$$\sum_{\mathbf{I}\in\mathscr{J}} a_{\mathbf{I}}\mathrm{d}x_{\mathbf{I}} = 0.$$

Then for each strictly increasing k-tuple $\mathbf{J}$ we have

$$0 = \Big(\sum_{\mathbf{I}\in\mathscr{J}} a_{\mathbf{I}}\mathrm{d}x_{\mathbf{I}}\Big)(e_{\mathbf{J}}) = a_{\mathbf{J}},$$

which shows that the family $\{\mathrm{d}x_{\mathbf{I}}\}$ is linearly independent. On the other hand, given $f \in \Lambda^k(\mathbb{R}^n)$, we consider

$$g := \sum_{\mathbf{I}\in\mathscr{J}} f(e_{\mathbf{I}})\mathrm{d}x_{\mathbf{I}}.$$

Then $g \in \Lambda^k(\mathbb{R}^n)$ and $g(e_{\mathbf{J}}) = f(e_{\mathbf{J}})$ for each strictly increasing k-tuple $\mathbf{J}$, which proves that $f = g$. □

It follows from this that $\Lambda^k(\mathbb{R}^n)$ is a vector space whose dimension is $\binom{n}{k}$ if $1 \le k \le n$ and 0 if $k > n$.

Definition 6.1.3. If $1 < j < n$, we denote by $\mathrm{d}x_1 \wedge \cdots \wedge \widehat{\mathrm{d}x_j} \wedge \cdots \wedge \mathrm{d}x_n$ the differential $(n-1)$-form obtained as the exterior product of the $n-1$ differential 1-forms $\{\mathrm{d}x_1, \mathrm{d}x_2, \ldots, \mathrm{d}x_{j-1}, \mathrm{d}x_{j+1}, \ldots, \mathrm{d}x_n\}$, in the order in which they appear. Analogously, $\widehat{\mathrm{d}x_1} \wedge \mathrm{d}x_2 \wedge \cdots \wedge \cdots \wedge \mathrm{d}x_n$ is the exterior product of $\{\mathrm{d}x_2, \ldots, \mathrm{d}x_n\}$, and $\mathrm{d}x_1 \wedge \cdots \wedge \mathrm{d}x_{n-1} \wedge \widehat{\mathrm{d}x_n}$ is the exterior product of $\{\mathrm{d}x_1, \mathrm{d}x_2, \ldots, \mathrm{d}x_{n-1}\}$.

Proposition 6.1.3. *(1) There is a linear isomorphism from* $\mathbb{R}^n$ *to* $\Lambda^1(\mathbb{R}^n)$, *given by*

$$\mathbb{R}^n \to \Lambda^1(\mathbb{R}^n)\ ,\ \boldsymbol{a} = (a_1,\dots,a_n) \mapsto \sum_{j=1}^{n} a_j \cdot \mathrm{d}x_j.$$

(2) There is a linear isomorphism $\boldsymbol{B} : \mathbb{R}^n \to \Lambda^{n-1}(\mathbb{R}^n)$ *defined by*

$$\boldsymbol{a} = (a_1,\dots,a_n) \mapsto \sum_{j=1}^{n} (-1)^j a_j \cdot \mathrm{d}x_1 \wedge \cdots \wedge \widehat{\mathrm{d}x_j} \wedge \cdots \wedge \mathrm{d}x_n.$$

The first of the isomorphisms given above was important in the study of line integrals, while the second will prove essential when we study the divergence theorem.

Example 6.1.1. Let $\boldsymbol{a} = (a_1,a_2,a_3)$ be a vector in $\mathbb{R}^3$ and

$$\boldsymbol{\omega} = a_1\ \mathrm{d}y \wedge \mathrm{d}z\ +\ a_2\ \mathrm{d}z \wedge \mathrm{d}x\ +\ a_3\ \mathrm{d}x \wedge \mathrm{d}y \in \Lambda^2(\mathbb{R}^3).$$

Then for each pair of vectors

$$\boldsymbol{u} = (u_1,u_2,u_3), \quad \boldsymbol{v} = (v_1,v_2,v_3) \in \mathbb{R}^3,$$

we have

$$\boldsymbol{\omega}(\boldsymbol{u},\boldsymbol{v}) = a_1 \begin{vmatrix} u_2 & u_3 \\ v_2 & v_3 \end{vmatrix} - a_2 \begin{vmatrix} u_1 & u_3 \\ v_1 & v_3 \end{vmatrix} + a_3 \begin{vmatrix} u_1 & u_2 \\ v_1 & v_2 \end{vmatrix}.$$

That is,

$$\boldsymbol{\omega}(\boldsymbol{u},\boldsymbol{v}) \ = \begin{vmatrix} a_1 & a_2 & a_3 \\ u_1 & u_2 & u_3 \\ v_1 & v_2 & v_3 \end{vmatrix} = \langle \boldsymbol{a}\ ,\ \boldsymbol{u} \times \boldsymbol{v} \rangle\,.$$

Observe that if $\boldsymbol{B} : \mathbb{R}^n \to \Lambda^{n-1}(\mathbb{R}^n)$ is the isomorphism given in Proposition 6.1.3, then for $\boldsymbol{v}^j = (v_1^j,\dots,v_n^j) \in \mathbb{R}^n$, $j = 1,\dots,n-1$, we have

$$\boldsymbol{B}(\boldsymbol{a})\left(\boldsymbol{v}^1,\boldsymbol{v}^2,\dots,\boldsymbol{v}^{n-1}\right) = \begin{vmatrix} a_1 & a_2 & \dots & a_n \\ v_1^1 & v_2^1 & \dots & v_n^1 \\ \dots & \dots & \dots & \dots \\ v_1^{n-1} & v_2^{n-1} & \dots & v_n^{n-1} \end{vmatrix},$$

which is a natural generalization of Example 6.1.1.

Definition 6.1.4. A *differential form* of *degree* k (briefly a differential k-form) on $\mathbb{R}^n$ is a mapping

$$\omega : U \subset \mathbb{R}^n \to \Lambda^k(\mathbb{R}^n).$$

By Proposition 6.1.2 we can write

$$\omega(\boldsymbol{x}) = \sum_{1\leq i_1<\cdots<i_k\leq n} f_{i_1,\ldots,i_k}(\boldsymbol{x})\mathrm{d}x_{i_1} \wedge \cdots \wedge \mathrm{d}x_{i_k}.$$

In shortened form,

$$\omega(\boldsymbol{x}) = \sum_{\mathbf{I}\in\mathscr{J}} f_{\mathbf{I}}(\boldsymbol{x})\mathrm{d}x_{\mathbf{I}},$$

where each $f_{\mathbf{I}} : U \subset \mathbb{R}^n \to \mathbb{R}$ is a unique mapping. A mapping ω is said to be *continuous* if each $f_{i_1,\ldots,i_k}(\boldsymbol{x})$ is continuous. If U is an open subset of $\mathbb{R}^n$ and each function $f_{i_1,\ldots,i_k}$ is differentiable (respectively of class C^p) on U, then ω is said to be *differentiable* (respectively of *class* C^p) on U.

It follows from Proposition 6.1.3 that *there exists a bijection between vector fields in U and 1-forms, as well as a bijection between vector fields in U and $(n-1)$-forms.* In particular, for $n = 3$ it turns out that a vector field on an open subset of $\mathbb{R}^3$ defines in a natural way a differential 1-form and also a differential 2-form on $\mathbb{R}^3$. It was this that allowed us in Chap. 2 to express the line integral of a vector field in terms of integration of 1-forms. It will also allow us in Chap. 7 to write the flux of a vector field across a surface as the integral of a differential form of degree 2. Notwithstanding this fact, for $1 < k < n-1$, it is not possible to obtain an isomorphism between $\mathbb{R}^n$ and $\Lambda^k(\mathbb{R}^n)$, since the differential forms of degree k do not define a vector field in a natural way. In spite of this, the integration of differential forms will be an essential tool in the integration of vector fields.

6.2 Exterior Product

The notation used to represent the elements of the basis of $\Lambda^k(\mathbb{R}^n)$ provides us with an *exterior product*, which is a natural way of combining differential forms of degrees k and s to create a differential form of degree $k+s$. This exterior product can be defined in a *basis-free* manner, as for example in the excellent book by Spivak [17]. However, we choose the basis-oriented approach to the exterior product, since it allows us to expressly deduce the properties of the exterior product that we need.

Definition 6.2.1. Let $\boldsymbol{\omega}(\boldsymbol{x}) = \sum_{\mathbf{I}\in\mathscr{J}} f_{\mathbf{I}}(\boldsymbol{x})\mathrm{d}x_{\mathbf{I}}$ be a k-form and $\boldsymbol{\varphi}(\boldsymbol{x}) = \sum_{\mathbf{J}\in\mathscr{J}} g_{\mathbf{J}}(\boldsymbol{x})\mathrm{d}x_{\mathbf{J}}$ an s-form on the set $U \subset \mathbb{R}^n$. The *exterior product* of $\boldsymbol{\omega}$ and $\boldsymbol{\varphi}$ is the $(k+s)$-form given by

$$\boldsymbol{\omega} \wedge \boldsymbol{\varphi} = \sum_{\mathbf{I},\mathbf{J}\in\mathscr{J}} f_{\mathbf{I}}\, g_{\mathbf{J}}\, \mathrm{d}x_{i_1} \wedge \cdots \wedge \mathrm{d}x_{i_k} \wedge \mathrm{d}x_{j_1} \wedge \cdots \wedge \mathrm{d}x_{j_s},$$

where the summation is over all strictly increasing tuples

$$\mathbf{I} = (i_1, \ldots, i_k), \quad 1 \le i_1 < \cdots < i_k \le n,$$

and

$$\mathbf{J} = (j_1, \ldots, j_s), \quad 1 \le j_1 < \cdots < j_s \le n.$$

Proposition 6.2.1. *Let $\boldsymbol{\omega}$ be a k-form, $\boldsymbol{\varphi}$ an s-form, and $\boldsymbol{\theta}$ an r-form on $U \subset \mathbb{R}^n$. Then:*

(1) $(\boldsymbol{\omega} \wedge \boldsymbol{\varphi}) \wedge \boldsymbol{\theta} = \boldsymbol{\omega} \wedge (\boldsymbol{\varphi} \wedge \boldsymbol{\theta})$,
(2) $\boldsymbol{\omega} \wedge (\boldsymbol{\varphi} + \boldsymbol{\theta}) = \boldsymbol{\omega} \wedge \boldsymbol{\varphi} + \boldsymbol{\omega} \wedge \boldsymbol{\theta}$ *if* $r = s$,
(3) $(\boldsymbol{\omega} \wedge \boldsymbol{\varphi}) = (-1)^{ks}(\boldsymbol{\varphi} \wedge \boldsymbol{\omega})$.

Proof. Leaving the other parts to the reader, we will prove (3) by way of an informal induction argument. In shortened form, we can write

$$\boldsymbol{\omega} \wedge \boldsymbol{\varphi} = \sum_{\mathbf{I}\in\mathscr{J}} \sum_{\mathbf{J}\in\mathscr{J}} f_{\mathbf{I}} g_{\mathbf{J}}\, \mathrm{d}x_{\mathbf{I}} \wedge \mathrm{d}x_{\mathbf{J}}$$

and also

$$\boldsymbol{\varphi} \wedge \boldsymbol{\omega} = \sum_{\mathbf{I}\in\mathscr{J}} \sum_{\mathbf{J}\in\mathscr{J}} f_{\mathbf{I}} g_{\mathbf{J}}\, \mathrm{d}x_{\mathbf{J}} \wedge \mathrm{d}x_{\mathbf{I}}.$$

Let $\mathbf{I} = (i_1, \ldots, i_k)$, $\mathbf{J} = (j_1, \ldots, j_s)$ be given. Observe that since $\mathrm{d}x_i \wedge \mathrm{d}x_j = -\mathrm{d}x_j \wedge \mathrm{d}x_i$,

$$\begin{aligned}
\mathrm{d}x_{\mathbf{I}} \wedge \mathrm{d}x_{\mathbf{J}} &= \mathrm{d}x_{i_1} \wedge \cdots \wedge \mathrm{d}x_{i_k} \wedge \mathrm{d}x_{j_1} \wedge \cdots \wedge \mathrm{d}x_{j_s} \\
&= (-1)\, \mathrm{d}x_{i_1} \wedge \cdots \wedge \mathrm{d}x_{i_{k-1}} \wedge \mathrm{d}x_{j_1} \wedge \mathrm{d}x_{i_k} \wedge \mathrm{d}x_{j_2} \wedge \cdots \wedge \mathrm{d}x_{j_s} \\
&= (-1)^s \mathrm{d}x_{i_1} \wedge \cdots \wedge \mathrm{d}x_{i_{k-1}} \wedge \mathrm{d}x_{j_1} \wedge \cdots \wedge \mathrm{d}x_{j_s} \wedge \mathrm{d}x_{i_k}.
\end{aligned}$$

Repeating this process k times, we get

$$\begin{aligned} dx_{\mathbf{I}} \wedge dx_{\mathbf{J}} &= (-1)^{ks} dx_{j_1} \wedge \cdots \wedge dx_{j_s} \wedge dx_{i_1} \wedge \cdots \wedge dx_{i_k} \\ &= (-1)^{ks} dx_{\mathbf{J}} \wedge dx_{\mathbf{I}}. \end{aligned}$$

We denote by $(\mathbb{R}^n)^* = \mathscr{L}(\mathbb{R}^n, \mathbb{R})$ the dual space of $\mathbb{R}^n$. In the next result we identify a linear form $\varphi \in \Lambda^1(\mathbb{R}^n)$ with the constant degree-1 differential form $\boldsymbol{\omega} : \mathbb{R}^n \to \Lambda^1(\mathbb{R}^n)$, $\boldsymbol{\omega}(\boldsymbol{x}) = \varphi$, for all $\boldsymbol{x} \in \mathbb{R}^n$.

Corollary 6.2.1. *Let $\varphi_1, \ldots, \varphi_k \in \Lambda^1(\mathbb{R}^n)$ be given. Then for each $(\boldsymbol{v}_1, \ldots, \boldsymbol{v}_k) \in (\mathbb{R}^n)^k$, and each $x \in \mathbb{R}^n$*

$$(\varphi_1 \wedge \cdots \wedge \varphi_k)(\boldsymbol{x})(\boldsymbol{v}_1, \ldots, \boldsymbol{v}_k) = \det\big(\varphi_i(\boldsymbol{v}_j)\big).$$

(Observe that $\varphi_1 \wedge \cdots \wedge \varphi_k$ is a k-form constant in $\boldsymbol{x}$.)

Proof. We fix $\boldsymbol{v}_1, \ldots, \boldsymbol{v}_k \in \mathbb{R}^n$ and consider the mappings

$$S, T : (\mathbb{R}^n)^* \times \cdots \times (\mathbb{R}^n)^* \to \mathbb{R}$$

defined by

$$S(\varphi_1, \ldots, \varphi_k) := \det\big(\varphi_i(\boldsymbol{v}_j)\big)$$

and

$$T(\varphi_1, \ldots, \varphi_k) := (\varphi_1 \wedge \cdots \wedge \varphi_k)(\boldsymbol{v}_1, \ldots, \boldsymbol{v}_k).$$

Both S and T are alternating k-linear mappings. Moreover, it follows from Definition 6.1.2 that

$$S(dx_{i_1}, \ldots, dx_{i_k}) = T(dx_{i_1}, \ldots, dx_{i_k})$$

for each $1 \leq i_1 < \cdots < i_k \leq n$. From the fact that $\{dx_j\}$ is a basis of $(\mathbb{R}^n)^*$, we deduce $S = T$. □

6.3 Exterior Differentiation

Already we have seen in Chap. 2 how one can obtain a differential form of degree 1 (the differential of g) from a differentiable function g. We are going to generalize this mechanism for differential forms of degree k. In addition, we will see that exterior differentiation can be viewed as a generalization of the operations of divergence and curl that were studied in Chap. 1.

Definition 6.3.1. Let $\boldsymbol{\omega} : U \subset \mathbb{R}^n \to \Lambda^k(\mathbb{R}^n)$ be the differentiable differential form of degree k on the open set U given by

$$\boldsymbol{\omega}(\boldsymbol{x}) = \sum_{1 \leq i_1 < \cdots < i_k \leq n} f_{i_1,\ldots,i_k}(\boldsymbol{x}) \wedge \mathrm{d}x_{i_1} \wedge \cdots \wedge \mathrm{d}x_{i_k}.$$

The exterior differential of $\boldsymbol{\omega}$, $\mathrm{d}\boldsymbol{\omega} : U \subset \mathbb{R}^n \to \Lambda^{k+1}(\mathbb{R}^n)$, is defined as

$$\mathrm{d}\boldsymbol{\omega}(\boldsymbol{x}) := \sum_{1 \leq i_1 < \cdots < i_k \leq n} \mathrm{d}f_{i_1,\ldots,i_k}(\boldsymbol{x}) \wedge \mathrm{d}x_{i_1} \wedge \cdots \wedge \mathrm{d}x_{i_k}.$$

It should be clear that $\mathrm{d}\boldsymbol{\omega}$ is a differential form of degree $k+1$.

Example 6.3.1. The exterior differential of a differential form of degree 1 *on* $\mathbb{R}^2$.

Let

$$\boldsymbol{F} = (P, Q)$$

be a vector field of class C^1 on an open set U in $\mathbb{R}^2$ and let us consider the associated differential form of degree 1,

$$\boldsymbol{\omega}(x,y) = P(x,y)\,\mathrm{d}x + Q(x,y)\,\mathrm{d}y.$$

Then

$$\begin{aligned} \mathrm{d}\boldsymbol{\omega} = dP \wedge \mathrm{d}x + dQ \wedge \mathrm{d}y &= \left(\frac{\partial P}{\partial x}\,\mathrm{d}x + \frac{\partial P}{\partial y}\,\mathrm{d}y\right) \wedge \mathrm{d}x \\ &\quad + \left(\frac{\partial Q}{\partial x}\,\mathrm{d}x + \frac{\partial Q}{\partial y}\,\mathrm{d}y\right) \wedge \mathrm{d}y \\ &= \frac{\partial P}{\partial y}\,\mathrm{d}y \wedge \mathrm{d}x + \frac{\partial Q}{\partial x}\,\mathrm{d}x \wedge \mathrm{d}y \\ &= \left(\frac{\partial Q}{\partial x} - \frac{\partial P}{\partial y}\right) \mathrm{d}x \wedge \mathrm{d}y. \end{aligned}$$

We have used that

$$\mathrm{d}x \wedge \mathrm{d}x = \mathrm{d}y \wedge \mathrm{d}y = 0$$

and

$$\mathrm{d}y \wedge \mathrm{d}x = -\mathrm{d}x \wedge \mathrm{d}y.$$

We observe that the differential form $\mathrm{d}\boldsymbol{\omega}$ is determined by the same function

$$\left(\frac{\partial Q}{\partial x} - \frac{\partial P}{\partial y}\right)$$

that appears in Green's Theorem (see Sect. 2.6).

We recall that if $\boldsymbol{F} = (f_1, f_2, f_3) : U \subset \mathbb{R}^3 \to \mathbb{R}^3$ is a vector field of class C^1 on an open set U, then the divergence of $\boldsymbol{F}$ is the scalar function

$$\operatorname{Div} \boldsymbol{F} := \frac{\partial f_1}{\partial x} + \frac{\partial f_2}{\partial y} + \frac{\partial f_3}{\partial z}$$

(Definition 1.2.10) and the curl of $\boldsymbol{F}$ is the vector field defined by the formal expression

$$\operatorname{Curl} \boldsymbol{F} := \boldsymbol{\nabla} \times \boldsymbol{F} = \begin{vmatrix} \boldsymbol{e}_1 & \boldsymbol{e}_2 & \boldsymbol{e}_3 \\ \dfrac{\partial}{\partial x} & \dfrac{\partial}{\partial y} & \dfrac{\partial}{\partial z} \\ f_1 & f_2 & f_3 \end{vmatrix}.$$

That is,

$$\operatorname{Curl} \boldsymbol{F} = \left(\frac{\partial f_3}{\partial y} - \frac{\partial f_2}{\partial z}, \frac{\partial f_1}{\partial z} - \frac{\partial f_3}{\partial x}, \frac{\partial f_2}{\partial x} - \frac{\partial f_1}{\partial y}\right)$$

(Definition 1.2.11). We also recall that, according to Proposition 6.1.2, the differential forms of degree 2

$$\{\mathrm{d}y \wedge \mathrm{d}z\ ,\ \mathrm{d}z \wedge \mathrm{d}x\ ,\ \mathrm{d}x \wedge \mathrm{d}y\}$$

are a basis of $\Lambda^2(\mathbb{R}^3)$.

Example 6.3.2. The exterior differential of a differential form of degree 1 *on* $\mathbb{R}^3$.

Let

$$\boldsymbol{F} = (f_1, f_2, f_3)$$

be a vector field of class C^1 in an open set U of $\mathbb{R}^3$. We intend to calculate the exterior differential of the associated differential form of degree 1,

$$\boldsymbol{\omega} = f_1\,\mathrm{d}x + f_2\,\mathrm{d}y + f_3\,\mathrm{d}z.$$

We obtain

$$\begin{aligned} d\omega &= df_1 \wedge dx + df_2 \wedge dy + df_3 \wedge dz \\ &= \left(\frac{\partial f_1}{\partial x}\, dx + \frac{\partial f_1}{\partial y}\, dy + \frac{\partial f_1}{\partial z}\, dz\right) \wedge dx \\ &+ \left(\frac{\partial f_2}{\partial x}\, dx + \frac{\partial f_2}{\partial y}\, dy + \frac{\partial f_2}{\partial z}\, dz\right) \wedge dy \\ &+ \left(\frac{\partial f_3}{\partial x}\, dx + \frac{\partial f_3}{\partial y}\, dy + \frac{\partial f_3}{\partial z}\, dz\right) \wedge dz \end{aligned}$$

from whence it follows

$$d\omega = \left(\frac{\partial f_3}{\partial y} - \frac{\partial f_2}{\partial z}\right) dy \wedge dz + \left(\frac{\partial f_1}{\partial z} - \frac{\partial f_3}{\partial x}\right) dz \wedge dx + \left(\frac{\partial f_2}{\partial x} - \frac{\partial f_1}{\partial y}\right) dx \wedge dy.$$

Finally, $d\omega$ is the differential form of degree 2 associated to the vector field

$$\text{Curl}\, \boldsymbol{F} = \boldsymbol{\nabla} \times \boldsymbol{F}.$$

Example 6.3.3 (The exterior differential of a differential form of degree 2 on $\mathbb{R}^3$). Let

$$\boldsymbol{F} = (f_1, f_2, f_3)$$

be a vector field of class C^1 in an open set U of $\mathbb{R}^3$. We intend to calculate the exterior differential of the associated differential form of degree 2,

$$\omega = f_1\, dy \wedge dz + f_2\, dz \wedge dx + f_3\, dx \wedge dy.$$

We get

$$d\omega = \frac{\partial f_1}{\partial x}\, dx \wedge dy \wedge dz + \frac{\partial f_2}{\partial y}\, dy \wedge dz \wedge dx + \frac{\partial f_3}{\partial z}\, dz \wedge dx \wedge dy.$$

Using

$$dy \wedge dz \wedge dx = -dy \wedge dx \wedge dz = (-1)^2 dx \wedge dy \wedge dz,$$

we conclude that

$$d\omega = (\text{Div}\, f)\, dx \wedge dy \wedge dz.$$

Proposition 6.3.1. *Let ω and η be two k-forms, φ an s-form, and f a function, each of class C^1 on an open set U of $\mathbb{R}^n$. Then*

(1) $d(\omega + \eta) = d\omega + d\eta$,
(2) $d(f\omega) = df \wedge \omega + f d\omega$,

(3) $\mathrm{d}(\boldsymbol{\omega}\wedge\boldsymbol{\varphi})=\mathrm{d}\boldsymbol{\omega}\wedge\boldsymbol{\varphi}+(-1)^k\boldsymbol{\omega}\wedge\mathrm{d}\boldsymbol{\varphi}$,
(4) $\mathrm{d}(\mathrm{d}\boldsymbol{\omega})=\mathbf{0}$ *if* $\boldsymbol{\omega}$ *is of class* C^2 *on* U.

Proof. (1) This is a simple verification.
(2) It suffices to consider the case

$$\boldsymbol{\omega}(\boldsymbol{x})=g(\boldsymbol{x})\ \mathrm{d}x_{i_1}\wedge\cdots\wedge\mathrm{d}x_{i_k}\ ,\ 1\le i_1<\cdots<i_k\le n,$$

where $g:U\to\mathbb{R}$ is a function of class C^1 on U. Keeping in mind that $g(\boldsymbol{x})$, $f(\boldsymbol{x})\in\mathbb{R}$, and $\mathrm{d}f(\boldsymbol{x}),\mathrm{d}g(\boldsymbol{x})\in\Lambda^1(\mathbb{R}^n)$, we obtain

$$\begin{aligned}\mathrm{d}(f\boldsymbol{\omega})(\boldsymbol{x})&=\mathrm{d}(fg)(\boldsymbol{x})\wedge\mathrm{d}x_{i_1}\wedge\cdots\wedge\mathrm{d}x_{i_k}\\&=\Big(g(\boldsymbol{x})\ \mathrm{d}f(\boldsymbol{x})+f(\boldsymbol{x})\ \mathrm{d}g(\boldsymbol{x})\Big)\wedge\mathrm{d}x_{i_1}\wedge\cdots\wedge\mathrm{d}x_{i_k}\\&=\mathrm{d}f(\boldsymbol{x})\wedge\Big(g(\boldsymbol{x})\mathrm{d}x_{i_1}\wedge\cdots\wedge\mathrm{d}x_{i_k}\Big)+f(\boldsymbol{x})\Big(\mathrm{d}g(\boldsymbol{x})\wedge\mathrm{d}x_{i_1}\wedge\cdots\wedge\mathrm{d}x_{i_k}\Big)\\&=(\mathrm{d}f\wedge\boldsymbol{\omega})(\boldsymbol{x})+f(\boldsymbol{x})(\mathrm{d}\boldsymbol{\omega})(\boldsymbol{x}),\end{aligned}$$

for every $\boldsymbol{x}\in U$.
(3) We will use some shorthand notation already introduced. It suffices to consider the case

$$\boldsymbol{\omega}(\boldsymbol{x})=f(\boldsymbol{x})\ \mathrm{d}x_{\mathbf{I}}\ ,\ \boldsymbol{\varphi}(\boldsymbol{x})=g(\boldsymbol{x})\ \mathrm{d}x_{\mathbf{J}},$$

where

$$\mathbf{I}\ =\ (i_1,\ldots,i_k)\ \text{ and }\ \mathbf{J}\ =\ (j_1,\ldots,j_s)$$

for $i_1<\cdots<i_k$, $j_1<\cdots<j_s$ and $f,g:U\to\mathbb{R}$ of class C^1 on U. Then

$$(\boldsymbol{\omega}\wedge\boldsymbol{\varphi})(\boldsymbol{x})=f(\boldsymbol{x})g(\boldsymbol{x})\ \mathrm{d}x_{\mathbf{I}}\wedge\mathrm{d}x_{\mathbf{J}},$$

and by the formula of the differential of a product,

$$\begin{aligned}\mathrm{d}(\boldsymbol{\omega}\wedge\boldsymbol{\varphi})(\boldsymbol{x})&=\mathrm{d}(fg)(\boldsymbol{x})\wedge\mathrm{d}x_{\mathbf{I}}\wedge\mathrm{d}x_{\mathbf{J}}\\&=\Big(g(\boldsymbol{x})\mathrm{d}f(\boldsymbol{x})+f(\boldsymbol{x})\mathrm{d}g(\boldsymbol{x})\Big)\wedge\mathrm{d}x_{\mathbf{I}}\wedge\mathrm{d}x_{\mathbf{J}}\\&=\mathrm{d}f(\boldsymbol{x})\wedge\mathrm{d}x_{\mathbf{I}}\wedge(g(\boldsymbol{x})\ \mathrm{d}x_{\mathbf{J}})+f(\boldsymbol{x})\mathrm{d}g(\boldsymbol{x})\wedge\mathrm{d}x_{\mathbf{I}}\wedge\mathrm{d}x_{\mathbf{J}}.\end{aligned}$$

We have

$$\mathrm{d}f(\boldsymbol{x})\wedge\mathrm{d}x_{\mathbf{I}}=\mathrm{d}\boldsymbol{\omega}(\boldsymbol{x}),$$

for each $\boldsymbol{x} \in U$. Also

$$\begin{aligned}
\mathrm{d}g(\boldsymbol{x}) \wedge \mathrm{d}x_{\mathbf{I}} &= \sum_{l=1}^{n} \frac{\partial g}{\partial x_l}(\boldsymbol{x})\mathrm{d}x_l \wedge \mathrm{d}x_{\mathbf{I}} \\
&= \sum_{l=1}^{n} (-1)^k \mathrm{d}x_{\mathbf{I}} \wedge \frac{\partial g}{\partial x_l}(\boldsymbol{x})\mathrm{d}x_l \\
&= (-1)^k \mathrm{d}x_{\mathbf{I}} \wedge \mathrm{d}g(\boldsymbol{x}).
\end{aligned}$$

Observe that we have used the fact that the permutation $(l, i_1, \ldots, i_k) \mapsto (i_1, \ldots, i_k, l)$ has signature k. Now we can conclude that

$$\mathrm{d}(\boldsymbol{\omega} \wedge \boldsymbol{\varphi}) = \mathrm{d}\boldsymbol{\omega} \wedge \boldsymbol{\varphi} + (-1)^k \boldsymbol{\omega} \wedge \mathrm{d}\boldsymbol{\varphi}.$$

(4) Suppose $\boldsymbol{\omega}(\boldsymbol{x}) = f(\boldsymbol{x})\mathrm{d}x_{\mathbf{I}}$, $\mathbf{I} = (i_1, \ldots, i_k)$ where $i_1 < \cdots < i_k$ and $f : U \subset \mathbb{R}^n \to \mathbb{R}$ is of class C^2 on U. Then

$$\mathrm{d}\boldsymbol{\omega}(\boldsymbol{x}) = \mathrm{d}f(\boldsymbol{x}) \wedge \mathrm{d}x_{\mathbf{I}} = \sum_{j=1}^{n} \frac{\partial f}{\partial x_j}(\boldsymbol{x})\, \mathrm{d}x_j \wedge \mathrm{d}x_{\mathbf{I}}$$

and therefore

$$\mathrm{d}(\mathrm{d}\boldsymbol{\omega})(\boldsymbol{x}) = \sum_{j=1}^{n} \Big(\sum_{k=1}^{n} \frac{\partial^2 f}{\partial x_k\, \partial x_j}(\boldsymbol{x})\, \mathrm{d}x_k \Big) \wedge \mathrm{d}x_j \wedge \mathrm{d}x_{\mathbf{I}},$$

for every $\boldsymbol{x} \in U$. Now, using

$$\mathrm{d}x_j \wedge \mathrm{d}x_j = \mathbf{0} \ \text{ and } \ \mathrm{d}x_k \wedge \mathrm{d}x_j = -\mathrm{d}x_j \wedge \mathrm{d}x_k,$$

we are left with

$$\mathrm{d}(\mathrm{d}\boldsymbol{\omega})(\boldsymbol{x}) = \sum_{1 \le j < k \le n} \Big(\frac{\partial^2 f}{\partial x_j\, \partial x_k}(\boldsymbol{x}) - \frac{\partial^2 f}{\partial x_k\, \partial x_j}(\boldsymbol{x}) \Big)\, \mathrm{d}x_j \wedge \mathrm{d}x_k \wedge \mathrm{d}x_{\mathbf{I}}$$

for all $\boldsymbol{x} \in U$, and it follows from Schwarz's theorem that $\mathrm{d}(\mathrm{d}\boldsymbol{\omega})(\boldsymbol{x}) = \mathbf{0}$ for all $\boldsymbol{x} \in U$. □

Corollary 6.3.1. *Let $\boldsymbol{F} := (f_1, f_2, f_3)$ be a vector field of class C^2 in an open set U of $\mathbb{R}^3$. Then*

$$\mathrm{Div}\ (\boldsymbol{\nabla} \times \boldsymbol{F}) = 0 \text{ in } U.$$

Proof. If

$$\boldsymbol{\omega} = f_1 \, \mathrm{d}x + f_2 \, \mathrm{d}y + f_3 \, \mathrm{d}z$$

is the differential form of degree 1 associated to the vector field $\boldsymbol{F}$, then $\boldsymbol{\nabla} \times \boldsymbol{F}$ is the vector field associated to the differential form (of degree 2) $\mathrm{d}\boldsymbol{\omega}$. Now the conclusion follows from

$$\mathbf{0} = \mathrm{d}(\mathrm{d}\boldsymbol{\omega}) = \mathrm{Div}\,(\boldsymbol{\nabla} \times \boldsymbol{F}) \, \mathrm{d}x \wedge \mathrm{d}y \wedge \mathrm{d}z. \qquad \square$$

6.4 Change of Variables: Pullback

Definition 6.4.1. Let $\boldsymbol{T} : U \subset \mathbb{R}^m \to G \subset \mathbb{R}^n$ be a mapping of class C^1 on the set U and let $\boldsymbol{\omega} : G \subset \mathbb{R}^n \to \Lambda^k(\mathbb{R}^n)$ be a differential k-form on the set G. The *pullback* of $\boldsymbol{\omega}$ under $\boldsymbol{T}$ is the differential k-form $\boldsymbol{T}^*\boldsymbol{\omega} : U \subset \mathbb{R}^m \to \Lambda^k(\mathbb{R}^m)$ defined by

$$\boldsymbol{T}^*\boldsymbol{\omega}(\boldsymbol{u})(\boldsymbol{v}_1,\dots,\boldsymbol{v}_k) := \boldsymbol{\omega}(\boldsymbol{T}(\boldsymbol{u}))\Big(\mathrm{d}\boldsymbol{T}(\boldsymbol{u})(\boldsymbol{v}_1),\dots,\mathrm{d}\boldsymbol{T}(\boldsymbol{u})(\boldsymbol{v}_k)\Big),$$

for every $\boldsymbol{u} \in U$ and every $(\boldsymbol{v}_1,\dots,\boldsymbol{v}_k) \in (\mathbb{R}^m)^k$.

Observe that the set G in this definition does not need to be open. If $\boldsymbol{f} : G \to \mathbb{R}^n$ is a function, we put $\boldsymbol{T}^*\boldsymbol{f} := \boldsymbol{f} \circ \boldsymbol{T}$. Also we will refer to $\boldsymbol{f}$ as a differential form of degree 0, or a 0-form.

Example 6.4.1 and subsequent results show that the operation of pullback is, in fact, a *change of variables*.

Example 6.4.1. Let $\boldsymbol{T} : U \subset \mathbb{R}^m \to G \subset \mathbb{R}^n$ be a mapping of class C^1, $\boldsymbol{T} = (T_1,\dots,T_n)$. For each $1 \le i \le n$ we have $\boldsymbol{T}^*\mathrm{d}x_i = \mathrm{d}\,T_i$.

For $\boldsymbol{u} \in U \subset \mathbb{R}^m$ we write $\boldsymbol{x} = \boldsymbol{T}(\boldsymbol{u})$, so $x_i = T_i(\boldsymbol{u})$. By definition,

$$(\boldsymbol{T}^*\mathrm{d}x_i)(\boldsymbol{u})(\boldsymbol{v}) = \mathrm{d}x_i(\boldsymbol{T}(\boldsymbol{u}))\Big(\mathrm{d}\boldsymbol{T}(\boldsymbol{u})(\boldsymbol{v})\Big) = \mathrm{d}x_i\Big(\mathrm{d}\boldsymbol{T}(\boldsymbol{u})(\boldsymbol{v})\Big)$$

for each $\boldsymbol{v} \in \mathbb{R}^m$. But

$$\mathrm{d}\boldsymbol{T}(\boldsymbol{u})(\boldsymbol{v}) = \big(\mathrm{d}T_1(\boldsymbol{u})(\boldsymbol{v}),\dots,\mathrm{d}T_n(\boldsymbol{u})(\boldsymbol{v})\big),$$

and hence

$$\mathrm{d}x_i\Big(\mathrm{d}\boldsymbol{T}(\boldsymbol{u})(\boldsymbol{v})\Big) = \mathrm{d}T_i(\boldsymbol{u})(\boldsymbol{v}).$$

This proves that $\boldsymbol{T}^*(\mathrm{d}x_i) = \mathrm{d}T_i$.

Corollary 6.4.1. *Let $\boldsymbol{T} : U \subset \mathbb{R}^m \to G \subset \mathbb{R}^n$ be a mapping of class C^1 on the open set U and let $\boldsymbol{\omega} : G \subset \mathbb{R}^n \to \Lambda^k(\mathbb{R}^n)$ be a differential form of degree k on the open set G. If*

$$\boldsymbol{\omega}(\boldsymbol{x}) = \sum_{1 \le i_1 < \cdots < i_k \le n} f_{i_1,\dots,i_k}(\boldsymbol{x}) \wedge \mathrm{d}x_{i_1} \wedge \cdots \wedge \mathrm{d}x_{i_k},$$

for every $x \in G$, then

$$\boldsymbol{T}^*(\boldsymbol{\omega})(\boldsymbol{u}) = \sum_{1 \le i_1 < \cdots < i_k \le n} f_{i_1,\dots,i_k} \circ \boldsymbol{T}(\boldsymbol{u}) \wedge \mathrm{d}T_{i_1} \wedge \cdots \wedge \mathrm{d}T_{i_k},$$

for every $u \in U$.

Proof. Consider the differential 1-form $\varphi_j = \mathrm{d}x_{i_j}$, $j = 1,\dots,k$. Clearly $\varphi_j(\boldsymbol{x}) = \mathrm{d}x_{i_j}$ for every $\boldsymbol{x} \in G$. Now,

$$T^*\left(\varphi_1 \wedge \cdots \wedge \varphi_k\right)(u)\Big(v_1,\ldots,v_k\Big)$$

by corollary 6.2.1, coincides with

$$\begin{aligned}
&\left(\varphi_1 \wedge \cdots \wedge \varphi_k\right)(T(u))\Big(\mathrm{d}T(u)(v_1),\ldots,\mathrm{d}T(u)(v_k)\Big) \\
&= \det\big(\varphi_i(T(u))(\mathrm{d}T(u)(v_j))\big)_{i,j=1,\ldots,k} \\
&= \big(T^*(\varphi_1) \wedge \cdots \wedge T^*(\varphi_k)\big)(u)(v_1,\ldots,v_k) \\
&= \big(\mathrm{d}T_{i_1} \wedge \cdots \wedge \mathrm{d}T_{i_k}\big)(u)(v_1,\ldots,v_k)
\end{aligned}$$

for every $u \in U$ and $(v_1,\ldots,v_k) \in (\mathbb{R}^m)^k$. □

If T and ω are of class C^p, $1 \leq p \leq \infty$, then $T^*(\omega)$ is a differential k-form of class C^{p-1}.

Example 6.4.2 (Polar coordinates). Let $T : \mathbb{R}^2_{(\rho,\theta)} \to \mathbb{R}^2_{(x,y)}$ be the mapping defined by

$$T(\rho,\theta) := (\rho\cos\theta, \rho\sin\theta).$$

Denote by (ρ,θ) the coordinates of an arbitrary point in the domain of T and by (x,y) the coordinates of an arbitrary point in the range of T, so that

$$\{\mathrm{d}\rho, \mathrm{d}\theta\}$$

is a basis of $\Lambda^1(\mathbb{R}^2_{(\rho,\theta)})$ and

$$\{\mathrm{d}x, \mathrm{d}y\}$$

is a basis of $\Lambda^1(\mathbb{R}^2_{(x,y)})$. According to Example 6.4.1, since $T_1(\rho,\theta) = \rho\cos\theta$ and $T_2(\rho,\theta) = \rho\sin\theta$,

$$T^*(\mathrm{d}x)(\rho,\theta) = \mathrm{d}T_1(\rho,\theta) = \cos\theta\mathrm{d}\rho - \rho\sin\theta\mathrm{d}\theta$$

and

$$T^*(\mathrm{d}y)(\rho,\theta) = \mathrm{d}T_2(\rho,\theta) = \sin\theta\mathrm{d}\rho + \rho\cos\theta\mathrm{d}\theta.$$

We observe that if we put $x = \rho\cos\theta$, $y = \rho\sin\theta$ and we perform a *formal calculation*

$$\begin{cases} \mathrm{d}x = \dfrac{\partial x}{\partial\rho}\mathrm{d}\rho + \dfrac{\partial x}{\partial\theta}\mathrm{d}\theta, \\[2ex] \mathrm{d}y = \dfrac{\partial y}{\partial\rho}\mathrm{d}\rho + \dfrac{\partial y}{\partial\theta}\mathrm{d}\theta, \end{cases}$$

we obtain

$$\begin{cases} dx = \cos\theta d\rho - \rho\sin\theta d\theta, \\ dy = \sin\theta d\rho + \rho\cos d\theta. \end{cases}$$

Thus, in changing coordinates, we are effectively calculating the *pullback* $\boldsymbol{T}^*(dx)$ and $\boldsymbol{T}^*(dy)$.

Prompted by this, and other similar examples, at times we will informally refer to the pullback as a *change of variables*, despite the fact that the pullback is also defined in the case that $\boldsymbol{T}$ is not a transformation of coordinates.

The basic properties of the pullback are the following (compare with Edwards [9]).

Proposition 6.4.1. *Let $\boldsymbol{T}: U \subset \mathbb{R}^m \to G \subset \mathbb{R}^n$ be a mapping of class C^1 on the open set U and let $\boldsymbol{\omega}$ and $\boldsymbol{\omega}_j$ be differential forms of degree k on the open set G in $\mathbb{R}^n$. Further, let $\boldsymbol{\varphi}$ be a differential form of degree s and f a function of class C^1 on G. Then:*

(1) If $\boldsymbol{S}: G \subset \mathbb{R}^n \to V \subset \mathbb{R}^p$ is of class C^1 and $\boldsymbol{\theta}: V \subset \mathbb{R}^p \to \Lambda^k(\mathbb{R}^p)$ is a differential form of degree k on V, then

$$(\boldsymbol{S} \circ \boldsymbol{T})^*(\boldsymbol{\theta}) = \boldsymbol{T}^*(\boldsymbol{S}^*(\boldsymbol{\theta})),$$

(2) $\boldsymbol{T}^*(f\boldsymbol{\omega}) = \boldsymbol{T}^*(f)\boldsymbol{T}^*(\boldsymbol{\omega})$,
(3) $\boldsymbol{T}^*(a\boldsymbol{\omega}) = a\boldsymbol{T}^*(\boldsymbol{\omega})$, $a \in \mathbb{R}$,
(4) $\boldsymbol{T}^*(\boldsymbol{\omega}_1 + \boldsymbol{\omega}_2) = \boldsymbol{T}^*(\boldsymbol{\omega}_1) + \boldsymbol{T}^*(\boldsymbol{\omega}_2)$,
(5) $\boldsymbol{T}^*(\boldsymbol{\omega} \wedge \boldsymbol{\varphi}) = \boldsymbol{T}^*(\boldsymbol{\omega}) \wedge \boldsymbol{T}^*(\boldsymbol{\varphi})$.

Proof. (1) is a consequence of Definition 6.4.1 and the chain rule, while (2), (3), and (4) are immediate from the definition.

For (5), write

$$\boldsymbol{\omega} = \sum_{\mathbf{I} \in \mathscr{J}} f_{\mathbf{I}}\, dx_{\mathbf{I}}\,,\ \boldsymbol{\varphi} = \sum_{\mathbf{J} \in \mathscr{J}} g_{\mathbf{J}}\, dx_{\mathbf{J}}.$$

Here $\mathbf{I} = (i_1, \ldots, i_k)$ and $\mathbf{J} = (j_1, \ldots, j_s)$, where

$$1 \le i_1 < \cdots < i_k \le n \text{ and } 1 \le j_1 < \cdots j_s \le n.$$

Then

$$\boldsymbol{\omega} \wedge \boldsymbol{\varphi} = \sum_{\mathbf{I},\mathbf{J}} f_{\mathbf{I}} \cdot g_{\mathbf{J}}\, dx_{\mathbf{I}} \wedge dx_{\mathbf{J}},$$

and it follows from properties (2) and (4) that

$$\boldsymbol{T}^*(\boldsymbol{\omega} \wedge \boldsymbol{\varphi}) = \sum_{\mathbf{I},\mathbf{J}} (f_{\mathbf{I}} \circ \boldsymbol{T})(g_{\mathbf{J}} \circ \boldsymbol{T})\, \boldsymbol{T}^*(dx_{\mathbf{I}} \wedge dx_{\mathbf{J}}).$$

As we already proved in Corollary 6.4.1,

$$T^*(\mathrm{d}x_{\mathbf{I}} \wedge \mathrm{d}x_{\mathbf{J}}) = (\mathrm{d}T_{i_1} \wedge \cdots \wedge \mathrm{d}T_{i_k}) \wedge (\mathrm{d}T_{j_1} \wedge \cdots \wedge \mathrm{d}T_{j_s}).$$

Applying Corollary 6.4.1 again, we get

$$\sum_{\mathbf{I} \in \mathscr{J}} (f_{\mathbf{I}} \circ T)\, \mathrm{d}T_{i_1} \wedge \cdots \wedge \mathrm{d}T_{i_k} = T^*(\omega)$$

and

$$\sum_{\mathbf{J} \in \mathscr{J}} (g_{\mathbf{J}} \circ T)\, \mathrm{d}T_{j_1} \wedge \cdots \wedge \mathrm{d}T_{j_s} = T^*(\varphi).$$

Thus

$$\begin{aligned} T^*(\omega) \wedge T^*(\varphi) &= \sum_{\mathbf{I},\mathbf{J}} (f_{\mathbf{I}} \circ T)(g_{\mathbf{J}} \circ T)\, T^*(\mathrm{d}x_{\mathbf{I}} \wedge \mathrm{d}x_{\mathbf{J}}) \\ &= T^*(\omega \wedge \varphi). \end{aligned}$$

□

Proposition 6.4.2. *Let $T : U \subset \mathbb{R}^m \to G \subset \mathbb{R}^n$ be a mapping of class C^2 on the open set U and let ω be a differential form of degree k and class C^1 on the open set G of $\mathbb{R}^n$. Then $T^*(\mathrm{d}\omega) = \mathrm{d}(T^*\omega)$.*

Proof. We first prove the result for differential forms of degree 0 (functions). To this end, let

$$f : G \subset \mathbb{R}^n \to \mathbb{R}$$

be a function of class C^1 and write $x_i = T_i(u)$. After applying Proposition 6.4.1 (2), Example 6.4.1, and then the chain rule, we obtain

$$\begin{aligned} T^*(\mathrm{d}f) = T^*\left(\sum_{i=1}^{n} \frac{\partial f}{\partial x_i}\mathrm{d}x_i\right) &= \sum_{i=1}^{n} T^*\left(\frac{\partial f}{\partial x_i}\right)\mathrm{d}T_i \\ &= \sum_{i=1}^{n}\sum_{k=1}^{m} T^*\left(\frac{\partial f}{\partial x_i}\right)\frac{\partial T_i}{\partial u_k}\mathrm{d}u_k \\ &= \sum_{k=1}^{m}\left(\sum_{i=1}^{n}\left(\frac{\partial f}{\partial x_i} \circ T\right)\frac{\partial T_i}{\partial u_k}\right)\mathrm{d}u_k \\ &= \sum_{k=1}^{m} \frac{\partial (f \circ T)}{\partial u_k}\mathrm{d}u_k = \mathrm{d}(f \circ T) = \mathrm{d}(T^* f). \end{aligned}$$

We now proceed by induction on k. Assume that the statement of the proposition holds for differential forms of degree $k-1$ $(k \geq 1)$ and consider

$$\boldsymbol{\omega}(\boldsymbol{x}) = f(\boldsymbol{x})\,\mathrm{d}x_{i_1} \wedge \cdots \wedge \mathrm{d}x_{i_{k-1}} \wedge \mathrm{d}x_{i_k}.$$

We take

$$\boldsymbol{\eta}(\boldsymbol{x}) := f(\boldsymbol{x})\,\mathrm{d}x_{i_1} \wedge \cdots \wedge \mathrm{d}x_{i_{k-1}},$$

which is a differential form of degree $k-1$ and satisfies

$$\boldsymbol{\omega} = \boldsymbol{\eta} \wedge \mathrm{d}x_{i_k}.$$

Then, by Proposition 6.3.1 (4) we have $\mathrm{d}(\mathrm{d}x_{i_k}) = \mathbf{0}$ and we deduce from Proposition 6.3.1 (3) that

$$\mathrm{d}\boldsymbol{\omega} = \mathrm{d}\boldsymbol{\eta} \wedge \mathrm{d}x_{i_k} + (-1)^{k-1}\boldsymbol{\eta} \wedge \mathrm{d}(\mathrm{d}x_{i_k}) = \mathrm{d}\boldsymbol{\eta} \wedge \mathrm{d}x_{i_k}.$$

Applying Proposition 6.4.1 (5) yields

$$\begin{aligned} \boldsymbol{T}^*(\mathrm{d}\boldsymbol{\omega}) = \boldsymbol{T}^*(\mathrm{d}\boldsymbol{\eta} \wedge \mathrm{d}x_{i_k}) &= \boldsymbol{T}^*(\mathrm{d}\boldsymbol{\eta}) \wedge \boldsymbol{T}^*(\mathrm{d}x_{i_k}) \\ &= \mathrm{d}(\boldsymbol{T}^*\boldsymbol{\eta}) \wedge \mathrm{d}T_{i_k} \end{aligned}$$

by the induction hypothesis. Also from Proposition 6.4.1 (5) and Example 6.4.1, we obtain

$$\begin{aligned} \mathrm{d}(\boldsymbol{T}^*\boldsymbol{\omega}) = \mathrm{d}\big(\boldsymbol{T}^*(\boldsymbol{\eta} \wedge \mathrm{d}x_{i_k})\big) &= \mathrm{d}\big(\boldsymbol{T}^*\boldsymbol{\eta} \wedge \boldsymbol{T}^*(\mathrm{d}x_{i_k})\big) \\ &= \mathrm{d}\big(\boldsymbol{T}^*\boldsymbol{\eta} \wedge \mathrm{d}T_{i_k}\big). \end{aligned}$$

Now Proposition 6.3.1 (parts (3) and (4)) allows us to conclude, since $\boldsymbol{T}$ is of class C^2 in U, that

$$\begin{aligned} \mathrm{d}(\boldsymbol{T}^*\boldsymbol{\omega}) &= \mathrm{d}\big(\boldsymbol{T}^*\boldsymbol{\eta} \wedge \mathrm{d}T_{i_k}\big) \\ &= \mathrm{d}(\boldsymbol{T}^*\boldsymbol{\eta}) \wedge \mathrm{d}T_{i_k} + (-1)^{k-1}\boldsymbol{T}^*\boldsymbol{\eta} \wedge \mathrm{d}(\mathrm{d}T_{i_k}) \\ &= \mathrm{d}(\boldsymbol{T}^*\boldsymbol{\eta}) \wedge \mathrm{d}T_{i_k}. \end{aligned}$$

Consequently $\boldsymbol{T}^*(\mathrm{d}\boldsymbol{\omega}) = \mathrm{d}(\boldsymbol{T}^*\boldsymbol{\omega})$. □

Example 6.4.3. In this example we will show that the definition provided in Chap. 2 of the integral of a differential 1-form ω along a path $\boldsymbol{\gamma}$ (or equivalently the integral of its associated vector field $\boldsymbol{F}$ along $\boldsymbol{\gamma}$) can be written in terms of the pullback.

Let $\boldsymbol{\gamma} : [a,b] \to \mathbb{R}^n$ be a path of class C^1 whose arc is contained in an open set U and let $\boldsymbol{F} = (f_1, \dots, f_n)$ be a continuous vector field on U with associated differential form of degree 1:

$$\boldsymbol{\omega} = \sum_{j=1}^{n} f_j \cdot \mathrm{d}x_j.$$

We consider the restriction of γ to the open interval (a,b). Then $\boldsymbol{\gamma}^*(\boldsymbol{\omega})$ is a differential form of degree 1 on the interval (a,b), and hence for $t \in (a,b)$,

$$\boldsymbol{\gamma}^*(\boldsymbol{\omega})(t) = h(t) \cdot \mathrm{d}t,$$

where $h : (a,b) \to \mathbb{R}$ is a continuous function. We claim that

$$\int_{\boldsymbol{\gamma}} \boldsymbol{\omega} = \int_{\boldsymbol{\gamma}} \boldsymbol{F} = \int_a^b h(t)\, \mathrm{d}t.$$

In fact,

$$\begin{aligned} \boldsymbol{\gamma}^*(\boldsymbol{\omega})(t) &= \sum_{j=1}^{n} (f_j \circ \boldsymbol{\gamma})(t) \boldsymbol{\gamma}^*(\mathrm{d}x_j) \\ &= \sum_{j=1}^{n} (f_j \circ \boldsymbol{\gamma})(t) \gamma_j'(t) \cdot \mathrm{d}t \\ &= \langle (\boldsymbol{F} \circ \boldsymbol{\gamma})(t), \boldsymbol{\gamma}'(t) \rangle \cdot \mathrm{d}t. \end{aligned}$$

Thus $h(t) = \langle (\boldsymbol{F} \circ \boldsymbol{\gamma})(t), \boldsymbol{\gamma}'(t) \rangle$ for every $t \in (a,b)$. But by definition,

$$\int_{\boldsymbol{\gamma}} \boldsymbol{F} = \int_a^b \langle (\boldsymbol{F} \circ \boldsymbol{\gamma})(t), \boldsymbol{\gamma}'(t) \rangle\, \mathrm{d}t.$$

The reader should be aware that the integrability of the function h above, and its integral, does not depend on the values taken by h at $x = a$ and $x = b$ or even on whether the function is defined at these points.

Example 6.4.4. Let S be a plane in $\mathbb{R}^3$ that contains the point $\boldsymbol{P}$ and let M denote the underlying vector subspace, that is, $M = S - \boldsymbol{P}$. We consider a linear isometry $\boldsymbol{L} : \mathbb{R}^2 \to M$ and assume that the surface S is oriented according to the parameterization

$$\boldsymbol{T} : \mathbb{R}^2 \to S\ ,\ (s,t) \mapsto \boldsymbol{P} + \boldsymbol{L}(s,t).$$

This means that the orientation of S is determined by the unit normal vector

$$\boldsymbol{N} := \boldsymbol{L}(\boldsymbol{e}_1) \times \boldsymbol{L}(\boldsymbol{e}_2).$$

For a given continuous vector field $\boldsymbol{F} = (f_1, f_2, f_3)$ in $\mathbb{R}^3$ with associated differential form

$$\boldsymbol{\omega} = f_1 \, \mathrm{d}y \wedge \mathrm{d}z + f_2 \, \mathrm{d}z \wedge \mathrm{d}x + f_3 \, \mathrm{d}x \wedge \mathrm{d}y$$

of degree 2, we are going to evaluate $\boldsymbol{T}^*(\boldsymbol{\omega})$.

Since $\boldsymbol{T}^*(\boldsymbol{\omega})$ is a differential form of degree 2 on $\mathbb{R}^2$, then, denoting by (s,t) the coordinates of an arbitrary point,

$$\boldsymbol{T}^*(\boldsymbol{\omega}) = H \cdot \mathrm{d}s \wedge \mathrm{d}t.$$

The function $H : \mathbb{R}^2 \to \mathbb{R}$ can be evaluated from the fact that $\mathrm{d}\boldsymbol{T}(s,t) = \boldsymbol{L}$, for every $(s,t) \in \mathbb{R}^2$ and

$$\begin{aligned} H(s,t) &= \boldsymbol{T}^*(\boldsymbol{\omega})(s,t)\,(\boldsymbol{e}_1, \boldsymbol{e}_2) \\ &= \boldsymbol{\omega}(\boldsymbol{T}(s,t))\,(\mathrm{d}\boldsymbol{T}(s,t)(\boldsymbol{e}_1), \mathrm{d}\boldsymbol{T}(s,t)(\boldsymbol{e}_2)) \\ &= \boldsymbol{\omega}(\boldsymbol{T}(s,t))\,(\boldsymbol{L}(\boldsymbol{e}_1), \boldsymbol{L}(\boldsymbol{e}_2)) \\ &= \langle \boldsymbol{F}(\boldsymbol{T}(s,t)) \,,\, \boldsymbol{L}(\boldsymbol{e}_1) \times \boldsymbol{L}(\boldsymbol{e}_2) \rangle \,. \end{aligned}$$

This means that

$$\boldsymbol{T}^*(\boldsymbol{\omega}) = (\langle \boldsymbol{F}, \boldsymbol{N} \rangle \circ \boldsymbol{T}) \cdot \mathrm{d}s \wedge \mathrm{d}t.$$

We have proved that the *pullback* of $\boldsymbol{\omega}$ is given by the *component of the vector field $\boldsymbol{F}$ in the direction of a normal vector to the plane.*

Recall the following notation that is commonly used in the study of functions of several variables. If

$$\boldsymbol{T} : U \subset \mathbb{R}^k \to \mathbb{R}^n$$

is a C^1 mapping in U, with $k \le n$, we let

$$\frac{\partial(T_{i_1}, \ldots, T_{i_k})}{\partial(u_1, \ldots, u_k)}(\boldsymbol{u}) = \det \begin{pmatrix} \boldsymbol{\nabla} T_{i_1}(\boldsymbol{u}) \\ \vdots \\ \boldsymbol{\nabla} T_{i_k}(\boldsymbol{u}) \end{pmatrix} = \begin{vmatrix} \frac{\partial T_{i_1}}{\partial u_1}(\boldsymbol{u}) & \ldots & \frac{\partial T_{i_1}}{\partial u_k}(\boldsymbol{u}) \\ \ldots & \ldots & \ldots \\ \frac{\partial T_{i_k}}{\partial u_1}(\boldsymbol{u}) & \ldots & \frac{\partial T_{i_k}}{\partial u_k}(\boldsymbol{u}) \end{vmatrix}.$$

Theorem 6.4.1. *Let $\boldsymbol{T} : U \subset \mathbb{R}^k \to G \subset \mathbb{R}^n$ be a mapping of class C^1 ($\boldsymbol{x} = \boldsymbol{T}(\boldsymbol{u})$) in an open set U and let $\boldsymbol{\omega} : G \subset \mathbb{R}^n \to \Lambda^k(\mathbb{R}^n)$ be the degree-k differential form*

$$\boldsymbol{\omega}(\boldsymbol{x}) = f(\boldsymbol{x}) \, \mathrm{d}x_{i_1} \wedge \cdots \wedge \mathrm{d}x_{i_k} \,, \ 1 \le i_1 < \cdots < i_k \le n.$$

Then:

(1) $(\boldsymbol{T}^*\boldsymbol{\omega})(\boldsymbol{u}) = f(\boldsymbol{T}(\boldsymbol{u})) \cdot \dfrac{\partial(T_{i_1}, \ldots, T_{i_k})}{\partial(u_1, \ldots, u_k)}(\boldsymbol{u}) \cdot \mathrm{d}u_1 \wedge \cdots \wedge \mathrm{d}u_k.$

(2) If $k = n$ and $\boldsymbol{\omega}(\boldsymbol{x}) = f(\boldsymbol{x}) \cdot \mathrm{d}x_1 \wedge \cdots \wedge \mathrm{d}x_n$, we have

$$(\boldsymbol{T}^* \boldsymbol{\omega})(\boldsymbol{u}) = f(\boldsymbol{T}(\boldsymbol{u})) \cdot J\boldsymbol{T}(\boldsymbol{u}) \cdot \mathrm{d}u_1 \wedge \cdots \wedge \mathrm{d}u_k,$$

for every $u \in U$.

Proof. It suffices to prove (1). Recall that $\Lambda^k(\mathbb{R}^k)$ is a 1-dimensional vector space and a basis is given by

$$\{\mathrm{d}u_1 \wedge \cdots \wedge \mathrm{d}u_k\}.$$

By Corollary 6.4.1,

$$(\boldsymbol{T}^* \boldsymbol{\omega})(\boldsymbol{u}) = f(\boldsymbol{T}(\boldsymbol{u})) \cdot \left(\mathrm{d}T_{i_1} \wedge \cdots \wedge \mathrm{d}T_{i_k}\right)(\boldsymbol{u})$$

for every $\boldsymbol{u} \in U$, and hence by Corollary 6.2.1,

$$\begin{aligned}
(\boldsymbol{T}^* \boldsymbol{\omega})(\boldsymbol{u})\Big(\boldsymbol{v}_1, \ldots, \boldsymbol{v}_k\Big) &= f(\boldsymbol{T}(\boldsymbol{u})) \begin{vmatrix} \mathrm{d}T_{i_1}(\boldsymbol{u})(\boldsymbol{v}_1) & \ldots & \mathrm{d}T_{i_1}(\boldsymbol{u})(\boldsymbol{v}_k) \\ \ldots & \ldots & \ldots \\ \mathrm{d}T_{i_k}(\boldsymbol{u})(\boldsymbol{v}_1) & \ldots & \mathrm{d}T_{i_k}(\boldsymbol{u})(\boldsymbol{v}_k) \end{vmatrix} \\
&= f(\boldsymbol{T}(\boldsymbol{u})) \begin{vmatrix} \langle \boldsymbol{\nabla} T_{i_1}(\boldsymbol{u}), \boldsymbol{v}_1 \rangle & \ldots & \langle \boldsymbol{\nabla} T_{i_1}(\boldsymbol{u}), \boldsymbol{v}_k \rangle \\ \ldots & \ldots & \ldots \\ \big\langle \boldsymbol{\nabla} T_{i_k}(\boldsymbol{u}), \boldsymbol{v}_1 \big\rangle & \ldots & \big\langle \boldsymbol{\nabla} T_{i_k}(\boldsymbol{u}), \boldsymbol{v}_k \big\rangle \end{vmatrix} \\
&= f(\boldsymbol{T}(\boldsymbol{u})) \det \begin{pmatrix} \boldsymbol{\nabla} T_{i_1}(\boldsymbol{u}) \\ \vdots \\ \boldsymbol{\nabla} T_{i_k}(\boldsymbol{u}) \end{pmatrix} \det(\boldsymbol{v}_1, \ldots, \boldsymbol{v}_k) \\
&= f(\boldsymbol{T}(\boldsymbol{u})) \cdot \frac{\partial(T_{i_1}, \ldots, T_{i_k})}{\partial(u_1, \ldots, u_k)}(\boldsymbol{u}) \cdot \mathrm{d}u_1 \wedge \cdots \wedge \mathrm{d}u_k \Big(\boldsymbol{v}_1, \ldots, \boldsymbol{v}_k\Big)
\end{aligned}$$

for every $\boldsymbol{u} \in U$ and $(\boldsymbol{v}_1, \ldots, \boldsymbol{v}_k) \in (\mathbb{R}^k)^k$. □

Corollary 6.4.2. *Let $\boldsymbol{T} : U \subset \mathbb{R}^k \to G \subset \mathbb{R}^n$ $(k \leq n)$ be a mapping of class C^1 on the open set U and*

$$\boldsymbol{\omega}(\boldsymbol{x}) = \sum_{1 \leq i_1 < \cdots < i_k \leq n} f_{i_1, \ldots, i_k}(\boldsymbol{x})\, \mathrm{d}x_{i_1} \wedge \cdots \wedge \mathrm{d}x_{i_k}$$

a differential form of degree k on G. Then

$$(\boldsymbol{T}^* \boldsymbol{\omega})(\boldsymbol{u}) = \left(\sum_{1 \leq i_1 < \cdots < i_k \leq n} f_{i_1, \ldots, i_k}(\boldsymbol{T}(\boldsymbol{u})) \frac{\partial(T_{i_1}, \ldots, T_{i_k})}{\partial(u_1, \ldots, u_k)}(\boldsymbol{u}) \right) \mathrm{d}u_1 \wedge \cdots \wedge \mathrm{d}u_k.$$

6.5 Appendix: On Green's Theorem

The perplexed student might question the purpose of the rather nonintuitive algebraic machinery developed up to this point and even venture to ask whether it can be avoided. To a limited extent, it can—it is true that in the case of two variables, in proving the following result relating to Green's theorem, the theory of differential forms can be avoided. However, the calculations that are then necessary are extraordinarily difficult. We expect to convince the student of the utility of the algebraic language introduced and also of the close relationship between the operation of pullback and that of changing variables. Let us begin by recalling the change of variables theorem in the setting of Lebesgue integration.

Theorem 6.5.1. *Let $\boldsymbol{T} : U \subset \mathbb{R}^n \to \boldsymbol{T}(U) \subset \mathbb{R}^n$ be a bijective mapping between two open sets U and $\boldsymbol{T}(U)$ such that $\boldsymbol{T}$ and $\boldsymbol{T}^{-1}$ are both of class C^1. Let $f : \boldsymbol{T}(U) \to \mathbb{R}$. The mapping f is Lebesgue integrable on $\boldsymbol{T}(U)$ if and only if $(f \circ \boldsymbol{T})|J\boldsymbol{T}|$ is Lebesgue integrable on U, and in that case,*

$$\int_{\boldsymbol{T}(U)} f = \int_U (f \circ \boldsymbol{T})|J\boldsymbol{T}|,$$

where $J\boldsymbol{T}(\boldsymbol{x})$ is the determinant of the matrix of $\mathrm{d}\boldsymbol{T}(\boldsymbol{x})$, $\boldsymbol{x} \in U$.

A very useful variant for measurable sets is the following theorem.

Theorem 6.5.2. *Let $\boldsymbol{T} : U \subset \mathbb{R}^n \to \boldsymbol{T}(U) \subset \mathbb{R}^n$ be a bijective mapping between two open sets U and $\boldsymbol{T}(U)$ such that $\boldsymbol{T}$ and $\boldsymbol{T}^{-1}$ are both of class C^1. Let K be a measurable subset of U and let $f : \boldsymbol{T}(K) \to \mathbb{R}$ be integrable on the measurable set $\boldsymbol{T}(K)$. Then $(f \circ \boldsymbol{T})|J\boldsymbol{T}|$ is integrable on K, and*

$$\int_{\boldsymbol{T}(K)} f = \int_K (f \circ \boldsymbol{T})|J\boldsymbol{T}|.$$

Let us observe that part of the above theorem asserts that if K is measurable, then $T(K)$ is measurable too.

Example 6.5.1 (Invariance of Green's theorem under transformation of coordinates). Let U and V be open sets in $\mathbb{R}^2$ and $\boldsymbol{T} : U \to V$ a transformation of class C^2 on U with strictly positive Jacobian. Let $(x,y) = \boldsymbol{T}(s,t)$. Suppose that K is a compact subset of U whose (topological) boundary coincides with the arc defined by a closed piecewise C^1 path $\boldsymbol{\gamma}$ and assume also that Green's formula

$$\int_{\boldsymbol{\gamma}} \eta = \iint_K \left(\frac{\partial f_2}{\partial s} - \frac{\partial f_1}{\partial t} \right) \mathrm{d}(s,t)$$

holds whenever $\boldsymbol{\eta} = f_1 \mathrm{d}s + f_2 \mathrm{d}t$ is a differential form of degree 1 and class C^1 in the open set U. Then Green's formula also holds in the region $\boldsymbol{T}(K)$ whose boundary is $\boldsymbol{T} \circ \boldsymbol{\gamma}$ (see Example 6.3.1). Or equivalently,

$$\int_{\boldsymbol{T}\circ\boldsymbol{\gamma}} \boldsymbol{\omega} = \iint_{\boldsymbol{T}(K)} \left(\frac{\partial Q}{\partial x} - \frac{\partial P}{\partial y} \right) \mathrm{d}(x,y)$$

for every differential form $\boldsymbol{\omega} = P\mathrm{d}x + Q\mathrm{d}y$ of degree 1 and class C^1 in V.

In fact, we consider $\boldsymbol{\eta} := \boldsymbol{T}^* \boldsymbol{\omega} = f_1 \mathrm{d}s + f_2 \mathrm{d}t$, which is a differential form of class C^1 in U. According to Example 6.4.3, $\boldsymbol{\gamma}^*(\boldsymbol{\eta}) = h(t) \cdot \mathrm{d}t$ for some continuous function $h : (a,b) \to \mathbb{R}$. Then from Proposition 6.4.1, we deduce that

$$(\boldsymbol{T} \circ \boldsymbol{\gamma})^* \boldsymbol{\omega} = \boldsymbol{\gamma}^*(\boldsymbol{T}^* \boldsymbol{\omega}) = \boldsymbol{\gamma}^*(\boldsymbol{\eta}) = h(t) \cdot \mathrm{d}t,$$

and another application of Example 6.4.3 gives

$$\int_{\boldsymbol{T}\circ\boldsymbol{\gamma}} \boldsymbol{\omega} = \int_a^b h(t)\, \mathrm{d}t = \int_{\boldsymbol{\gamma}} \boldsymbol{\eta}. \tag{6.1}$$

Now we put $G = \left(\frac{\partial Q}{\partial x} - \frac{\partial P}{\partial y} \right)$, so that $\mathrm{d}\boldsymbol{\omega} = G \cdot \mathrm{d}x \wedge \mathrm{d}y$, and we write $\mathrm{d}\eta = H \cdot \mathrm{d}s \wedge \mathrm{d}t$ with $H = (\frac{\partial f_2}{\partial s} - \frac{\partial f_1}{\partial t})$. By hypothesis,

$$\int_{\boldsymbol{\gamma}} \boldsymbol{\eta} = \iint_K H(s,t)\, \mathrm{d}(s,t),$$

and our objective is to prove that

$$\int_{\boldsymbol{T}\circ\boldsymbol{\gamma}} \boldsymbol{\omega} = \iint_{\boldsymbol{T}(K)} G(x,y)\, \mathrm{d}(x,y).$$

According to Theorem 6.4.1,

$$\mathrm{d}\eta = \boldsymbol{T}^*(\mathrm{d}\boldsymbol{\omega}) = (G \circ \boldsymbol{T}) J\boldsymbol{T} \mathrm{d}s \wedge \mathrm{d}t,$$

which means that

$$H(s,t) = G(\boldsymbol{T}(s,t)) J\boldsymbol{T}(s,t).$$

Finally, since the Jacobian of $\boldsymbol{T}$ is positive, we deduce from the change of variables theorem that

$$\begin{aligned}\int_{\boldsymbol{T}\circ\boldsymbol{\gamma}} \omega = \int_{\boldsymbol{\gamma}} \eta &= \iint_K G(\boldsymbol{T}(s,t))J\boldsymbol{T}(s,t)\mathrm{d}(s,t)\\ &= \iint_K G(\boldsymbol{T}(s,t))|J\boldsymbol{T}(s,t)|\mathrm{d}(s,t)\\ &= \iint_{\boldsymbol{T}(K)} G(x,y)\mathrm{d}(x,y).\end{aligned}$$

The example that follows is based on Example 6.4.4.

Example 6.5.2. Let S be a plane in $\mathbb{R}^3$ that contains the point $\boldsymbol{P}$, and let M denote the corresponding vector subspace. Let $\boldsymbol{L}:\mathbb{R}^2\to M$ be a linear isometry and assume that the surface S is oriented according to the parameterization

$$\boldsymbol{T}:\mathbb{R}^2\to S\ ,\ (s,t)\mapsto \boldsymbol{P}+\boldsymbol{L}(s,t).$$

This means that the orientation is determined by the unit normal vector

$$\boldsymbol{N} := \boldsymbol{L}(\boldsymbol{e}_1)\times\boldsymbol{L}(\boldsymbol{e}_2).$$

If $\boldsymbol{C}_\varepsilon$ denotes the circle in $\mathbb{R}^2$ centered at the origin and with radius $\varepsilon>0$, oriented counterclockwise, then $\boldsymbol{\gamma}_\varepsilon := \boldsymbol{C}_\varepsilon\circ\boldsymbol{T}$ is a circle in the plane S, centered at $\boldsymbol{P}$ and with radius ε. Let $\boldsymbol{F}$ be a vector field of class C^1 in a neighborhood of $\boldsymbol{P}$. We want to estimate

$$\lim_{\varepsilon\to 0^+}\frac{1}{\pi\varepsilon^2}\int_{\boldsymbol{\gamma}_\varepsilon}\boldsymbol{F},$$

which is the work per unit area done by the vector field $\boldsymbol{F}$ in the plane S at the point $\boldsymbol{P}$. To this end, let

$$\boldsymbol{\omega} = f_1\,\mathrm{d}x + f_2\,\mathrm{d}y + f_3\,\mathrm{d}z$$

be the differential form of degree 1 associated to $\boldsymbol{F}$. Put $\boldsymbol{\nu} := \boldsymbol{T}^*(\boldsymbol{\omega})$, which is a differential form of degree 1, $\boldsymbol{\nu} = P\cdot\mathrm{d}s + Q\cdot\mathrm{d}t$, on $\mathbb{R}^2$, so that by (6.1),

$$\int_{\boldsymbol{\gamma}_\varepsilon}\boldsymbol{F} = \int_{\boldsymbol{C}_\varepsilon}\boldsymbol{\nu}.$$

Moreover, since $\mathrm{d}\boldsymbol{\nu} = \boldsymbol{T}^*(\mathrm{d}\boldsymbol{\omega})$ (see Proposition 6.4.1) and $\mathrm{d}\boldsymbol{\omega}$ is the differential form of degree 2 associated to the curl $\boldsymbol{\nabla}\times\boldsymbol{F}$ (by Example 6.3.2), it follows from Example 6.4.4 that

$$\mathrm{d}\boldsymbol{\nu} = \langle\boldsymbol{\nabla}\times\boldsymbol{F},\boldsymbol{N}\rangle\circ\boldsymbol{T}\cdot\mathrm{d}s\wedge\mathrm{d}t,$$

and this implies

$$\frac{\partial Q}{\partial s} - \frac{\partial P}{\partial t} = \langle \boldsymbol{\nabla} \times \boldsymbol{F}, \boldsymbol{N} \rangle \circ \boldsymbol{T}.$$

Green's formula allows us to conclude that

$$\int_{\boldsymbol{\gamma}_\varepsilon} \boldsymbol{F} = \int_{C_\varepsilon} \nu = \int_{D_\varepsilon} \big(\langle \boldsymbol{\nabla} \times \boldsymbol{F}, \boldsymbol{N} \rangle \circ \boldsymbol{T} \big)(s,t)\, \mathrm{d}(s,t),$$

where D_ε is the disk of radius ε centered at the origin. It is easy to deduce

$$\lim_{\varepsilon \to 0} \frac{1}{\pi \varepsilon^2} \int_{\boldsymbol{\gamma}_\varepsilon} \boldsymbol{F} = \langle \boldsymbol{\nabla} \times \boldsymbol{F}, \boldsymbol{N} \rangle (\boldsymbol{P}).$$

6.6 Appendix: Simply Connected Open Sets

We saw in Chap. 2 that if $\boldsymbol{F}$ is a vector field of class C^1 on an open set $U \subset \mathbb{R}^3$, then a necessary condition that $\boldsymbol{F}$ be conservative is that its curl be zero. We intend to examine the class of open sets for which the converse holds, those open sets for which every field with vanishing curl is conservative. For simplicity, we will formulate the results in terms of differential forms of degree 1 in $\mathbb{R}^n$.

Definition 6.6.1. A differential 1-form $\boldsymbol{\omega}$ of class C^1 on an open set $U \subset \mathbb{R}^n$ is said to be *closed* if $\mathrm{d}\boldsymbol{\omega} = \mathbf{0}$ in U.

Since if $g : U \subset \mathbb{R}^n \to \mathbb{R}$ is a mapping of class C^2 on the open set U, then $\mathrm{d}(dg) = \mathbf{0}$, every exact differential 1-form of class C^1 (see Definition 2.5.1) is necessarily closed. Also, if

$$\boldsymbol{\omega} = \sum_{j=1}^{n} f_j \cdot \mathrm{d}x_j,$$

then the condition $\mathrm{d}\boldsymbol{\omega} = \mathbf{0}$ in U is equivalent to

$$\frac{\partial f_j}{\partial x_k}(\boldsymbol{x}) = \frac{\partial f_k}{\partial x_j}(\boldsymbol{x})$$

for every $\boldsymbol{x} \in U$ and for each selection of the indices j and k in $\{1,\dots,n\}$. Consequently, Theorem 2.5.6 can be reformulated as follows:

Theorem 6.6.1. *Let $U \subset \mathbb{R}^n$ be an open set that is starlike with respect to a point $\boldsymbol{a} \in U$. If $\boldsymbol{\omega}$ is a closed differential 1-form of class C^1 in U, then $\boldsymbol{\omega}$ is exact.*

That is, for a starlike open set, the exact differential 1-forms coincide with the closed differential 1-forms. The objective of this section is to study the most general class of open sets for which this equivalence remains true. In view of the existing relation between differential 1-forms and vector fields, it is clear that what we are asking for is the most general class of open sets $U \subset \mathbb{R}^n$ with the property that whenever

$$\boldsymbol{F} = (f_1, f_2, \dots, f_n)$$

is a vector field of class C^1 in U, then $\boldsymbol{F}$ is conservative if and only if

$$\frac{\partial f_j}{\partial x_k}(\boldsymbol{x}) = \frac{\partial f_k}{\partial x_j}(\boldsymbol{x})$$

for every $\boldsymbol{x} \in U$ and for every choice of the indices $1 \leq j,k \leq n$.

Definition 6.6.2. Let $\boldsymbol{\alpha}, \boldsymbol{\beta} : [a,b] \to U \subset \mathbb{R}^n$ be two continuous paths with $\boldsymbol{\alpha}(a) = \boldsymbol{\beta}(a)$ and $\boldsymbol{\alpha}(b) = \boldsymbol{\beta}(b)$. Then $\boldsymbol{\alpha}$ is said to be homotopic to $\boldsymbol{\beta}$ in U if there exists a continuous function

$$\boldsymbol{H} : [a,b] \times [0,1] \to U$$

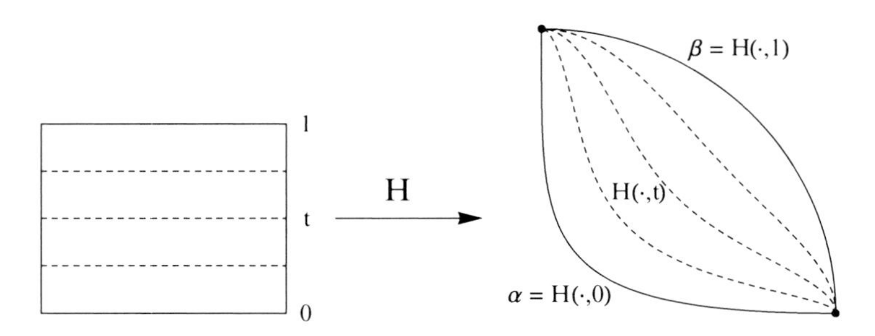

Fig. 6.1 Homotopy between paths α and β

with the following properties (Fig. 6.1):
(a) $\boldsymbol{H}(s,0)=\boldsymbol{\alpha}(s)$ y $\boldsymbol{H}(s,1)=\boldsymbol{\beta}(s)$
(b) $\boldsymbol{H}(a,t)=\boldsymbol{\alpha}(a)=\boldsymbol{\beta}(a)$ and $\boldsymbol{H}(b,t)=\boldsymbol{\alpha}(b)=\boldsymbol{\beta}(b)$.

The mapping $\boldsymbol{H}$ is called a homotopy. A homotopy can be considered a family of curves parameterized by $t\in[0,1]$. More precisely, for each $t\in[0,1]$,

$$\boldsymbol{H}(\cdot,t):[a,b]\to U$$

is a continuous path with initial point $\boldsymbol{\alpha}(a)=\boldsymbol{\beta}(a)$ and with final point $\boldsymbol{\alpha}(b)=\boldsymbol{\beta}(b)$. Moreover, for $t=0$, the path $\boldsymbol{H}(\cdot,0)$ coincides with $\boldsymbol{\alpha}$, while the path $\boldsymbol{H}(\cdot,1)$ that we obtain for $t=1$ coincides with $\boldsymbol{\beta}$. We can envisage this as $\boldsymbol{H}$ continuously deforming the path $\boldsymbol{\alpha}$ into the path $\boldsymbol{\beta}$, *without leaving the open set* U while keeping the initial and final points fixed.

Proposition 6.6.1. *Let* $\boldsymbol{\alpha},\boldsymbol{\beta},\boldsymbol{\gamma}:[a,b]\to U\subset\mathbb{R}^n$ *be three continuous paths.*
(1) *If* $\boldsymbol{\alpha}$ *is homotopic to* $\boldsymbol{\beta}$, *then* $\boldsymbol{\beta}$ *is homotopic to* $\boldsymbol{\alpha}$.
(2) *If* $\boldsymbol{\alpha}$ *is homotopic to* $\boldsymbol{\beta}$ *and* $\boldsymbol{\beta}$ *is homotopic to* $\boldsymbol{\gamma}$, *then* $\boldsymbol{\alpha}$ *is homotopic to* $\boldsymbol{\gamma}$.

Proof. (1) If $\boldsymbol{H}:[a,b]\times[0,1]\to U$ is a homotopy from $\boldsymbol{\alpha}$ to $\boldsymbol{\beta}$ (Definition 6.6.2), then $\tilde{\boldsymbol{H}}:[a,b]\times[0,1]\to U$, $\tilde{\boldsymbol{H}}(s,t)=\boldsymbol{H}(s,1-t)$, is a homotopy from $\boldsymbol{\beta}$ to $\boldsymbol{\alpha}$.

(2) If $\boldsymbol{H}_1:[a,b]\times[0,1]\to U$ is a homotopy from $\boldsymbol{\alpha}$ to $\boldsymbol{\beta}$ and $\boldsymbol{H}_2:[a,b]\times[0,1]\to U$ is a homotopy from $\boldsymbol{\beta}$ to $\boldsymbol{\gamma}$, then $\boldsymbol{H}:[a,b]\times[0,1]\to U$ defined by

$$\boldsymbol{H}(s,t):=\begin{cases}\boldsymbol{H}_1(s,2t), & 0\le t\le\frac{1}{2},\\ \boldsymbol{H}_2(s,2t-1), & \frac{1}{2}\le t\le 1,\end{cases}$$

is a homotopy from $\boldsymbol{\alpha}$ to $\boldsymbol{\gamma}$.

□

Example 6.6.1. Let $U\subset\mathbb{R}^n$ be an open set and $\boldsymbol{\alpha}:[a,b]\to U\subset\mathbb{R}^n$ a continuous path. Then $\boldsymbol{\alpha}$ is homotopic to a polygonal arc contained in U.

In fact, for each $s \in [a,b]$ take $r_s > 0$ such that $B(\boldsymbol{\alpha}(s), 2r_s) \subset U$. The family

$$\{B(\boldsymbol{\alpha}(s), r_s) : \ s \in [a,b]\}$$

is an open cover of the compact set $\boldsymbol{\alpha}([a,b])$; hence there are finitely many points $\boldsymbol{\xi}_1, \ldots, \boldsymbol{\xi}_k$ such that

$$\boldsymbol{\alpha}([a,b]) \subset \bigcup_{i=1}^{k} B(\boldsymbol{\alpha}(\boldsymbol{\xi}_i), r_{\boldsymbol{\xi}_i}).$$

Let ε be the smallest of the radii $r_{\boldsymbol{\xi}_j}$, $1 \le j \le k$. By the Heine–Cantor theorem, $\boldsymbol{\alpha}$ is uniformly continuous, and hence there exists $\delta > 0$ such that $|\boldsymbol{\alpha}(s) - \boldsymbol{\alpha}(t)| \le \varepsilon$ whenever $s,t \in [a,b]$ and $|s-t| \le \delta$. Finally, we take a partition $P := \{a = s_0 < s_1 < \cdots < s_m = b\}$ with the property that $\|P\| < \delta$ and consider the polygonal arc, denoted by $\boldsymbol{\beta}$, with vertices $\boldsymbol{\alpha}(a), \boldsymbol{\alpha}(s_1), \ldots, \boldsymbol{\alpha}(s_{m-1}), \boldsymbol{\alpha}(b)$, that is,

$$\boldsymbol{\beta} = \bigcup_{j=1}^{m} \boldsymbol{\beta}_j, \quad \text{where } \boldsymbol{\beta}_j : [s_{j-1}, s_j] \to \mathbb{R}^n$$

is the segment from $\boldsymbol{\alpha}(s_{j-1})$ to $\boldsymbol{\alpha}(s_j)$ defined by

$$\boldsymbol{\beta}_j(s) := \boldsymbol{\alpha}(s_{j-1}) + \frac{s - s_{j-1}}{s_j - s_{j-1}} (\boldsymbol{\alpha}(s_j) - \boldsymbol{\alpha}(s_{j-1})).$$

For each $j = 1, \ldots, m$ there is $i \in \{1, \ldots, k\}$ such that $\boldsymbol{\alpha}(s_j) \in B(\boldsymbol{\alpha}(\boldsymbol{\xi}_i), r_{\boldsymbol{\xi}_i})$. This implies, by construction of the partition, that $\boldsymbol{\alpha}([s_{j-1}, s_j])$ is contained in $B(\boldsymbol{\alpha}(\boldsymbol{\xi}_i), 2r_{\boldsymbol{\xi}_i}) \subset U$, and also $\boldsymbol{\beta}_j([s_{j-1}, s_j])$ is contained in $B(\boldsymbol{\alpha}(\boldsymbol{\xi}_i), 2r_{\boldsymbol{\xi}_i})$. Since the ball is a convex set, we can define

$$\boldsymbol{H}_j : [s_{j-1}, s_j] \times [0,1] \to B(\boldsymbol{\alpha}(\boldsymbol{\xi}_i), 2r_{\boldsymbol{\xi}_i}) \subset U$$

by

$$\boldsymbol{H}_j(s,t) = (1-t)\boldsymbol{\alpha}(s) + t\boldsymbol{\beta}_j(s).$$

Finally, we define

$$\boldsymbol{H} : [a,b] \times [0,1] \to U$$

to coincide with $\boldsymbol{H}_j$ in $[s_{j-1}, s_j] \times [0,1]$, for each $j = 1, \ldots, m$. Then $\boldsymbol{H}$ is a homotopy on U from $\boldsymbol{\alpha}$ to the polygonal arc $\boldsymbol{\beta}$.

Definition 6.6.3. Let $\boldsymbol{\alpha}, \boldsymbol{\beta} : [a,b] \to U \subset \mathbb{R}^n$ be two continuous closed paths. These are said to be *strictly homotopic* in U if there exists a continuous function

$$\boldsymbol{H} : [a,b] \times [0,1] \to U$$

with the following properties:
(a) $\boldsymbol{H}(s,0) = \boldsymbol{\alpha}(s)$ and $\boldsymbol{H}(s,1) = \boldsymbol{\beta}(s)$.
(b) $\boldsymbol{H}(a,t) = \boldsymbol{H}(b,t)$ for every $0 \leq t \leq 1$.

The mapping $\boldsymbol{H}$ is called a strict homotopy. Condition (b) means that each path $\boldsymbol{H}(\cdot,t)$ is closed. Obviously, two closed paths that are homotopic are also strictly homotopic.

The next result tells us that the integral of a closed differential form of degree 1 along two homotopic paths (piecewise of class C^1), or two strictly homotopic paths, coincide.

Theorem 6.6.2. *Let $\boldsymbol{\alpha}, \boldsymbol{\beta} : [a,b] \to U \subset \mathbb{R}^n$ be two paths that are piecewise of class C^1 and satisfy* (a) *or* (b):
(a) $\boldsymbol{\alpha}(a) = \boldsymbol{\beta}(a)$, $\boldsymbol{\alpha}(b) = \boldsymbol{\beta}(b)$ *and $\boldsymbol{\alpha}, \boldsymbol{\beta}$ are homotopic in U.*
(b) *$\boldsymbol{\alpha}$ and $\boldsymbol{\beta}$ are strictly homotopic closed paths.*
If $\boldsymbol{\omega}$ is a closed differential form of degree 1 in U, then

$$\int_{\boldsymbol{\alpha}} \boldsymbol{\omega} = \int_{\boldsymbol{\beta}} \boldsymbol{\omega}.$$

Proof. Since

$$\mathrm{d}\boldsymbol{\omega} = \boldsymbol{0},$$

it follows from Theorem 6.6.1 that $\boldsymbol{\omega}$ is exact on each ball contained in U. Let

$$\boldsymbol{H} : [a,b] \times [0,1] \to U$$

be a homotopy according to Definition 6.6.2. For every $\boldsymbol{x} \in R := [a,b] \times [0,1]$ take $r_{\boldsymbol{x}} > 0$ such that $B(\boldsymbol{H}(\boldsymbol{x}), 2r_{\boldsymbol{x}}) \subset U$. The family

$$\{B(\boldsymbol{H}(\boldsymbol{x}), r_{\boldsymbol{x}}) : \boldsymbol{x} \in R\}$$

is an open cover of the compact set $\boldsymbol{H}(R)$, and so there are finitely many points $\boldsymbol{x}_1, \ldots, \boldsymbol{x}_k \in R$ with

$$\boldsymbol{H}(R) \subset \bigcup_{i=1}^{k} B(\boldsymbol{H}(\boldsymbol{x}_i), r_{\boldsymbol{x}_i}).$$

Let ε be the minimum of the radii $r_{\boldsymbol{x}_i}$, $i = 1, \ldots, k$. By the Heine-Cantor theorem, $\boldsymbol{H}$ is uniformly continuous and there is $\delta > 0$ with

$$\| \boldsymbol{H}(\boldsymbol{x}) - \boldsymbol{H}(\boldsymbol{y}) \| \leq \varepsilon,$$

whenever $\boldsymbol{x}, \boldsymbol{y} \in R$, $\| \boldsymbol{x} - \boldsymbol{y} \| \leq \delta$. Observe that if A is a subset of R with diameter less than δ, then $\boldsymbol{H}(A)$ is contained in some ball

$$B_i := B(\boldsymbol{H}(\boldsymbol{x}_i), 2r_{\boldsymbol{x}_i}) \subset U.$$

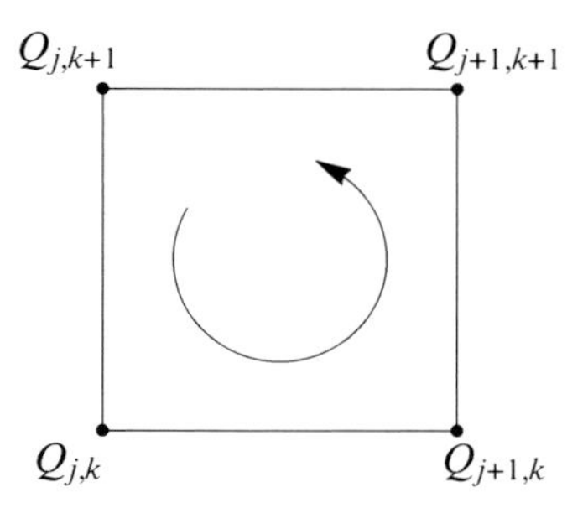

Fig. 6.2 Grid used in the proof of Theorem 6.6.2

In fact, it suffices to fix a point $\boldsymbol{x}_0 \in A$ and select the index $i = 1, \ldots, k$ such that $\boldsymbol{H}(\boldsymbol{x}_0) \in B(\boldsymbol{H}(\boldsymbol{x}_i), r_{\boldsymbol{x}_i})$. If $\boldsymbol{x} \in A$, then $\| \boldsymbol{x} - \boldsymbol{x}_0 \| < \delta$ implies $\| \boldsymbol{H}(\boldsymbol{x}) - \boldsymbol{H}(\boldsymbol{x}_0) \| < \varepsilon \leq r_{\boldsymbol{x}_i}$, and consequently $\boldsymbol{H}(\boldsymbol{x}) \in B_i$.

Now using appropriate partitions of the intervals $[a,b]$ and $[0,1]$, we divide the rectangle R into subrectangles $R_{j,k}$ with diameter less than δ. Consequently, for each subrectangle $R_{j,k}$ there exists an open ball B_i such that

$$\boldsymbol{H}(R_{j,k}) \subset B_i \subset U.$$

Denote by

$$\{\boldsymbol{Q}_{j,k} = (s_j, t_k) \ : \ 1 \leq j \leq m \, ; 1 \leq k \leq p\}$$

the vertices of the grid, so that (Fig. 6.2)

$$\boldsymbol{Q}_{1,1} = (a,0) \ , \ \boldsymbol{Q}_{1,p} = (a,1) \ , \ \boldsymbol{Q}_{m,1} = (b,0) \ , \ \boldsymbol{Q}_{m,p} = (b,1).$$

Next we consider the following paths contained in U, which are associated to the segments determined by the previous grid. For each $1 < k < p$ and $1 \leq j \leq m-1$ denote by

$$\varphi_{j,k}$$

the segment from the point $\boldsymbol{H}(\boldsymbol{Q}_{j,k})$ to the point $\boldsymbol{H}(\boldsymbol{Q}_{j+1,k})$. By the way in which we have selected the grid, this segment is contained in an open ball that is contained in U. For $k = 1$ and $k = p$ we proceed differently. For $k = 1$,

$$\boldsymbol{\varphi}_{j,1}$$

is the restriction of the path $\boldsymbol{\alpha}$ to the interval $[s_j, s_{j+1}]$, while for $k = p$,

$$\boldsymbol{\varphi}_{j,p}$$

is the restriction of the path $\boldsymbol{\beta}$ to the interval $[s_j, s_{j+1}]$. Then

$$\bigcup_{j=1}^{m-1} \boldsymbol{\varphi}_{j,1} \sim \boldsymbol{\alpha}$$

and

$$\bigcup_{j=1}^{m-1} \boldsymbol{\varphi}_{j,p} \sim \boldsymbol{\beta}.$$

Also, for every $1 < k < p$,

$$\bigcup_{j=1}^{m-1} \boldsymbol{\varphi}_{j,k}$$

is a polygonal arc contained in U. Now we proceed to associate a path contained in U to each one of the vertical segments determined by the grid of the rectangle R. Fix $1 \leq j \leq m$, and for each $1 \leq k < p$, denote by

$$\boldsymbol{\psi}_{j,k}$$

the segment from the point $\boldsymbol{H}(\boldsymbol{Q}_{j,k})$ to the point $\boldsymbol{H}(\boldsymbol{Q}_{j,k+1})$. In this way

$$\bigcup_{k=1}^{p-1} \boldsymbol{\psi}_{j,k}$$

is a polygonal arc in U from the point $\boldsymbol{H}(\boldsymbol{Q}_{j,1}) = \boldsymbol{\alpha}(s_j)$ to the point $\boldsymbol{H}(\boldsymbol{Q}_{j,p}) = \boldsymbol{\beta}(s_j)$. Given $1 \leq j < m$ and $1 \leq k < p$, the four piecewise C^1 paths

$$\boldsymbol{\varphi}_{j,k}, \quad \boldsymbol{\varphi}_{j,k+1}, \quad \boldsymbol{\psi}_{j,k}, \quad \boldsymbol{\psi}_{j+1,k}$$

are contained in a ball B_i that is contained in U. By Theorem 6.6.1, $\boldsymbol{\omega}$ is an exact differential 1-form in B_i, while

$$\boldsymbol{\varphi}_{j,k} \cup \boldsymbol{\psi}_{j+1,k} \cup (-\boldsymbol{\varphi}_{j,k+1}) \cup (-\boldsymbol{\psi}_{j,k})$$

is a closed path contained in B_i that is piecewise of class C^1. Hence

$$\int_{\boldsymbol{\varphi}_{j,k}} \boldsymbol{\omega} - \int_{\boldsymbol{\varphi}_{j,k+1}} \boldsymbol{\omega} + \int_{\boldsymbol{\psi}_{j+1,k}} \boldsymbol{\omega} - \int_{\boldsymbol{\psi}_{j,k}} \boldsymbol{\omega} = 0.$$

Consequently, summing for all the values of $1 \leq j < m$ and $1 \leq k < p$, we obtain

$$\sum_{j=1}^{m-1} \left(\int_{\boldsymbol{\varphi}_{j,1}} \boldsymbol{\omega} - \int_{\boldsymbol{\varphi}_{j,p}} \boldsymbol{\omega} \right) + \sum_{k=1}^{p-1} \left(\int_{\boldsymbol{\psi}_{m,k}} \boldsymbol{\omega} - \int_{\boldsymbol{\psi}_{1,k}} \boldsymbol{\omega} \right) = 0.$$

Moreover,

$$\bigcup_{j=1}^{m-1} \boldsymbol{\varphi}_{j,1} \sim \boldsymbol{\alpha}, \quad \bigcup_{j=1}^{m-1} \boldsymbol{\varphi}_{j,p} \sim \boldsymbol{\beta}.$$

We first suppose that we are in case (a). Then the paths $\boldsymbol{\psi}_{1,k}$ and $\boldsymbol{\psi}_{m,k}$ are constants, and we can conclude that

$$\int_{\boldsymbol{\alpha}} \boldsymbol{\omega} - \int_{\boldsymbol{\beta}} \boldsymbol{\omega} = 0.$$

Finally, suppose that we have case (b). Then the condition $\boldsymbol{H}(a,t_k) = \boldsymbol{H}(b,t_k)$ implies that

$$\bigcup_{k=1}^{p-1} \boldsymbol{\psi}_{1,k} \sim \bigcup_{k=1}^{p-1} \boldsymbol{\psi}_{m,k}.$$

We again conclude that

$$\int_{\boldsymbol{\alpha}} \boldsymbol{\omega} - \int_{\boldsymbol{\beta}} \boldsymbol{\omega} = 0.$$

□

Definition 6.6.4. An open set $U \subset \mathbb{R}^n$ is said to be simply connected if every closed path $\boldsymbol{\alpha} : [a,b] \to U$ is strictly homotopic to a constant path.

An immediate consequence of Theorem 6.6.2 is the following.

Theorem 6.6.3. *Let $U \subset \mathbb{R}^n$ be a simply connected open set and suppose that $\boldsymbol{\omega}$ is a closed differential form of degree 1 and class C^1. Then $\boldsymbol{\omega}$ is exact in U.*

Proof. If $\boldsymbol{\alpha}$ is a closed path that is piecewise of class C^1 and contained in U, then

$$\int_{\boldsymbol{\alpha}} \boldsymbol{\omega} = 0$$

by Theorem 6.6.2, since $\boldsymbol{\alpha}$ is strictly homotopic to a constant path. Without loss of generality we can assume that U is connected (for if not, we can repeat the argument for each connected component). We fix $\boldsymbol{x}_0 \in U$ and define

$$f : U \to \mathbb{R}, \quad f(\boldsymbol{x}) = \int_{\boldsymbol{\gamma}_{\boldsymbol{x}}} \boldsymbol{\omega},$$

where $\boldsymbol{\gamma}_{\boldsymbol{x}}$ is an arbitrary path of class C^1 contained in U whose initial point is $\boldsymbol{x}_0$ and whose final point is $\boldsymbol{x}$. From the previous considerations, the definition of $f(\boldsymbol{x})$ is independent of the chosen path. Moreover (see the proof of Theorem 2.5.3), f is differentiable in U and $\mathrm{d}f = \boldsymbol{\omega}$. □

Example 6.6.2. (a) Every open convex set is simply connected.
(b) $\mathbb{R}^2 \setminus \{(0,0)\}$ is not simply connected.
(c) $\mathbb{R}^3 \setminus \{(0,0,0)\}$ is simply connected.

(a) Let U be an open convex set and let $\boldsymbol{\alpha} : [a,b] \to U$ be a closed path. We let $\boldsymbol{x}_0 := \boldsymbol{\alpha}(a) = \boldsymbol{\alpha}(b)$ and define

$$\boldsymbol{H} : [a,b] \times [0,1] \to U$$

by

$$\boldsymbol{H}(s,t) = (1-t)\boldsymbol{\alpha}(s) + t\boldsymbol{x}_0.$$

Since U is convex, $\boldsymbol{H}(s,t) \in U$ for every $(s,t) \in [a,b] \times [0,1]$. Moreover, $\boldsymbol{H}$ is a homotopy (in the sense of Definition 6.6.2) between the path $\boldsymbol{\alpha}$ and the constant path $\boldsymbol{\beta} : [a,b] \to U$ defined as $\boldsymbol{\beta}(s) = \boldsymbol{x}_0$.

(b) If $U := \mathbb{R}^2 \setminus \{(0,0)\}$ were simply connected, then according to Theorem 6.6.3,

$$\int_{\boldsymbol{\alpha}} \boldsymbol{\omega} = 0$$

for each closed path $\boldsymbol{\alpha}$ contained in U and for every closed differential form $\boldsymbol{\omega}$ of degree 1. However, as we saw in Example 2.5.3, this is not true.

(c) From Proposition 6.6.1 and Example 6.6.1 it suffices to show that each closed polygonal arc $\boldsymbol{\alpha} : [a,b] \to \mathbb{R}^3 \setminus \{(0,0,0)\}$ is strictly homotopic to a constant function. We proceed in two steps: first we move the polygonal arc away from the origin and later we contract it to a point. Since the polygonal arc is a finite union of segments and we are dealing with three dimensions, we can select a point $\boldsymbol{v} \in \mathbb{R}^3$ such that the straight line that passes through the origin in the direction $\boldsymbol{v}$ does not intersect $\boldsymbol{\alpha}([a,b])$ and $\|\boldsymbol{v}\| > \|\boldsymbol{\alpha}(t)\|$ for every $t \in [a,b]$. Then the translated arc, $\boldsymbol{v} + \boldsymbol{\alpha}([a,b])$, is contained in an open half-space U that does not contain the origin. It is the existence of this vector that differentiates this example from that of part (b). We consider the strict homotopy

$$\boldsymbol{H}_1 : [a,b] \times [0,1] \to \mathbb{R}^3 \setminus \{(0,0,0)\}$$

defined by

$$\boldsymbol{H}_1(s,t) = \boldsymbol{\alpha}(s) + t\boldsymbol{v}.$$

Since U is convex, we can find a (strict) homotopy

$$\boldsymbol{H}_2 : [a,b] \times [0,1] \to U \subset \mathbb{R}^3 \setminus \{(0,0,0)\}$$

between the closed path $\boldsymbol{v} + \boldsymbol{\alpha}$ and a constant path. Finally, as in Proposition 6.6.1, define

$$\boldsymbol{H} : [a,b] \times [0,1] \to \mathbb{R}^3 \setminus \{(0,0,0)\}$$

by

$$H(s,t) := \begin{cases} H_1(s,2t), & 0 \le t \le \frac{1}{2}, \\ H_2(s,2t-1), & \frac{1}{2} \le t \le 1. \end{cases}$$

In this way we obtain a strict homotopy in $\mathbb{R}^3 \setminus \{(0,0,0)\}$ that transforms the polygonal arc $\boldsymbol{\alpha}$ into a constant path.

From Theorem 6.6.3 it follows that a vector field $\boldsymbol{F} = (f_1, \ldots, f_n)$ in a simply connected open set $U \subset \mathbb{R}^n$ is conservative if and only if the differential 1-form $\omega = \sum_{j=1}^{n} f_j \mathrm{d}x_j$ is closed. In particular, we have the following corollaries.

Corollary 6.6.1. *Let $U \subset \mathbb{R}^2$ be a simply connected open set. A vector field $\boldsymbol{F} = (f_1, f_2)$ of class C^1 in U is conservative if and only if*

$$\frac{\partial f_1}{\partial y}(x,y) = \frac{\partial f_2}{\partial x}(x,y)$$

for every $(x,y) \in U$.

Corollary 6.6.2. *Let $U \subset \mathbb{R}^3$ be a simply connected open set. A vector field $\boldsymbol{F} = (f_1, f_2, f_3)$ of class C^1 in U is conservative if and only if*

$$(\nabla \times \boldsymbol{F})(x,y,z) = \boldsymbol{0}$$

for every $(x,y,z) \in U$.

6.7 Exercises

Exercise 6.7.1. For

$$\boldsymbol{\omega} = \mathrm{d}y \wedge \mathrm{d}z - 2\, \mathrm{d}z \wedge \mathrm{d}x + 3\, \mathrm{d}x \wedge \mathrm{d}y \in \Lambda^2(\mathbb{R}^3)$$

find $\boldsymbol{\omega}(\boldsymbol{u},\boldsymbol{v})$, where $\boldsymbol{u} = (1,-1,0)$ and $\boldsymbol{v} = (2,1,1)$.

Exercise 6.7.2. Let $\boldsymbol{F} = (F_1,F_2,F_3)$ and $\boldsymbol{G} = (G_1,G_2,G_3)$ be vector fields on an open set $U \subset \mathbb{R}^3$ and let $\boldsymbol{\omega}$, $\boldsymbol{\varphi}$ be the associated differential forms of degree 1. Find the relation between the vector field $\boldsymbol{F} \times \boldsymbol{G}$ and the vector field associated to the differential form of degree 2, $\boldsymbol{\omega} \wedge \boldsymbol{\varphi}$.

Exercise 6.7.3. We consider in $\mathbb{R}^{2n}$ the differential form of degree 2

$$\boldsymbol{\omega} = \mathrm{d}x_1 \wedge \mathrm{d}x_2 + \mathrm{d}x_3 \wedge \mathrm{d}x_4 + \cdots + \mathrm{d}x_{2n-1} \wedge \mathrm{d}x_{2n}.$$

Evaluate the exterior product of n copies of $\boldsymbol{\omega}$.

Exercise 6.7.4. Let $\boldsymbol{\omega}$ be the differential form of degree 1 on an open set U in $\mathbb{R}^3$ associated to the vector field $\boldsymbol{F}$ and $\boldsymbol{\nu}$ the differential form of degree 2 in U associated to the vector field $\boldsymbol{G}$. Find the relation between the scalar product $\langle \boldsymbol{F}, \boldsymbol{G} \rangle$ and the degree-3 differential form $\boldsymbol{\omega} \wedge \boldsymbol{\nu}$.

Exercise 6.7.5. Consider the following differential forms in $\mathbb{R}^3$:
(1) $\boldsymbol{\omega} = \mathrm{d}x - z\mathrm{d}y$.
(2) $\boldsymbol{\nu} = (x^2 + y^2 + z^2)\, \mathrm{d}x \wedge \mathrm{d}z + (xyz)\, \mathrm{d}y \wedge \mathrm{d}z$.
Evaluate

$$\mathrm{d}\boldsymbol{\omega}\ ,\ \boldsymbol{\omega} \wedge \mathrm{d}\boldsymbol{\omega}\ ,\ \mathrm{d}\boldsymbol{\nu}\ ,\ \boldsymbol{\omega} \wedge \boldsymbol{\nu}.$$

Exercise 6.7.6. For the differential form

$$\boldsymbol{\omega} = \mathrm{d}x + (x^2 + y^2)\, \mathrm{d}y - \sin x\, \mathrm{d}z,$$

find the associated vector field $\boldsymbol{F}$ and show that

$$\boldsymbol{\nabla} \times \boldsymbol{F}$$

is the vector field associated to the differential form $\mathrm{d}\boldsymbol{\omega}$.

Exercise 6.7.7. Deduce from the previous exercises and the properties of the exterior differential that if $\boldsymbol{F}$ and $\boldsymbol{G}$ are vector fields of class C^1 on an open set $U \subset \mathbb{R}^3$, then

$$\mathrm{Div}(\boldsymbol{F} \times \boldsymbol{G}) = \langle \boldsymbol{G},\ \boldsymbol{\nabla} \times \boldsymbol{F} \rangle - \langle \boldsymbol{F},\ \boldsymbol{\nabla} \times \boldsymbol{G} \rangle\,.$$

Exercise 6.7.8. Let

$$\boldsymbol{\omega} = xz\,\mathrm{d}y - y\,\mathrm{d}x\ ;\ \boldsymbol{\nu} = x^3\,\mathrm{d}z + \mathrm{d}x$$

and

$$\boldsymbol{\varphi}(s,t) = (\cos s, \sin s, t)$$

be given. Evaluate each of

$$\boldsymbol{\varphi}^*(\boldsymbol{\omega}), \quad \boldsymbol{\varphi}^*(\mathrm{d}\boldsymbol{\omega}), \quad \boldsymbol{\varphi}^*(\boldsymbol{\omega}\wedge\boldsymbol{\nu}), \quad \boldsymbol{\varphi}^*(\boldsymbol{\omega}\wedge\mathrm{d}\boldsymbol{\nu}).$$

Chapter 7
Integration on Surfaces

We intend to study the integration of a differential k-form over a regular k-surface of class C^1 in $\mathbb{R}^n$. To begin with, in Sect. 7.1, we undertake the integration over a portion of the surface that is contained in a coordinate neighborhood. Where possible, we will express the obtained results in terms of integration of vector fields. For example, we study the integral of a vector field on a portion of a regular surface in $\mathbb{R}^3$ and also the integral over a portion of a hypersurface in $\mathbb{R}^n$. In Sect. 7.3 we study the integration of differential k-forms on regular k-surfaces admitting a finite atlas. We discuss the need for the surface to be orientable so that the defined integral makes sense in this more general situation. Although the requirement of having a finite atlas seems rather restrictive, all compact surfaces fall into this category, as do almost all the surfaces that one might naturally encounter (including many that are not compact). Thus, in the context of *vector calculus*, where applications play a key role, this restriction is not as great as it first appears. The advantage of this approach is that we don't have to consider convergence of infinite series and we can also avoid the intensive use of Lebesgue integration theory required in working with a general atlas. Addressing these difficulties might produce a more satisfactory theory from the point of view of *vector analysis*, but the time and space invested would yield virtually nothing in the way of applications, and it is these on which we wish to focus. Partitions of unity will be indispensable in the presentation of the general theorem of Stokes's (Chap. 9), but we think that is not a good idea to utilize them at this point to define the integral (as is done, for example, in Fleming [10]). We prefer instead to present the definition of integral in terms of divisions of the surface, because this is what is done in practice when one wishes to solve a concrete integral.

A. Galbis and M. Maestre, *Vector Analysis Versus Vector Calculus*, Universitext,
DOI 10.1007/978-1-4614-2200-6_7,

7.1 Integration of differential k-forms in $\mathbb{R}^n$

We first introduce the integral of a k-form on a coordinate neighborhood of an oriented regular k-surface.

The following definition is a generalization of the integral of a differential form of degree 1 along a curve (Definition 2.3.2).

Definition 7.1.1. Let M be an orientable regular k-surface of class C^1 and (A,φ) a coordinate system of M that is compatible with the orientation. Let

$$\omega : M \subset \mathbb{R}^n \to \Lambda^k(\mathbb{R}^n)$$

be a continuous differential k-form in M. For a measurable subset $\Delta \subset A$ we define

$$\int_{(\Delta,\varphi)} \omega := \int_{\varphi(\Delta)} \omega := \int_{\Delta} \omega(\varphi(t)) \left(\frac{\partial \varphi}{\partial t_1}(t), \ldots, \frac{\partial \varphi}{\partial t_k}(t) \right) \mathrm{d}t,$$

whenever this integral exists.

Already we have seen in Example 6.4.3 a particular case of the theorem that follows.

Theorem 7.1.1. *Under the conditions of Definition 7.1.1, let*

$$(\varphi^*\omega)(t) = g(t) \cdot \mathrm{d}t_1 \wedge \cdots \wedge \mathrm{d}t_k$$

for $t \in A$. Then

$$\int_{\varphi(\Delta)} \omega = \int_{\Delta} g(t)\mathrm{d}t.$$

Proof. Let

$$\omega(x) = \sum_{1 \le i_1 < \cdots < i_k \le n} f_{i_1,\ldots,i_k}(x) \cdot \mathrm{d}x_{i_1} \wedge \cdots \wedge \mathrm{d}x_{i_k}.$$

From Definitions 6.1.2 and 7.1.1,

$$\begin{aligned}
\int_{\varphi(\Delta)} \omega &= \int_{\Delta} \omega(\varphi(t)) \left(\frac{\partial \varphi}{\partial t_1}(t), \ldots, \frac{\partial \varphi}{\partial t_k}(t) \right) \mathrm{d}t \\
&= \int_{\Delta} \sum_{1 \le i_1 < \cdots < i_k \le n} f_{i_1,\ldots,i_k}(\varphi(t))(\mathrm{d}x_{i_1} \wedge \cdots \wedge \mathrm{d}x_{i_k}) \left(\frac{\partial \varphi}{\partial t_1}(t), \ldots, \frac{\partial \varphi}{\partial t_k}(t) \right) \mathrm{d}t \\
&= \int_{\Delta} \sum_{1 \le i_1 < \cdots < i_k \le n} f_{i_1,\ldots,i_k}(\varphi(t)) \frac{\partial(\varphi_{i_1}, \ldots, \varphi_{i_k})}{\partial(t_1, \ldots, t_k)}(t)\mathrm{d}t.
\end{aligned}$$

From Corollary 6.4.2 with $T = \varphi : A \subset \mathbb{R}^k \to M \subset \mathbb{R}^n$, we have

$$(\varphi^*\omega)(t) = g(t) \cdot \mathrm{d}t_1 \wedge \cdots \wedge \mathrm{d}t_k,$$

where

$$g(\boldsymbol{t}) = \sum_{1 \leq i_1 < \cdots < i_k \leq n} f_{i_1,\ldots,i_k}(\boldsymbol{\varphi}(\boldsymbol{t})) \frac{\partial(\varphi_{i_1},\ldots,\varphi_{i_k})}{\partial(t_1,\ldots,t_k)}(\boldsymbol{t}).$$

Hence

$$\int_{\boldsymbol{\varphi}(\Delta)} \omega = \int_\Delta g(\boldsymbol{t})\,\mathrm{d}\boldsymbol{t}.$$

An alternative proof is possible if we apply first the pullback

$$\begin{aligned}
\int_{\boldsymbol{\varphi}(\Delta)} \boldsymbol{\omega} &= \int_\Delta \omega(\boldsymbol{\varphi}(\boldsymbol{t})) \left(\frac{\partial \boldsymbol{\varphi}}{\partial t_1}(\boldsymbol{t}),\ldots,\frac{\partial \boldsymbol{\varphi}}{\partial t_k}(\boldsymbol{t})\right) \mathrm{d}\boldsymbol{t} \\
&= \int_\Delta \omega(\boldsymbol{\varphi}(\boldsymbol{t}))\big(\mathrm{d}\boldsymbol{\varphi}(\boldsymbol{t})(\boldsymbol{e}_1),\ldots,\mathrm{d}\boldsymbol{\varphi}(\boldsymbol{t})(\boldsymbol{e}_k)\big)\mathrm{d}\boldsymbol{t} \\
&= \int_\Delta (\boldsymbol{\varphi}^* \boldsymbol{\omega})(\boldsymbol{t})(\boldsymbol{e}_1,\ldots,\boldsymbol{e}_k)\mathrm{d}\boldsymbol{t} \\
&= \int_\Delta g(\boldsymbol{t})\mathrm{d}t_1 \wedge \cdots \wedge \mathrm{d}t_k(\boldsymbol{e}_1,\ldots,\boldsymbol{e}_k)\,\mathrm{d}\boldsymbol{t} = \int_\Delta g(\boldsymbol{t})\,\mathrm{d}\boldsymbol{t}.
\end{aligned}$$

□

Proposition 7.1.1. *Definition 7.1.1 is independent of the chosen coordinate system (provided that it is compatible with the orientation of M).*

Proof. Suppose that $(A_1,\boldsymbol{\varphi}_1)$ and $(A_2,\boldsymbol{\varphi}_2)$ are two coordinate systems of M (each compatible with the orientation) and let $\Delta_1 \subset A_1$, $\Delta_2 \subset A_2$ be measurable subsets such that $\boldsymbol{\varphi}_1(\Delta_1) = \boldsymbol{\varphi}_2(\Delta_2)$. If we put

$$(\boldsymbol{\varphi}_1^* \boldsymbol{\omega})(\boldsymbol{t}) = g(\boldsymbol{t}) \cdot \mathrm{d}t_1 \wedge \cdots \wedge \mathrm{d}t_k$$

and

$$(\boldsymbol{\varphi}_2^* \boldsymbol{\omega})(\boldsymbol{u}) = h(\boldsymbol{u}) \cdot \mathrm{d}u_1 \wedge \cdots \wedge \mathrm{d}u_k,$$

then we have to show that

$$\int_{\Delta_1} g(\boldsymbol{t})\mathrm{d}\boldsymbol{t} = \int_{\Delta_2} h(\boldsymbol{u})\mathrm{d}\boldsymbol{u}.$$

To this end, put

$$D := \boldsymbol{\varphi}_1(A_1) \cap \boldsymbol{\varphi}_2(A_2)$$

and consider the open sets $B_1 = \boldsymbol{\varphi}_1^{-1}(D)$ and $B_2 = \boldsymbol{\varphi}_2^{-1}(D)$ in $\mathbb{R}^k$ and the invertible C^1 mapping

$$\boldsymbol{T} : B_2 \to B_1, \;\; \boldsymbol{T}(\boldsymbol{u}) := \left(\boldsymbol{\varphi}_1^{-1} \circ \boldsymbol{\varphi}_2\right)(\boldsymbol{u}), \;\; \boldsymbol{u} \in B_2.$$

Then $\boldsymbol{T}(\Delta_2) = \Delta_1$ and $\boldsymbol{\varphi}_1 \circ \boldsymbol{T} = \boldsymbol{\varphi}_2$. Consequently, from Proposition 6.4.1 and Theorem 6.4.1,

$$\boldsymbol{\varphi}_2^* \boldsymbol{\omega}(\boldsymbol{u}) = (\boldsymbol{\varphi}_1 \circ \boldsymbol{\varphi}_1^{-1} \circ \boldsymbol{\varphi}_2)^* \boldsymbol{\omega}(\boldsymbol{u}) = (\boldsymbol{\varphi}_1 \circ \boldsymbol{T})^* \boldsymbol{\omega}(\boldsymbol{u})$$

$$= \boldsymbol{T}^* (\boldsymbol{\varphi}_1^* \boldsymbol{\omega})(\boldsymbol{u}) = g(\boldsymbol{T}(\boldsymbol{u})) \cdot J\boldsymbol{T}(\boldsymbol{u}) \cdot \mathrm{d}u_1 \wedge \cdots \wedge \mathrm{d}u_k,$$

which proves that

$$h(\boldsymbol{u}) = g(\boldsymbol{T}(\boldsymbol{u})) \cdot J\boldsymbol{T}(\boldsymbol{u}).$$

By hypothesis, the two coordinate systems are compatible with the orientation of M, which means that the Jacobian $J\boldsymbol{T}(\boldsymbol{u})$ is positive at each point $\boldsymbol{u} \in B_2$ (see Proposition 5.2.1). It follows from the change of variables theorem, Theorem 6.5.2, applied to $\Delta_1 = \boldsymbol{T}(\Delta_2)$

$$\int_{\Delta_1} g(\boldsymbol{t})\, \mathrm{d}\boldsymbol{t} = \int_{\Delta_2} g(\boldsymbol{T}(\boldsymbol{u}))\, |J\boldsymbol{T}(\boldsymbol{u})|\, \mathrm{d}\boldsymbol{u}$$

$$= \int_{\Delta_2} g(\boldsymbol{T}(\boldsymbol{u}))\, J\boldsymbol{T}(\boldsymbol{u}) \mathrm{d}u$$

$$= \int_{\Delta_2} h(\boldsymbol{u})\, \mathrm{d}\boldsymbol{u},$$

as was required. □

The reader will be aware that this proposition, by itself, justifies all the orientation machinery introduced and developed in Chap. 5.

Example 7.1.1. Let $\boldsymbol{\omega} = f \mathrm{d}t_1 \wedge \cdots \wedge \mathrm{d}t_k$ be a continuous differential k-form

$$\boldsymbol{\omega} : W \subset \mathbb{R}^k \to \Lambda^k(\mathbb{R}^k)$$

with compact support contained in the open set W. Then

$$\int_{(W, \mathrm{id}_{\mathbb{R}^k})} \boldsymbol{\omega} = \int_W f(\boldsymbol{t})\, \mathrm{d}\boldsymbol{t}.$$

Indeed, every continuous function with compact support is Lebesgue integrable; hence the integrals exist, and

$$\int_{(W, \mathrm{Id}_{\mathbb{R}^k})} \boldsymbol{\omega} = \int_W \boldsymbol{\omega}\left(\mathrm{Id}(\boldsymbol{t})\right) \left(\frac{\partial \mathrm{Id}}{\partial t_1}(\boldsymbol{t}), \ldots, \frac{\partial \mathrm{Id}}{\partial t_k}(\boldsymbol{t})\right) \mathrm{d}\boldsymbol{t}$$

$$= \int_W f(\boldsymbol{t}) \mathrm{d}t_1 \wedge \cdots \wedge \mathrm{d}t_k (\boldsymbol{e}_1, \ldots, \boldsymbol{e}_k) \mathrm{d}\boldsymbol{t}$$

$$= \int_W f(\boldsymbol{t})\, \mathrm{d}\boldsymbol{t}.$$

At this point, we change our way of thinking very slightly, and proceed from the point of view of *vector analysis*.

Let

$$\boldsymbol{F} = (f_1, f_2, f_3)$$

be a continuous vector field on an open set $U \subset \mathbb{R}^3$ and M an oriented regular surface of class C^1 contained in U. We consider the degree-2 differential form

$$\boldsymbol{\omega} := f_1 \cdot \mathrm{d}y \wedge \mathrm{d}z + f_2 \cdot \mathrm{d}z \wedge \mathrm{d}x + f_3 \cdot \mathrm{d}x \wedge \mathrm{d}y$$

associated to the vector field $\boldsymbol{F}$ and a coordinate system $(A, \boldsymbol{\varphi})$ of M compatible with the orientation. For each measurable subset $\Delta \subset A$, we intend to express

$$\int_{\boldsymbol{\varphi}(\Delta)} \boldsymbol{\omega}$$

in terms of the vector field $\boldsymbol{F}$. In order to do this, we first evaluate

$$\boldsymbol{\varphi}^* \boldsymbol{\omega} : A \subset \mathbb{R}^2 \to \Lambda^2(\mathbb{R}^2).$$

Denoting by (s,t) the coordinates of an arbitrary point, it turns out that

$$(\boldsymbol{\varphi}^* \boldsymbol{\omega})(s,t) = f(s,t) \cdot \mathrm{d}s \wedge \mathrm{d}t,$$

where

$$\begin{aligned} f(s,t) &= \boldsymbol{\varphi}^*(\boldsymbol{\omega})(s,t)\Big(\boldsymbol{e}_1, \boldsymbol{e}_2\Big) \\ &= \boldsymbol{\omega}(\boldsymbol{\varphi}(s,t))\Big(\frac{\partial \boldsymbol{\varphi}}{\partial s}(s,t), \frac{\partial \boldsymbol{\varphi}}{\partial t}(s,t)\Big) \\ &= \Big\langle \boldsymbol{F}(\boldsymbol{\varphi}(s,t)) \, , \, \frac{\partial \boldsymbol{\varphi}}{\partial s}(s,t) \times \frac{\partial \boldsymbol{\varphi}}{\partial t}(s,t) \Big\rangle \end{aligned}$$

for every point $(s,t) \in A$ (the last identity following from Example 6.1.1). According to Theorem 7.1.1, we can conclude that

$$\int_{\boldsymbol{\varphi}(\Delta)} \boldsymbol{\omega} = \iint_{\Delta} \Big\langle \boldsymbol{F}(\boldsymbol{\varphi}(s,t)), \frac{\partial \boldsymbol{\varphi}}{\partial s}(s,t) \times \frac{\partial \boldsymbol{\varphi}}{\partial t}(s,t) \Big\rangle \, \mathrm{d}(s,t). \tag{7.1}$$

Also, if we put $\boldsymbol{\varphi}(s,t) = (x,y,z)$, then the integral can be written as

$$\int_{\boldsymbol{\varphi}(\Delta)} \boldsymbol{\omega} = \iint_{\Delta} \begin{vmatrix} f_1 & f_2 & f_3 \\ \frac{\partial x}{\partial s} & \frac{\partial y}{\partial s} & \frac{\partial z}{\partial s} \\ \frac{\partial x}{\partial t} & \frac{\partial y}{\partial t} & \frac{\partial z}{\partial t} \end{vmatrix} \, \mathrm{d}(s,t). \tag{7.2}$$

The previous expression coincides with that obtained after Definition 4.3.1 in evaluating the flux of the vector field $\boldsymbol{F}$ across a regular surface with a unique coordinate system.

The reader can treat the following section as optional at this point; it is not required until one undertakes the divergence theorem. It is based, in part, on Edwards [9, Chap. V, Sect. 7] and also on exercises contained in Spivak [17].

7.2 Integration of Vector Fields in $\mathbb{R}^n$

We intend to extend the study of the previous section to the case in which we integrate a differential form of degree $n-1$ on an regular $(n-1)$-surface in $\mathbb{R}^n$.

Definition 7.2.1. Let $\boldsymbol{v}_1, \ldots, \boldsymbol{v}_{n-1}$ be linearly independent vectors in $\mathbb{R}^n$. The *vector product* (also called *cross product*)

$$\boldsymbol{z} := \boldsymbol{v}_1 \times \cdots \times \boldsymbol{v}_{n-1}$$

is defined as the unique vector $\boldsymbol{z} \in \mathbb{R}^n$ with the property that

$$\langle \boldsymbol{z}, \boldsymbol{w} \rangle = \det \begin{bmatrix} \boldsymbol{w} \\ \boldsymbol{v}_1 \\ \cdots \\ \boldsymbol{v}_{n-1} \end{bmatrix}$$

for every $\boldsymbol{w} \in \mathbb{R}^n$.

We observe that the mapping

$$T : \mathbb{R}^n \to \mathbb{R} \ , \ T(\boldsymbol{w}) := \det \begin{bmatrix} \boldsymbol{w} \\ \boldsymbol{v}_1 \\ \cdots \\ \boldsymbol{v}_{n-1} \end{bmatrix}$$

is a linear form, and consequently, the components of the vector $\boldsymbol{z}$ are given by

$$z_j = T(\boldsymbol{e}_j).$$

In the case that $\boldsymbol{v}_1, \ldots, \boldsymbol{v}_{n-1}$ are linearly dependent, we define $\boldsymbol{v}_1 \times \cdots \times \boldsymbol{v}_{n-1} := \boldsymbol{0}$. If we put $\boldsymbol{v}_i := (v_{i1}, \ldots, v_{in})$, then

$$\boldsymbol{v}_1 \times \cdots \times \boldsymbol{v}_{n-1} = \begin{vmatrix} \boldsymbol{e}_1 & \ldots & \boldsymbol{e}_j & \ldots & \boldsymbol{e}_n \\ v_{11} & \ldots & v_{1j} & \ldots & v_{1n} \\ \vdots & \vdots & \vdots & \vdots & \vdots \\ v_{i1} & \ldots & v_{ij} & \ldots & v_{in} \\ \vdots & \vdots & \vdots & \vdots & \vdots \\ v_{n-1,1} & \ldots & v_{n-1,j} & \ldots & v_{n-1,n} \end{vmatrix} . \tag{7.3}$$

Of course, this is a *formal* expression, since the entries $\{\boldsymbol{e}_i\}$ are not numbers but vectors in the canonical basis of $\mathbb{R}^n$. We interpret the formal expression by understanding that the coordinates of the vector product are obtained after expanding the determinant along the first row. Thus, the jth coordinate is the product of $(-1)^{j-1}$ and the determinant of the numerical matrix that we obtain on deleting the first row and the jth column. Hence, for every $\boldsymbol{a} = (a_1, \ldots, a_n) \in \mathbb{R}^n$,

$$\langle \boldsymbol{a}, \boldsymbol{v}_1 \times \cdots \times \boldsymbol{v}_{n-1} \rangle = \begin{vmatrix} a_1 & \ldots & a_j & \ldots & a_n \\ v_{11} & \ldots & v_{1j} & \ldots & v_{1n} \\ \vdots & \vdots & \vdots & \vdots & \vdots \\ v_{i1} & \ldots & v_{ij} & \ldots & v_{in} \\ \vdots & \vdots & \vdots & \vdots & \vdots \\ v_{n-1,1} & \ldots & v_{n-1,j} & \ldots & v_{n-1,n} \end{vmatrix}. \tag{7.4}$$

Proposition 7.2.1. *The vector $\boldsymbol{v}_1 \times \cdots \times \boldsymbol{v}_{n-1}$ is orthogonal to each of the vectors $\boldsymbol{v}_1, \ldots, \boldsymbol{v}_{n-1}$.*

Proof. It suffices to observe that

$$\langle \boldsymbol{v}_1 \times \cdots \times \boldsymbol{v}_{n-1}, \boldsymbol{v}_i \rangle$$

is the determinant of a matrix with two equal rows: both row 1 and row $i+1$ consist of the coordinates of the vector $\boldsymbol{v}_i$. □

The next result generalizes Proposition 5.1.2 to include the vector product of $n-1$ vectors in $\mathbb{R}^n$.

Proposition 7.2.2. *Let $\boldsymbol{v}_1, \ldots, \boldsymbol{v}_{n-1}$ be linearly independent vectors in $\mathbb{R}^n$. Then $\{\boldsymbol{v}_1 \times \cdots \times \boldsymbol{v}_{n-1}, \boldsymbol{v}_1, \ldots, \boldsymbol{v}_{n-1}\}$ is a positively oriented basis of $\mathbb{R}^n$ (i.e., has the same orientation that the canonical basis of $\mathbb{R}^n$).*

Proof. If $\boldsymbol{v}_1, \ldots, \boldsymbol{v}_{n-1}$ are linearly independent vectors in $\mathbb{R}^n$, then by Proposition 7.2.1, $\{\boldsymbol{v}_1 \times \cdots \times \boldsymbol{v}_{n-1}, \boldsymbol{v}_1, \ldots, \boldsymbol{v}_{n-1}\}$ is a basis of $\mathbb{R}^n$. The matrix of the linear mapping $\boldsymbol{L} : \mathbb{R}^n \to \mathbb{R}^n$ determined by $\boldsymbol{L}(\boldsymbol{e}_1) = \boldsymbol{v}_1 \times \cdots \times \boldsymbol{v}_{n-1}$ and $\boldsymbol{L}(\boldsymbol{e}_j) = \boldsymbol{v}_{j-1}$ for $j = 2, \ldots, n$, is the transpose of the matrix in (7.4), where we take $(a_1, \ldots, a_n) := \boldsymbol{v}_1 \times \cdots \times \boldsymbol{v}_{n-1}$. Hence,

$$\det(\boldsymbol{L}(\boldsymbol{e}_j)) = \|\boldsymbol{v}_1 \times \cdots \times \boldsymbol{v}_{n-1}\|^2 > 0,$$

and so, by definition, $\{\boldsymbol{v}_1 \times \cdots \times \boldsymbol{v}_{n-1}, \boldsymbol{v}_1, \ldots, \boldsymbol{v}_{n-1}\}$ is a positively oriented basis in $\mathbb{R}^n$. □

The notation used in the next proposition was introduced in Definition 6.1.3. The result itself should be compared with Example 6.1.1.

Proposition 7.2.3. *(1) Let $\boldsymbol{v}_1, \ldots, \boldsymbol{v}_{n-1}$ be vectors in $\mathbb{R}^n$. Then the jth coordinate of $\boldsymbol{v}_1 \times \cdots \times \boldsymbol{v}_{n-1}$ is given by*

$$(-1)^{j-1} \mathrm{d}x_1 \wedge \cdots \wedge \widehat{\mathrm{d}x_j} \wedge \cdots \wedge \mathrm{d}x_n (\boldsymbol{v}_1, \ldots, \boldsymbol{v}_{n-1}).$$

(2) Let $\boldsymbol{a} = (a_1, a_2, \ldots, a_n)$ be a vector in $\mathbb{R}^n$ and

$$\omega := \sum_{j=1}^{n} (-1)^{j-1} a_j \cdot \mathrm{d}x_1 \wedge \cdots \wedge \widehat{\mathrm{d}x_j} \wedge \cdots \wedge \mathrm{d}x_n$$

the associated element of $\Lambda^{n-1}(\mathbb{R}^n)$. Then

$$\omega(\boldsymbol{v}_1, \ldots, \boldsymbol{v}_{n-1}) = \langle \boldsymbol{a}, \boldsymbol{v}_1 \times \cdots \times \boldsymbol{v}_{n-1} \rangle .$$

Proof. From Definition 6.1.2, we have that

$$\mathrm{d}x_1 \wedge \cdots \wedge \widehat{\mathrm{d}x_j} \wedge \cdots \wedge \mathrm{d}x_n (\boldsymbol{v}_1, \ldots, \boldsymbol{v}_{n-1})$$

is the determinant of the matrix obtained by deleting the first row and the jth column in the matrix (7.3), from which we deduce (1). On the other hand, (2) is an immediate consequence of (1) and the definition of the scalar product. □

Definition 7.2.2. Let M be an orientable regular $(n-1)$-surface of class C^1 in $\mathbb{R}^n$ and $(A, \boldsymbol{\varphi})$ a coordinate system of M compatible with the orientation. For every $\boldsymbol{x} := \boldsymbol{\varphi}(\boldsymbol{t}) \in M$ we define the unit vector $\boldsymbol{N}(\boldsymbol{x})$ (orthogonal to the tangent space $T_{\boldsymbol{x}}M$) by

$$\boldsymbol{N}(\boldsymbol{x}) = \frac{\frac{\partial \boldsymbol{\varphi}}{\partial t_1}(\boldsymbol{t}) \times \cdots \times \frac{\partial \boldsymbol{\varphi}}{\partial t_{n-1}}(\boldsymbol{t})}{\left\| \frac{\partial \boldsymbol{\varphi}}{\partial t_1}(\boldsymbol{t}) \times \cdots \times \frac{\partial \boldsymbol{\varphi}}{\partial t_{n-1}}(\boldsymbol{t}) \right\|}.$$

Let

$$\boldsymbol{F} := (f_1, f_2, \ldots, f_n)$$

be a continuous vector field on M and

$$\omega := \sum_{j=1}^{n} (-1)^{j-1} f_j \cdot \mathrm{d}x_1 \wedge \cdots \wedge \widehat{\mathrm{d}x_j} \wedge \cdots \wedge \mathrm{d}x_n$$

the associated differential form of degree $(n-1)$. For each measurable subset $\Delta \subset A$, we intend to express the integral

$$\int_{\boldsymbol{\varphi}(\Delta)} \omega$$

in terms of the vector field $\boldsymbol{F}$. In order to do this, we first evaluate the differential form

$$\boldsymbol{\varphi}^* \omega : A \subset \mathbb{R}^{n-1} \to \Lambda^{n-1}(\mathbb{R}^{n-1})$$

of degree $n-1$ in $\mathbb{R}^{n-1}$. For every $\boldsymbol{t} := (t_1, \ldots, t_{n-1}) \in A$,

$$(\boldsymbol{\varphi}^* \omega)(\boldsymbol{t}) = f(\boldsymbol{t}) \cdot \mathrm{d}t_1 \wedge \cdots \wedge \mathrm{d}t_{n-1},$$

where

$$f(t) = \varphi^*(\omega)(t)(e_1, \dots, e_{n-1})$$
$$= \omega(\varphi(t))\left(\frac{\partial \varphi}{\partial t_1}(t), \dots, \frac{\partial \varphi}{\partial t_{n-1}}(t)\right).$$

It follows from Proposition 7.2.3 that

$$f(t) = \left\langle F(\varphi(t)), \frac{\partial \varphi}{\partial t_1}(t) \times \cdots \times \frac{\partial \varphi}{\partial t_{n-1}}(t) \right\rangle.$$

Consequently,

$$\int_{\varphi(\Delta)} \omega = \int_{\Delta} \langle F(\varphi(t)) \, , \, N(\varphi(t)) \rangle \cdot \left\| \frac{\partial \varphi}{\partial t_1}(t) \times \cdots \times \frac{\partial \varphi}{\partial t_{n-1}}(t) \right\| \, \mathrm{d}t.$$

This motivates the following definition.

Definition 7.2.3. Let M be an oriented regular $(n-1)$-surface of class C^1 in $\mathbb{R}^n$. Let $F := (f_1, f_2, \dots, f_n)$ be a continuous vector field in M, (A, φ) a coordinate system of M that is compatible with the orientation, and $\Delta \subset A$ a measurable subset. Then

$$\int_{\varphi(\Delta)} F \cdot N \, \mathrm{d}V_{n-1} := \int_{\Delta} \langle F(\varphi(t)) \, , \, N(\varphi(t)) \rangle \cdot \left\| \frac{\partial \varphi}{\partial t_1}(t) \times \cdots \times \frac{\partial \varphi}{\partial t_{n-1}}(t) \right\| \, \mathrm{d}t.$$

Here V_{n-1} stands for the $(n-1)$-dimensional volume element.

That is,

$$\int_{\varphi(\Delta)} F \cdot N \, \mathrm{d}V_{n-1} := \int_{\varphi(\Delta)} \sum_{j=1}^{n} (-1)^{j-1} f_j \cdot \mathrm{d}x_1 \wedge \cdots \wedge \widehat{\mathrm{d}x_j} \wedge \cdots \wedge \mathrm{d}x_n.$$

Of course, it is also possible to define the integral of a scalar function on a surface (see for instance Definition 7.3.3), but a vector-based approach better serves our purpose. For the scalar approach we recommend the book of Fleming [10].

7.3 Surfaces with a Finite Atlas

If we intersect a compact regular surface with an open set, we obtain a new regular surface that in general, is not compact but on the other hand can be covered with a finite number of coordinate neighborhoods. For example, a cone without its vertex is a noncompact regular surface that has a finite atlas. On the other hand, the boundary of the unit cube is a compact set that is not a regular surface in $\mathbb{R}^3$. One can obtain a regular surface from this object by eliminating the edges and vertices, but of course this means we lose the property of compactness. Nevertheless, the surface so obtained can be covered with six coordinate neighborhoods. These examples justify the development of a theory of integration for regular surfaces admitting a finite atlas. Such a theory will encompass compact surfaces and, in addition, allow the treatment of most interesting examples where compactness fails.

We recall (see Lemma 4.2.2) that if M is an oriented regular k-surface admitting a finite atlas

$$\{(A_j,\boldsymbol{\varphi}_j)\}_{j=1}^m$$

that is compatible with the orientation, then there exists a partition of M into subsets $\{Y_j\}$ such that $Y_j = \boldsymbol{\varphi}_j(\Delta_j)$ with Δ_j a measurable subset of A_j. Moreover, according to the proof of Lemma 4.2.2, we can take Δ_j to be the intersection of an open and a closed subset (for the relative topology) of A_j.

Under these conditions, the following definition makes sense, provided all the Lebesgue integrals involved are finite.

Definition 7.3.1. For a continuous differential k-form $\boldsymbol{\omega} : M \to \Lambda^k(\mathbb{R}^n)$, we define

$$\int_M \boldsymbol{\omega} := \sum_{j=1}^m \int_{\boldsymbol{\varphi}_j(\Delta_j)} \boldsymbol{\omega}.$$

Proposition 7.3.1. *Definition 7.3.1 is independent of the chosen atlas.*

Proof. Let $\{(A_j,\boldsymbol{\varphi}_j)\}_{j=1}^m$ and $\{(U_i,\boldsymbol{\psi}_i)\}_{i=1}^s$ be two atlases of M, each one compatible with its orientation, and let $\{Y_j\}$ and $\{X_i\}$ be two partitions of M such that $Y_j = \boldsymbol{\varphi}_j(\Delta_j)$ and $X_i = \boldsymbol{\psi}_i(\Lambda_i)$, where Δ_j is the intersection of an open and a closed subset (for the relative topology) of A_j, and Λ_i is the intersection of an open and a closed subset (for the relative topology) of U_i. Assume that each integral $\int_{\boldsymbol{\varphi}_j(\Delta_j)} \boldsymbol{\omega}$ is finite. For each $1 \le j \le m$, $1 \le i \le s$, put

$$\Delta_{i,j} = \boldsymbol{\varphi}_j^{-1}(Y_j \cap X_i), \quad \Lambda_{i,j} = \boldsymbol{\psi}_i^{-1}(Y_j \cap X_i).$$

Then $\Delta_{i,j}$ is a measurable subset of A_j, and $\Lambda_{i,j}$ is a measurable subset of U_i. Since for every i and j, $\Delta_{i,j}$ is a measurable subset of Δ_j, by the properties of the Lebesgue integral, we have that each integral $\int_{\boldsymbol{\varphi}_j(\Delta_{i,j})} \boldsymbol{\omega}$ is finite. According to Proposition 7.1.1,

$$\int_{\boldsymbol{\varphi}_j(\Delta_{i,j})} \boldsymbol{\omega} = \int_{\boldsymbol{\psi}_i(\Lambda_{i,j})} \boldsymbol{\omega}.$$

Since $\Delta_j = \cup_{i=1}^{s} \Delta_{i,j}$ and $\Lambda_i = \cup_{j=1}^{m} \Lambda_{i,j}$, and both of these are disjoint unions, we conclude that

$$\begin{aligned}\sum_{j=1}^{m} \int_{\boldsymbol{\varphi}_j(\Delta_j)} \boldsymbol{\omega} &= \sum_{j=1}^{m} \sum_{i=1}^{s} \int_{\boldsymbol{\varphi}_j(\Delta_{i,j})} \boldsymbol{\omega} \\ &= \sum_{i=1}^{s} \sum_{j=1}^{m} \int_{\boldsymbol{\psi}_i(\Lambda_{i,j})} \boldsymbol{\omega} \\ &= \sum_{i=1}^{s} \int_{\boldsymbol{\psi}(\Lambda_i)} \boldsymbol{\omega}.\end{aligned}$$

Observe that in addition, we have proven that each of the integrals in the above sum is finite too. □

Theorem 7.3.1. *Suppose that M is a compact and oriented regular k-surface and $\boldsymbol{\omega} : M \to \Lambda^k(\mathbb{R}^n)$ is a continuous differential k-form. Then the integral $\int_M \boldsymbol{\omega}$ exists.*

Proof. For every $\boldsymbol{x} \in M$, one can find a coordinate system $(A_{\boldsymbol{x}}, \boldsymbol{\varphi}_{\boldsymbol{x}})$ of M, compatible with the orientation, and an open set $V_{\boldsymbol{x}}$ in $\mathbb{R}^k$, contained in a compact subset of $A_{\boldsymbol{x}}$,

$$V_{\boldsymbol{x}} \subset \overline{V_{\boldsymbol{x}}} \subset A_{\boldsymbol{x}},$$

such that $x \in \boldsymbol{\varphi}_{\boldsymbol{x}}(V_{\boldsymbol{x}})$. For each $\boldsymbol{x} \in M$ there is an open set $U_{\boldsymbol{x}}$ in $\mathbb{R}^n$ such that

$$\boldsymbol{\varphi}_{\boldsymbol{x}}(V_{\boldsymbol{x}}) = U_{\boldsymbol{x}} \cap M.$$

Then

$$\{U_{\boldsymbol{x}} \ : \boldsymbol{x} \in M\}$$

is an open cover of M, and since M is compact, we can find $\{\boldsymbol{x}_1, \ldots, \boldsymbol{x}_m\}$ such that

$$M \subset \bigcup_{j=1}^{m} U_{\boldsymbol{x}_j}.$$

Hence, if we let $V_j := V_{\boldsymbol{x}_j}$, $A_j := A_{\boldsymbol{x}_j}$, and $\boldsymbol{\varphi}_j := \boldsymbol{\varphi}_{\boldsymbol{x}_j}$, then $\{(A_j, \boldsymbol{\varphi}_j)\}_{j=1}^{m}$ is a finite family of coordinate systems of M and $\{(V_j, \boldsymbol{\varphi}_j)\}_{j=1}^{m}$ is an atlas of M. If $\{\boldsymbol{\varphi}_j(\Delta_j)\}_{j=1}^{m}$ is a partition of M, where Δ_j is a measurable subset of V_j, then the integral

$$\int_{\boldsymbol{\varphi}_j(\Delta_j)} \boldsymbol{\omega}$$

is well defined because $\boldsymbol{\omega}$ is continuous and $\overline{\Delta_j}$ is a compact subset of A_j. The conclusion follows. □

Let us suppose that M is an oriented regular $(n-1)$-surface admitting a finite atlas

$$\{(A_j,\varphi_j)\}_{j=1}^m$$

(compatible with the orientation). We can decompose M into disjoint subsets $\{Y_j\}$ such that $Y_j = \varphi_j(\Delta_j)$, where Δ_j is a measurable subset of A_j. Now we can extend Definition 7.2.3. In what follows, V_{n-1} stands for the $(n-1)$-dimensional volume element.

Definition 7.3.2. Given a continuous vector field $\boldsymbol{F} = (f_1,\dots,f_n)$ in M, we define the flux of $\boldsymbol{F}$ across M as

$$\int_M \boldsymbol{F}\cdot\boldsymbol{N}\,\mathrm{d}V_{n-1} := \sum_{j=1}^m \int_{\varphi_j(\Delta_j)} \boldsymbol{F}\cdot\boldsymbol{N}\,\mathrm{d}V_{n-1},$$

whenever each of the integrals on the right-hand side exists and is finite.

As in Proposition 7.3.1, can be checked that this definition does not depend on the chosen partition.

Here, $\boldsymbol{N}$ represents the continuous vector field of unit normal vectors to M as in Definition 7.2.2.

We observe that

$$\int_M \boldsymbol{F}\cdot\boldsymbol{N}\,\mathrm{d}V_{n-1} = \int_M \boldsymbol{\omega}$$

for

$$\boldsymbol{\omega} = \sum_{j=1}^n (-1)^{j-1} f_j\cdot \mathrm{d}x_1\wedge\cdots\wedge\widehat{\mathrm{d}x_j}\wedge\cdots\wedge\mathrm{d}x_n.$$

Example 7.3.1. Let M be an oriented regular surface contained in the open set $U\subset\mathbb{R}^3$, $\boldsymbol{F} = (f_1,f_2,f_3)$ a continuous vector field in U, and

$$\boldsymbol{\omega} := f_1\cdot\mathrm{d}y\wedge\mathrm{d}z + f_2\cdot\mathrm{d}z\wedge\mathrm{d}x + f_3\cdot\mathrm{d}x\wedge\mathrm{d}y$$

the differential form of degree 2 associated to the vector field $\boldsymbol{F}$. Then

$$\int_M \boldsymbol{\omega} = \int_M \boldsymbol{F}\cdot\boldsymbol{N}\,\mathrm{d}V_2.$$

We also use the notation

$$\int_M \boldsymbol{\omega} = \int_M \langle \boldsymbol{F},\boldsymbol{N}\rangle\ \mathrm{d}S = \int_M \boldsymbol{F}\cdot\mathrm{d}\boldsymbol{S}.$$

Hence, if $M = \boldsymbol{\varphi}(A)$, where $(A,\boldsymbol{\varphi})$ is a coordinate system of M compatible with the orientation

$$\int_M \langle F,N\rangle \mathrm{d}\boldsymbol{S} = \int_A \langle F(\boldsymbol{\varphi}(s,t')), \frac{\partial\boldsymbol{\varphi}}{\partial s}(s,t)\times\frac{\partial\boldsymbol{\varphi}}{\partial t}(s,t)\rangle \mathrm{d}(s,t).$$

From the expression for the integral obtained in Sect. 7.2 and from the discussion relating to the area of a surface in Chap. 4, one is led to the next result.

Proposition 7.3.2. *Let M be an oriented regular surface contained in the open set $U \subset \mathbb{R}^3$ and let $\boldsymbol{F} = \boldsymbol{N}$ be the continuous vector field of unit normal vectors defined by the orientation of M. If M admits a finite atlas, then*

$$\text{area}\, M = \int_M \langle \boldsymbol{F}, \boldsymbol{N} \rangle \, \mathrm{d}S.$$

(This may be finite or $+\infty$.)

Proof. It suffices to show that if $(A, \boldsymbol{\varphi})$ is a coordinate system of M compatible with the orientation and $\Delta \subset A$ is a measurable subset, then

$$\text{area}\,(\Delta, \boldsymbol{\varphi}) = \int_{\boldsymbol{\varphi}(\Delta)} \langle \boldsymbol{F}, \boldsymbol{N} \rangle \, \mathrm{d}S.$$

We put $\boldsymbol{F} = (f_1, f_2, f_3)$ and

$$\boldsymbol{\omega} := f_1 \cdot \mathrm{d}y \wedge \mathrm{d}z + f_2 \cdot \mathrm{d}z \wedge \mathrm{d}x + f_3 \cdot \mathrm{d}x \wedge \mathrm{d}y.$$

Then, since

$$\boldsymbol{F}(\boldsymbol{\varphi}(s,t)) = \frac{\frac{\partial \boldsymbol{\varphi}}{\partial s}(s,t) \times \frac{\partial \boldsymbol{\varphi}}{\partial t}(s,t)}{\left\| \frac{\partial \boldsymbol{\varphi}}{\partial s}(s,t) \times \frac{\partial \boldsymbol{\varphi}}{\partial t}(s,t) \right\|},$$

we obtain from Definitions 4.2.1 and 4.2.2 that

$$\begin{aligned}
\int_{\boldsymbol{\varphi}(\Delta)} \langle \boldsymbol{F}, \boldsymbol{N} \rangle \, \mathrm{d}S &= \int_{\boldsymbol{\varphi}(\Delta)} \boldsymbol{\omega} \\
&\overset{(7.1)}{=} \iint_\Delta \left\langle \boldsymbol{F}(\boldsymbol{\varphi}(s,t)), \frac{\partial \boldsymbol{\varphi}}{\partial s}(s,t) \times \frac{\partial \boldsymbol{\varphi}}{\partial t}(s,t) \right\rangle \mathrm{d}(s,t) \\
&= \iint_\Delta \left\| \frac{\partial \boldsymbol{\varphi}}{\partial s}(s,t) \times \frac{\partial \boldsymbol{\varphi}}{\partial t}(s,t) \right\| \mathrm{d}(s,t) \\
&= \text{area}\,(\Delta, \boldsymbol{\varphi}).
\end{aligned}$$

□

Definition 7.3.3. Let M be an oriented regular surface in $\mathbb{R}^3$ and

$$f : M \subset \mathbb{R}^3 \to \mathbb{R}$$

a continuous function on M. If $(A, \boldsymbol{\varphi})$ is a coordinate system of M and $\Delta \subset A$ is a measurable subset, we consider the vector field

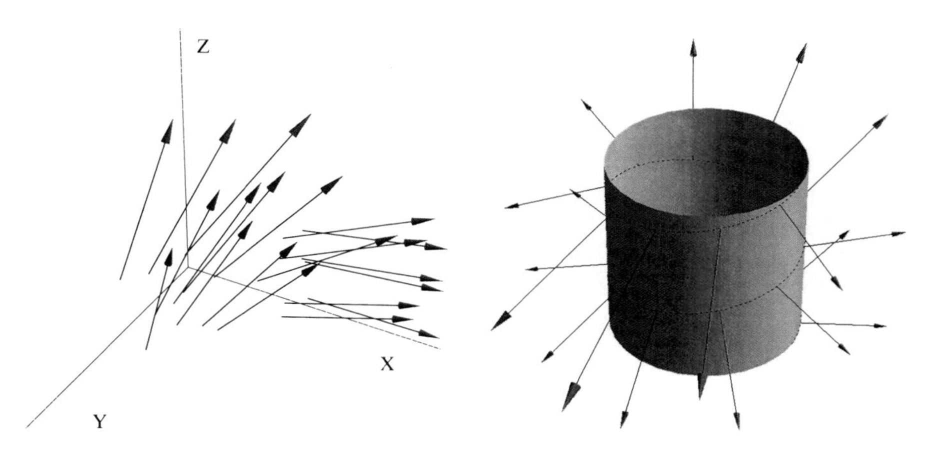

Fig. 7.1 Vector field of Example 7.3.2

$$F := f \cdot N$$

and define

$$\int_{\varphi(\Delta)} f \, \mathrm{d}S := \int_{\varphi(\Delta)} \langle F, N \rangle \, \mathrm{d}S,$$

whenever the integral on the right-hand side exists and is finite.

Thus,

$$\int_{\varphi(\Delta)} f \, \mathrm{d}S = \iint_{\Delta} f(\varphi(s,t)) \cdot \left\| \frac{\partial \varphi}{\partial s}(s,t) \times \frac{\partial \varphi}{\partial t}(s,t) \right\| \, \mathrm{d}(s,t).$$

This definition is independent of the chosen coordinate system and also of the orientation of the surface, since a change of orientation implies also a corresponding change in the vector field N. Hence, the last expression obtained can be used to define the surface integral of a scalar function even in the case that the surface M is not orientable.

Example 7.3.2. Consider the vector field $F(x,y,z) = (x, yz, z)$ and the cylinder

$$M := \{(x,y,z) \in \mathbb{R}^3 \; : \; x^2 + y^2 = 1 \; , \; 0 < z < 1\}.$$

Choose an orientation for M and evaluate the integral

$$\int_M F \cdot \mathrm{d}S$$

according to the chosen orientation (Fig. 7.1).

Solution: Put $U := \{(x,y,z) \in \mathbb{R}^3 \; : \; 0 < z < 1\}$, which is an open set, and

$$\Phi(x,y,z) = x^2 + y^2 - 1.$$

Since

$$M = \{(x,y,z) \in U \; : \Phi(x,y,z) = 0\}$$

and $\nabla \Phi(x,y,z) = (2x, 2y, 0) \neq (0,0,0)$ for every $(x,y,z) \in M$, it follows from Example 3.2.2 that M is a regular surface of class C^∞. Also, according to Example 3.4.3,

$$\boldsymbol{N}(x,y,z) := \frac{\nabla \Phi(x,y,z)}{\| \nabla \Phi(x,y,z) \|} = (x,y,0)$$

is a continuous vector field of unit normal vectors to M. We suppose that M is oriented according to this vector field $\boldsymbol{N}$ (see Theorem 5.2.2). To evaluate the integral we need an atlas of M, so consider the mapping $\boldsymbol{\varphi} : \mathbb{R}^2 \to \mathbb{R}^3$,

$$\boldsymbol{\varphi}(s,t) := (\cos s, \sin s, t),$$

and the open sets

$$A_1 := (0, 2\pi) \times (0,1), \quad A_2 := \left(-\frac{\pi}{2}, \frac{\pi}{2}\right) \times (0,1).$$

Since $\boldsymbol{\varphi}_{|A_1}$ and $\boldsymbol{\varphi}_{|A_2}$ are injective functions of class C^∞ and

$$\frac{\partial \boldsymbol{\varphi}}{\partial s}(s,t) \times \frac{\partial \boldsymbol{\varphi}}{\partial t}(s,t) = \begin{pmatrix} \boldsymbol{e}_1 & \boldsymbol{e}_2 & \boldsymbol{e}_3 \\ -\sin s & \cos s & 0 \\ 0 & 0 & 1 \end{pmatrix}$$

$$= (\cos s, \sin s, 0)$$

is nonzero, we conclude, using Corollary 3.3.1, that $(A_1, \boldsymbol{\varphi})$ and $(A_2, \boldsymbol{\varphi})$ are two coordinate systems of M and they form an atlas. Moreover, since

$$\frac{\partial \boldsymbol{\varphi}}{\partial s}(s,t) \times \frac{\partial \boldsymbol{\varphi}}{\partial t}(s,t) = \boldsymbol{N}(\boldsymbol{\varphi}(s,t)),$$

the two coordinate systems are compatible with the chosen orientation for M (see Definition 5.2.3). Finally, we observe that

$$M = \boldsymbol{\varphi}(A_1) \cup \boldsymbol{\varphi}(\{0\} \times (0,1))$$

is a partition of M. Keeping in mind that

$$\boldsymbol{F}(\boldsymbol{\varphi}(s,t)) = (\cos s, t \sin s, t)$$

and $\{0\} \times (0,1)$ is a null subset, we deduce from Definition 7.3.2 of the surface integral that

$$\int_M \boldsymbol{F} \cdot \mathrm{d}\boldsymbol{S} = \int_{\boldsymbol{\varphi}(A_1)} \mathbf{F} \cdot \mathrm{d}\boldsymbol{S}$$

$$= \iint_{A_1} \begin{vmatrix} \cos s & t\sin s & t \\ -\sin s & \cos s & 0 \\ 0 & 0 & 1 \end{vmatrix} \mathrm{d}(s,t)$$

$$= \iint_{A_1} (\cos^2 s + t\sin^2 s)\ \mathrm{d}(s,t)$$

$$= \int_0^{2\pi} \left(\cos^2 s + \frac{1}{2}\sin^2 s\right) \mathrm{d}s = \frac{3\pi}{2}.$$

To finish, we include some remarks relating to integration on *regular curves*, that is, regular 1-surfaces of class C^1. It can be shown that every regular curve is orientable.

Lemma 7.3.1. *Let M be an (orientable) regular curve in $\mathbb{R}^n$. Then there is a continuous function*

$$\boldsymbol{T} : M \to \mathbb{R}^n$$

such that for each $\boldsymbol{x} \in M$, $\boldsymbol{T}(\boldsymbol{x})$ is a unit tangent vector to M at $\boldsymbol{x}$. Moreover, the orientation $\theta_{\boldsymbol{x}}$ of the tangent line to M at the point $\boldsymbol{x}$ is precisely $\theta_{\boldsymbol{x}} = [\boldsymbol{T}(\boldsymbol{x})]$.

Proof. For every $\boldsymbol{x} \in M$ let $((a,b),\boldsymbol{\gamma})$ be a coordinate system of M, of class C^1, compatible with the orientation,

$$\boldsymbol{\gamma} : (a,b) \to \mathbb{R}^n,$$

such that $\boldsymbol{x} = \boldsymbol{\gamma}(t)$ for some $t \in (a,b)$. Now we define

$$\boldsymbol{T}(\boldsymbol{x}) := \frac{\boldsymbol{\gamma}'(t)}{\| \boldsymbol{\gamma}'(t) \|},$$

which is a unit tangent vector to M at $\boldsymbol{x}$. The definition of $\boldsymbol{T}(\boldsymbol{x})$ does not depend on the chosen coordinate system, as long as it is compatible with the orientation, because the tangent space to M at $\boldsymbol{x}$ is one-dimensional. Thus, the defined function

$$\boldsymbol{T} : M \to \mathbb{R}^n$$

has the required properties. □

Example 7.3.3. Let M be a regular curve in $\mathbb{R}^n$ of class C^1 and $\boldsymbol{F} = (f_1,\dots,f_n)$ a continuous vector field in M. We consider its associated differential form of degree 1

$$\boldsymbol{\omega} = \sum_{j=1}^{n} f_j \, \mathrm{d}x_j.$$

Take a coordinate system $\big((a,b),\boldsymbol{\varphi}\big)$,

$$\boldsymbol{\gamma} : (a,b) \to \mathbb{R}^n$$

of M compatible with the orientation. Now we evaluate the integral of the differential form $\boldsymbol{\omega}$ on the path $\boldsymbol{\varphi}$ (see Example 6.4.3)

$$\begin{aligned}\int_{\boldsymbol{\gamma}} \boldsymbol{\omega} = \int_{\boldsymbol{\gamma}} \boldsymbol{F} &= \int_a^b \big\langle \boldsymbol{F}(\boldsymbol{\gamma}(t)), \boldsymbol{\gamma}'(t) \big\rangle \, \mathrm{d}t \\ &= \int_a^b \langle \boldsymbol{F}(\boldsymbol{\gamma}(t)), \boldsymbol{T}(\boldsymbol{\gamma}(t)) \rangle \cdot \| \, \boldsymbol{\gamma}'(t) \, \| \, \mathrm{d}t.\end{aligned}$$

From now on, we may use any of the following expressions to represent the integral of the vector field $\boldsymbol{F}$ over the regular curve M:

$$\int_M \boldsymbol{F} \cdot \mathrm{d}s := \int_M \langle \boldsymbol{F}, \boldsymbol{T} \rangle \, \mathrm{d}s := \int_M \boldsymbol{\omega}.$$

7.4 Exercises

Exercise 7.4.1. Find $\int_M \omega$, where

$$\omega = x^2 \, dy \wedge dz - y \, dx \wedge dz$$

and M is the portion of the plane $x+y+z=2$ whose projection onto the xy-plane is the triangle with vertices $(0,0),(0,1),(1,1)$, oriented according to the normal vector $(1,1,1)$.

Exercise 7.4.2. Find the height of the center of gravity of the portion of the paraboloid $2z = x^2 + y^2$ that is under the plane $2z = 1$, assuming that the density (mass per unit area) is constant. Hint: The third coordinate of the center of gravity of a surface M with constant density is given by

$$\bar{z} = \frac{1}{\text{area}(M)} \int_M z \, dS$$

(see, for instance, Marsden–Tromba [14]).

Exercise 7.4.3. Integrate the vector field $\boldsymbol{F}(x,y,z) = (0,0,z)$ over the ellipsoid of Exercise 3.5.7, oriented according to the coordinate systems described in that exercise.

Exercise 7.4.4. Determine the flux of the vector field

$$\boldsymbol{F}(x,y,z) = (x,y,z)$$

across the surface $x^2 + y^2 + z^2 = R^2$.

Exercise 7.4.5. Find the flux of the vector field $\boldsymbol{F}(x,y,z) = (1,\ xy,\ z^2)$ through the square in the yz-plane defined by $0 < y < 2$, $-1 < z < 1$ and oriented according to the normal vector $(-1,0,0)$.

Chapter 8
Surfaces with Boundary

One of the objectives of this book is to obtain a rigorous proof of a version of Green's formula for compact subsets of $\mathbb{R}^2$ whose topological boundary is a regular curve of class C^2. These sets are typical examples of what we will call regular 2-surfaces with boundary in $\mathbb{R}^2$. The analogous three-dimensional example would consist of a compact set of $\mathbb{R}^3$ whose topological boundary is a regular surface of class C^2. The following example is perhaps instructive. Consider in $\mathbb{R}^3$ the portion of the cylinder $x^2 + y^2 = 1$ limited by the planes $z = 0$ and $z = 1$; that is,

$$M := \{(x,y,z) \in \mathbb{R}^3 : x^2 + y^2 = 1, \ 0 \leq z \leq 1\}.$$

The set M is not a regular surface according to Definition 3.1.1. However, we can decompose it as $M = M_1 \cup M_2$, where

$$M_1 = \{(x,y,z) \in \mathbb{R}^3 : x^2 + y^2 = 1, \ 0 < z < 1\}$$

is a regular surface and M_2, being the union of two circles, is a regular curve. This set M is an example of what is called a regular 2-surface with boundary in $\mathbb{R}^3$. Here, *boundary* refers to the curve M_2, which, in this case, does not coincide with the topological boundary of M. In general, a regular k-surface with boundary in $\mathbb{R}^n$ $(2 \leq k \leq n)$ is the union of a regular k-surface in $\mathbb{R}^n$ and a regular $(k-1)$-surface (called the boundary) that are "glued" together in a very particular way (see Definitions 8.2.1 and 8.2.2, and Theorem 8.2.1). The concept of regular k-surface with boundary in $\mathbb{R}^n$ is essential to the formulation of Stokes's theorem.

A. Galbis and M. Maestre, *Vector Analysis Versus Vector Calculus*, Universitext,
DOI 10.1007/978-1-4614-2200-6_8, © Springer Science+Business Media, LLC 2012

8.1 Functions of Class C^p in a Half-Space

The domain of parameters of a regular surface with boundary in $\mathbb{R}^3$ will be a subset of the half-plane formed by the points of $\mathbb{R}^2$ whose first coordinate is less than or equal to zero. In general, the domain of parameters of a regular k-surface with boundary in $\mathbb{R}^n$, with $n \geq k$, will be an open subset (for the relative topology) of the closed half-space $\mathbb{H}^k$ formed by the points whose first coordinate is less than or equal to zero. We will need to develop a differential calculus for functions defined on such a set before investigating further the notion of regular surface with boundary.

Definition 8.1.1. The half-space $\mathbb{H}^k$ is defined by

$$\mathbb{H}^k := \{(t_1,\ldots,t_k) \in \mathbb{R}^k \; : \; t_1 \leq 0\} \; = (-\infty,0] \times \mathbb{R}^{k-1}.$$

Since the half-space is not an open set, we must clarify what is meant by saying that a function defined on $\mathbb{H}^k$ is differentiable. We recall that a set $\mathbb{A} \subset \mathbb{H}^k$ is *open in* $\mathbb{H}^k$ *for the relative topology* if there exists an open set $A \subset \mathbb{R}^k$ such that $A \cap \mathbb{H}^k = \mathbb{A}$.

Definition 8.1.2. Let $\mathbb{A}$ be an open set in $\mathbb{H}^k$. The mapping

$$\boldsymbol{\varphi} : \mathbb{A} \subset \mathbb{H}^k \to \mathbb{R}^n$$

is said to be *differentiable* (C^p) if there exist an open set $A \subset \mathbb{R}^k$ and a differentiable (C^p) mapping

$$\boldsymbol{g} : A \subset \mathbb{R}^k \to \mathbb{R}^n$$

such that $A \cap \mathbb{H}^k = \mathbb{A}$ and $\boldsymbol{g}_{|\mathbb{A}} = \boldsymbol{\varphi}$.

Implicit in this definition is the fact that the domain $\mathbb{A}$ of $\boldsymbol{\varphi}$ is an open subset of $\mathbb{H}^k$ with the relative topology.

We next verify that the partial derivatives of $\boldsymbol{g}$ at $\boldsymbol{t} \in \mathbb{A} \subset \mathbb{H}^k$ are determined by the restriction $\boldsymbol{g}_{|\mathbb{A}} = \boldsymbol{\varphi}$. This allows us to refer to the partial derivatives of $\boldsymbol{\varphi} : \mathbb{A} \subset \mathbb{H}^k \to \mathbb{R}^n$ at $\boldsymbol{t} \in \mathbb{A}$ without any ambiguity. We fix $\boldsymbol{t} \in \mathbb{A} \subset \mathbb{H}^k$ and suppose

$$2 \leq j \leq k.$$

Since $A \subset \mathbb{R}^k$ is open, there exists $\delta > 0$ such that $\boldsymbol{t} + s\,\boldsymbol{e}_j \in A$ whenever $s \in \mathbb{R}$ and $|s| < \delta$. Since the first coordinate of $\boldsymbol{t} + s\,\boldsymbol{e}_j$ is $t_1 \leq 0$, we have

$$\boldsymbol{t} + s\,\boldsymbol{e}_j \in A \cap \mathbb{H}^k \; = \; \mathbb{A}$$

for every $|s| < \delta$. Now,

$$\begin{aligned} \frac{\partial \boldsymbol{g}}{\partial t_j}(\boldsymbol{t}) &= \lim_{s \to 0} \frac{\boldsymbol{g}(\boldsymbol{t} + s\boldsymbol{e}_j) - \boldsymbol{g}(\boldsymbol{t})}{s} \\ &= \lim_{s \to 0} \frac{\boldsymbol{\varphi}(\boldsymbol{t} + s\boldsymbol{e}_j) - \boldsymbol{\varphi}(\boldsymbol{t})}{s}. \end{aligned} \tag{8.1}$$

Notice that the last limit does not depend on the concrete extension $\boldsymbol{g}$ of $\boldsymbol{\varphi}$ that we are using.

Analysis of the partial derivative with respect to the first variable is slightly different. As before, there exists $\delta > 0$ such that $\boldsymbol{t} + s\boldsymbol{e}_1 \in A$ whenever $|s| < \delta$. However, the first coordinate of $\boldsymbol{t} + s\boldsymbol{e}_1$ is $t_1 + s$, and if $t_1 = 0$, we are guaranteed that $\boldsymbol{t} + s\boldsymbol{e}_1 \in \mathbb{A} \subset \mathbb{H}^k$ only if we restrict ourselves to negative values of s. So, for $-\delta < s < 0$, we have

$$\boldsymbol{t} + s\boldsymbol{e}_1 \in A \cap \mathbb{H}^k = \mathbb{A},$$

and we can write

$$\begin{aligned} \frac{\partial \boldsymbol{g}}{\partial t_1}(\boldsymbol{t}) &= \lim_{s \to 0} \frac{\boldsymbol{g}(\boldsymbol{t} + s\boldsymbol{e}_1) - \boldsymbol{g}(\boldsymbol{t})}{s} \\ &= \lim_{s \to 0} \frac{\boldsymbol{\varphi}(\boldsymbol{t} + s\boldsymbol{e}_1) - \boldsymbol{\varphi}(\boldsymbol{t})}{s} \end{aligned} \tag{8.2}$$

when $t_1 < 0$, and

$$\begin{aligned} \frac{\partial \boldsymbol{g}}{\partial t_1}(\boldsymbol{t}) &= \lim_{s \to 0^-} \frac{\boldsymbol{g}(\boldsymbol{t} + s\boldsymbol{e}_1) - \boldsymbol{g}(\boldsymbol{t})}{s} \\ &= \lim_{s \to 0^-} \frac{\boldsymbol{\varphi}(\boldsymbol{t} + s\,\boldsymbol{e}_1) - \boldsymbol{\varphi}(\boldsymbol{t})}{s} \end{aligned} \tag{8.3}$$

when $t_1 = 0$.

This discussion justifies the following definition.

Definition 8.1.3. Let $\mathbb{A}$ be an open set for the relative topology of $\mathbb{H}^k$ and $\boldsymbol{\varphi} : \mathbb{A} \subset \mathbb{H}^k \to \mathbb{R}^n$. The *partial derivatives* of $\boldsymbol{\varphi}$ at $\boldsymbol{t} \in \mathbb{A}$ are defined as follows (assuming existence of the corresponding limits):

If $2 \le j \le k$,

$$\frac{\partial \boldsymbol{\varphi}}{\partial t_j}(\boldsymbol{t}) := \lim_{s \to 0} \frac{\boldsymbol{\varphi}(\boldsymbol{t} + s\boldsymbol{e}_j) - \boldsymbol{\varphi}(\boldsymbol{t})}{s}.$$

If $t_1 < 0$,

$$\frac{\partial \boldsymbol{\varphi}}{\partial t_1}(\boldsymbol{t}) := \lim_{s \to 0} \frac{\boldsymbol{\varphi}(\boldsymbol{t} + s\boldsymbol{e}_1) - \boldsymbol{\varphi}(\boldsymbol{t})}{s}.$$

If $t_1 = 0$,

$$\frac{\partial \boldsymbol{\varphi}}{\partial t_1}(\boldsymbol{t}) := \lim_{s \to 0^-} \frac{\boldsymbol{\varphi}(\boldsymbol{t} + s\boldsymbol{e}_1) - \boldsymbol{\varphi}(\boldsymbol{t})}{s}.$$

It is obvious how the partial derivatives of higher order can be defined. According to the discussion prior to Definition 8.1.3, if $A \subset \mathbb{R}^k$ is an open set such that $A \cap \mathbb{H}^k = \mathbb{A}$ and

$$\boldsymbol{g} : A \subset \mathbb{R}^k \to \mathbb{R}^n$$

is a differentiable function on A with the property that $\boldsymbol{g}_{|\mathbb{A}} = \boldsymbol{\varphi}$, then the partial derivatives of $\boldsymbol{\varphi}$ exist at all points of $\mathbb{A}$ and coincide with the partial derivatives of $\boldsymbol{g}$.

Definition 8.1.4. Let $\boldsymbol{\varphi} : \mathbb{A} \subset \mathbb{H}^k \to \mathbb{R}^n$ be a differentiable function on $\mathbb{A}$ according to Definition 8.1.2. For every $\boldsymbol{t} \in \mathbb{A}$, *the differential* of $\boldsymbol{\varphi}$ at $\boldsymbol{t}$ is defined to be the linear mapping $\mathrm{d}\boldsymbol{\varphi}(\boldsymbol{t}) = \mathrm{d}\boldsymbol{g}(\boldsymbol{t}) : \mathbb{R}^k \to \mathbb{R}^n$, where $\boldsymbol{g}$ is the function of Definition 8.1.2.

It follows from the previous discussion that the differential does not depend on the choice of the function $\boldsymbol{g}$ associated to $\boldsymbol{\varphi}$.

Lemma 8.1.1. *Let $a > 0$ and suppose the function $f : (-a,0] \to \mathbb{R}$ has derivative of order p at each $s \in (-a,0)$ and left derivative of order p, $1 \leq p < \infty$, at $s = 0$. Then there exist numbers $\alpha_0, \alpha_1, \ldots, \alpha_p$ such that the function*

$$g(s) := \begin{cases} f(s), & s \leq 0 \\ \alpha_0 f(0) + \sum_{l=1}^{p} \alpha_l f(-ls), & s \geq 0, \end{cases}$$

is well defined and of class C^p on the open interval $(-a, \frac{a}{p})$.

Proof. We can select scalars $\alpha_0, \ldots, \alpha_p$ such that

$$\alpha_0 + \alpha_1 + \cdots + \alpha_p = 1$$

and

$$\sum_{l=1}^{p} l^m \alpha_l = (-1)^m$$

for every $m = 1, 2, \ldots, p$. Indeed, this linear system has a unique solution, since the determinant of the matrix of coefficients is a Vandermonde determinant different from zero. For $0 \leq s < \frac{a}{p}$ we have

$$g(s) = \alpha_0 f(0) + \sum_{l=1}^{p} \alpha_l f(-ls).$$

Then

$$g(0) = (\alpha_0 + \alpha_1 + \cdots + \alpha_p) f(0) = f(0),$$

and for every $m = 1, 2, \ldots, p$,

$$g^{(m)}(s) = \sum_{l=1}^{p} (-l)^m \alpha_l f^{(m)}(-ls),$$

whenever $0 < s < \frac{a}{p}$. Moreover, g has one-sided derivatives at the origin up to order p given by

$$g_+^{(m)}(0) = \left(\sum_{l=1}^{p} (-l)^m \alpha_l \right) f_-^{(m)}(0) = f_-^{(m)}(0).$$

The conclusion follows. □

We are going to show that if a function has continuous partial derivatives (up to order p) in $\mathbb{A}$, then there exists a function $\boldsymbol{g}$ defined on an open set of $\mathbb{R}^k$ that satisfies

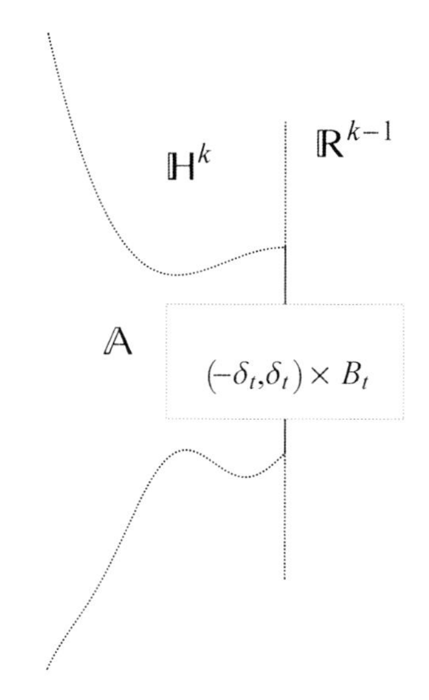

Fig. 8.1 Illustration of the proof of Proposition 8.1.1

Definition 8.1.2. Thus, the differentiability of a function (and the degree thereof) on $\mathbb{A}$ can be established *internally* without recourse to an open set containing $\mathbb{A}$.

Proposition 8.1.1. *Given $\boldsymbol{\varphi} : \mathbb{A} \subset \mathbb{H}^k \to \mathbb{R}^n$, and $1 \leq p < \infty$ the following conditions are equivalent:*

(1) $\boldsymbol{\varphi}$ is a function of class C^p on $\mathbb{A}$.
(2) $\boldsymbol{\varphi}$ has partial derivatives of order $\leq p$, which are continuous functions on $\mathbb{A}$.

Proof. It is obvious that (1) implies (2). Indeed, if $\boldsymbol{g}$ is as in Definition 8.1.2, then the partial derivatives of $\boldsymbol{\varphi}$ up to order p are the restrictions to $\mathbb{A}$ of the corresponding partial derivatives of $\boldsymbol{g}$ in the open set $A \subset \mathbb{R}^k$.

Suppose now that (2) holds. In order to deduce (1) we will construct a set A and a function $\boldsymbol{g}$ that satisfy the conditions in Definition 8.1.2. We first consider the set

$$\tilde{\mathbb{A}}_1 := \mathbb{A} \cap \left((-\infty, 0) \times \mathbb{R}^{k-1} \right),$$

which is open in $\mathbb{R}^k$. For every $\boldsymbol{t} \in \mathbb{A}$ with $t_1 = 0$, take $\delta_{\boldsymbol{t}} > 0$ and $B_{\boldsymbol{t}} \subset \mathbb{R}^{k-1}$ open such that

$$\boldsymbol{t} \in (-\delta_{\boldsymbol{t}}, 0] \times B_{\boldsymbol{t}} \subset \mathbb{A}.$$

We first analyze the case $p = 1$, that is, $\boldsymbol{\varphi}$ admits continuous partial derivatives of first order on $\mathbb{A}$. In this case, consider (Fig. 8.1)

$$\tilde{\mathbb{A}}_2 := \bigcup \{ (-\delta_{\boldsymbol{t}}, \delta_{\boldsymbol{t}}) \times B_{\boldsymbol{t}} \, : \, \boldsymbol{t} \in \mathbb{A}, t_1 = 0 \}$$

and

$$A := \tilde{\mathbb{A}}_1 \cup \tilde{\mathbb{A}}_2,$$

which is an open set in $\mathbb{R}^k$.

It is easy to check that $A \cap \mathbb{H}^k = \mathbb{A}$. Now define $\boldsymbol{g} : A \subset \mathbb{R}^k \to \mathbb{R}^n$ as follows. For $\boldsymbol{s} \in \tilde{\mathbb{A}}_1 \subset \mathbb{A}$ let $\boldsymbol{g}(\boldsymbol{s}) := \varphi(\boldsymbol{s})$, while if $\boldsymbol{s} \in \tilde{\mathbb{A}}_2$ (that is, $\boldsymbol{s} \in (-\delta_t, \delta_t) \times B_t$ for some $\boldsymbol{t} \in \mathbb{A}$ with $t_1 = 0$), then put

$$\boldsymbol{g}(\boldsymbol{s}) := \begin{cases} \boldsymbol{\varphi}(\boldsymbol{s}) & \text{if } s_1 \leq 0, \\ 2\boldsymbol{\varphi}(0, s_2, \ldots, s_k) - \boldsymbol{\varphi}(-s_1, s_2, \ldots, s_k) & \text{if } 0 < s_1 < \delta_t. \end{cases}$$

By this construction, $\boldsymbol{g} : A \subset \mathbb{R}^k \to \mathbb{R}^n$ is a mapping of class C^1 on A whose restriction to the set $\mathbb{A}$ coincides with $\boldsymbol{\varphi}$. In the general case, $p > 1$, we proceed essentially as before but with some slight variation. Again $\tilde{\mathbb{A}}_1 := \mathbb{A} \cap \big((-\infty, 0) \times \mathbb{R}^{k-1}\big)$, but now

$$\tilde{\mathbb{A}}_2 := \bigcup \Big\{ \Big(-\delta_t, \frac{\delta_t}{p} \Big) \times B_t \ : \ \boldsymbol{t} \in \mathbb{A}, t_1 = 0 \Big\}.$$

As before, $A := \tilde{\mathbb{A}}_1 \cup \tilde{\mathbb{A}}_2$ and $\boldsymbol{g} : A \subset \mathbb{R}^k \to \mathbb{R}^n$ is defined in such a way that $\boldsymbol{g}(\boldsymbol{s}) = \boldsymbol{\varphi}(\boldsymbol{s})$ whenever $\boldsymbol{s} \in \tilde{\mathbb{A}}_1$, but this time, if $\boldsymbol{s} \in (-\delta_t, \frac{\delta_t}{p}) \times B_t$ for some $\boldsymbol{t} \in \mathbb{A}$ with $t_1 = 0$ we let

$$\boldsymbol{g}(\boldsymbol{s}) := \begin{cases} \boldsymbol{\varphi}(\boldsymbol{s}) & \text{if } s_1 \leq 0, \\ \alpha_0 \boldsymbol{\varphi}(0, s_2, \ldots, s_p) + \sum_{l=1}^{p} \alpha_l \boldsymbol{\varphi}(-l s_1, s_2, \ldots, s_p) & \text{if } 0 < s_1 < \frac{\delta_t}{p}, \end{cases}$$

where $\alpha_0, \alpha_1, \ldots, \alpha_p$ are the scalars provided by Lemma 8.1.1. Their properties allow us to conclude that $\boldsymbol{g}$ is a function of class C^p on the open set A, which is an extension of $\boldsymbol{\varphi}$. □

Proposition 8.1.1 allows us to deduce that a function

$$\boldsymbol{\varphi} : \mathbb{A} \subset \mathbb{H}^k \to \mathbb{R}^n$$

is of class C^p if and only if each point of $\mathbb{A}$ has an open neighborhood for the relative topology in $\mathbb{H}^k$ in which $\boldsymbol{\varphi}$ is of class C^p. That is, the property of being of class C^p in an open set of the half-space, which was defined globally, is, in fact, a local property.

Corollary 8.1.1. *Given $\boldsymbol{\varphi} : \mathbb{A} \subset \mathbb{H}^k \to \mathbb{R}^n$, the following conditions are equivalent:*

1. *$\mathbb{A}$ is an open set in $\mathbb{H}^k$ and $\boldsymbol{\varphi}$ is a function of class C^p on $\mathbb{A}$;*
2. *for every $\boldsymbol{t}_0 \in \mathbb{A}$ there exists an open neighborhood U of $\boldsymbol{t}_0$ in $\mathbb{R}^k$ such that*

$$U \cap \mathbb{H}^k \subset \mathbb{A}$$

and

$$\boldsymbol{\varphi} : U \cap \mathbb{H}^k \subset \mathbb{H}^k \to \mathbb{R}^n$$

is a function of class C^p on $U \cap \mathbb{H}^k$.

8.2 Coordinate Systems in a Surface with Boundary

Having now dealt with the necessary technical preliminaries, we are in a position to formally define the main concept in this chapter: a regular k-surface with boundary in $\mathbb{R}^n$ (Fig. 8.2).

Definition 8.2.1. Let $2 \le k \le n$ and $M \subset \mathbb{R}^n$, $M \neq \varnothing$, be given. Then M is said to be a *regular k-surface with boundary* of class C^p if for every $\boldsymbol{x} \in M$ there exists an injective mapping

$$\boldsymbol{\varphi} : \mathbb{A} \subset \mathbb{H}^k \to M \subset \mathbb{R}^n$$

of class C^p in $\mathbb{A}$ such that $\boldsymbol{x} \in \boldsymbol{\varphi}(\mathbb{A})$ and the following hold

(SB1) There exists an open subset U of $\mathbb{R}^n$ such that $\boldsymbol{\varphi}(\mathbb{A}) = M \cap U$ and $\boldsymbol{\varphi} : \mathbb{A} \to \boldsymbol{\varphi}(\mathbb{A})$ is a homeomorphism.

(SB2) For every $\boldsymbol{t} \in \mathbb{A}$, $\mathrm{d}\boldsymbol{\varphi}(\boldsymbol{t}) : \mathbb{R}^k \to \mathbb{R}^n$ is injective.

The pair $(\mathbb{A}, \boldsymbol{\varphi})$ is called a *coordinate system.* An *atlas* of M is a family of coordinate systems

$$(\mathbb{A}_\alpha, \boldsymbol{\varphi}_\alpha)_{\alpha \in \Lambda}$$

such that

$$M = \bigcup_{\alpha \in \Lambda} \boldsymbol{\varphi}_\alpha(\mathbb{A}_\alpha).$$

In analogy to the definition of a regular k-surface, continuity of $\boldsymbol{\varphi}$ implies that condition (SB1) is equivalent to the following condition:

(SB1′) Given an open subset $\mathbb{B}$ of $\mathbb{A}$ for the relative topology, there exists an open subset V of $\mathbb{R}^n$ such that $\boldsymbol{\varphi}(\mathbb{B}) = M \cap V$.

Example 8.2.1. The first and most natural example of a regular k-surface with boundary of class C^∞ is the half-space $\mathbb{H}^k$ itself. Indeed we can take as coordinate system $(\mathbb{H}^k, \mathrm{Id}_{\mathbb{H}^k})$, where $\mathrm{Id}_{\mathbb{H}^k}$ is the restriction to $\mathbb{H}^k$ of the identity mapping on $\mathbb{R}^k$.

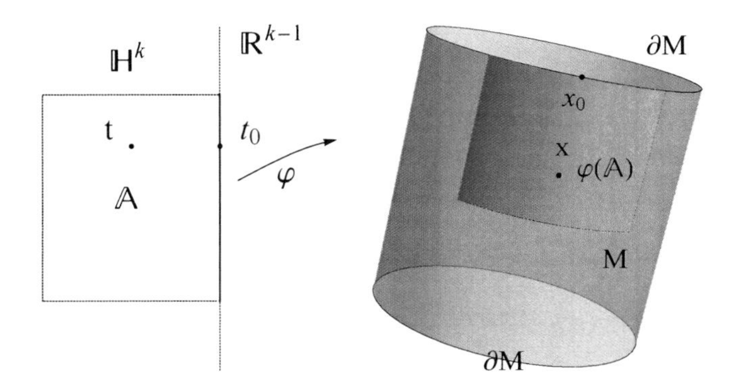

Fig. 8.2 Coordinate system of a regular surface with boundary

Remark 8.2.1. Recall that the injectivity of $\mathrm{d}\boldsymbol{\varphi}(\boldsymbol{t})$ is equivalent to the matrix

$$\left(\frac{\partial \varphi_i}{\partial t_j}(\boldsymbol{t})\right)_{i=1,\dots,n\ j=1,\dots,k}$$

being of rank k, that is, having a $k \times k$ submatrix with nonzero determinant.

Proposition 8.2.1. *Let $2 \le k \le n$ and let $M \subset \mathbb{R}^n$ be a regular k-surface with boundary of class C^p. If $(\mathbb{A}, \boldsymbol{\varphi})$ is a coordinate system of M, $\boldsymbol{\psi} : \mathbb{B} \subset \mathbb{H}^k \to M \subset \mathbb{R}^n$ is of class C^p, such that*

$$D := \boldsymbol{\varphi}(\mathbb{A}) \cap \boldsymbol{\psi}(\mathbb{B}) \neq \phi,$$

then $\boldsymbol{\psi}^{-1}(D)$ is an open set in $\mathbb{H}^k$ and

$$\boldsymbol{\varphi}^{-1} \circ \boldsymbol{\psi} : \boldsymbol{\psi}^{-1}(D) \subset \mathbb{H}^k \to \mathbb{R}^k$$

is a mapping of class C^p in $\boldsymbol{\psi}^{-1}(D)$. Moreover, if $(\mathbb{B}, \boldsymbol{\psi})$ is a coordinate system, then

$$J(\boldsymbol{\varphi}^{-1} \circ \boldsymbol{\psi})(s) \neq 0$$

at each point $\boldsymbol{s} \in \boldsymbol{\psi}^{-1}(D)$.

Proof. Let $\boldsymbol{g} : A \subset \mathbb{R}^k \to \mathbb{R}^n$ be the C^p extension of $\boldsymbol{\varphi}$ to an open set A of $\mathbb{R}^k$ given by Definition 8.1.2. We fix $\boldsymbol{s}_0 \in \boldsymbol{\psi}^{-1}(D)$. Then $\boldsymbol{s}_0 \in \mathbb{B}$ and there exists $\boldsymbol{t}_0 \in \mathbb{A}$ such that $\boldsymbol{\varphi}(\boldsymbol{t}_0) = \boldsymbol{\psi}(\boldsymbol{s}_0)$. Since $(\mathbb{A}, \boldsymbol{\varphi})$ is a coordinate system of M, we can assume, following a permutation of the coordinates, that

$$\frac{\partial(g_1, \dots, g_k)}{\partial(t_1, \dots, t_k)}(\boldsymbol{t}_0) \neq 0.$$

Hence $\tilde{\boldsymbol{g}} := (g_1, \dots, g_k)$ admits a local inverse of class C^p in an open neighborhood of $\boldsymbol{t}_0$. More precisely, there exists an open neighborhood $U \subset \mathbb{R}^k$ of $\boldsymbol{t}_0$ such that $V := \tilde{\boldsymbol{g}}(U)$ is open in $\mathbb{R}^k$ and $\tilde{\boldsymbol{g}} : U \to V$ has inverse $\boldsymbol{h} : V \to U$ of class C^p. Observe now that

$$\tilde{\boldsymbol{g}} \circ \boldsymbol{\varphi}^{-1}$$

transforms $\boldsymbol{\varphi}(\boldsymbol{t})$ in its first k coordinates. This suggests that we consider the mapping

$$\boldsymbol{\pi} : \mathbb{R}^n \to \mathbb{R}^k, \ \boldsymbol{\pi}(x_1, \dots, x_n) = (x_1, \dots, x_k),$$

that is, projection onto the first k coordinates. We take an open set $B \subset \mathbb{R}^k$ and a function $\boldsymbol{f} : B \subset \mathbb{R}^k \to \mathbb{R}^n$ of class C^p such that $B \cap \mathbb{H}^k = \mathbb{B}$ and $\boldsymbol{f} = \boldsymbol{\psi}$ in $\mathbb{B}$. Since $(\boldsymbol{\varphi}^{-1} \circ \boldsymbol{\psi})(\boldsymbol{s}_0) = \boldsymbol{t}_0 \in U$ and

$$(\boldsymbol{\pi} \circ \boldsymbol{f})(\boldsymbol{s}_0) = \boldsymbol{\pi}(\boldsymbol{\varphi}(\boldsymbol{t}_0)) = \tilde{\boldsymbol{g}}(\boldsymbol{t}_0) \in V,$$

there is an open neighborhood W of s_0 in $\mathbb{R}^k$, $W \subset B$, with the following properties for $D = \boldsymbol{\varphi}(\mathbb{A}) \cap \boldsymbol{\psi}(\mathbb{B})$:

1. $W \cap \mathbb{H}^k \subset \boldsymbol{\psi}^{-1}(D)$;
2. $(\boldsymbol{\varphi}^{-1} \circ \boldsymbol{\psi})(\boldsymbol{s}) \in U$ whenever $\boldsymbol{s} \in W \cap \mathbb{H}^k$;
3. $(\boldsymbol{\pi} \circ \boldsymbol{f})(\boldsymbol{s}) \in V$ for every $\boldsymbol{s} \in W$.

We note that

$$\tilde{\boldsymbol{g}} \circ \boldsymbol{\varphi}^{-1} \circ \boldsymbol{\psi}(\boldsymbol{s}) = \boldsymbol{\pi} \circ \boldsymbol{\psi}(\boldsymbol{s}) \tag{8.4}$$

for each $\boldsymbol{s} \in \boldsymbol{\psi}^{-1}(D)$, because for every $\boldsymbol{s} \in \boldsymbol{\psi}^{-1}(D)$ there is $\boldsymbol{t} \in \mathbb{A}$ such that $\boldsymbol{\psi}(\boldsymbol{s}) = \boldsymbol{\varphi}(\boldsymbol{t})$. Consequently,

$$\boldsymbol{\varphi}^{-1} \circ \boldsymbol{\psi} = \boldsymbol{h} \circ \boldsymbol{\pi} \circ \boldsymbol{f}$$

in the neighborhood $W \cap \mathbb{H}^k$ of s_0 in $\mathbb{H}^k$, which proves that $\boldsymbol{\varphi}^{-1} \circ \boldsymbol{\psi}$ is the restriction of a function of class C^p in a neighborhood of s_0 in $\mathbb{R}^k$. It follows from Corollary 8.1.1 that $\boldsymbol{\psi}^{-1}(D)$ is an open set in $\mathbb{H}^k$ and

$$\boldsymbol{\varphi}^{-1} \circ \boldsymbol{\psi} : \boldsymbol{\psi}^{-1}(D) \subset \mathbb{H}^k \to \mathbb{R}^k$$

is a mapping of class C^p in $\boldsymbol{\psi}^{-1}(D)$. In the case that $(\mathbb{B}, \boldsymbol{\psi})$ is also a coordinate system, we can exchange the roles of $\boldsymbol{\varphi}$ and $\boldsymbol{\psi}$ to conclude that

$$\boldsymbol{\psi}^{-1} \circ \boldsymbol{\varphi} : \boldsymbol{\varphi}^{-1}(D) \subset \mathbb{H}^k \to \boldsymbol{\psi}^{-1}(D)$$

is also a mapping of class C^p in $\boldsymbol{\varphi}^{-1}(D)$, inverse to $\boldsymbol{\varphi}^{-1} \circ \boldsymbol{\psi}$. If $\boldsymbol{\Phi}_j : W_j \subset \mathbb{R}^k \to \mathbb{R}^k$ $(j = 1, 2)$ are functions of class C^p on the open subsets W_j of $\mathbb{R}^k$ such that $W_1 \cap \mathbb{H}^k = \boldsymbol{\varphi}^{-1}(D)$, $W_2 \cap \mathbb{H}^k = \boldsymbol{\psi}^{-1}(D)$ and $\boldsymbol{\Phi}_1(\boldsymbol{t}) = \boldsymbol{\psi}^{-1} \circ \boldsymbol{\varphi}(\boldsymbol{t})$ for all $\boldsymbol{t} \in \boldsymbol{\varphi}^{-1}(D)$, $\boldsymbol{\Phi}_2(\boldsymbol{s}) = \boldsymbol{\varphi}^{-1} \circ \boldsymbol{\psi}(\boldsymbol{s})$ for every $\boldsymbol{s} \in \boldsymbol{\psi}^{-1}(D)$, then $\boldsymbol{\Phi}_2 \circ \boldsymbol{\Phi}_1(\boldsymbol{t}) = \boldsymbol{t}$ for every $\boldsymbol{t} \in \boldsymbol{\varphi}^{-1}(D)$. As in the discussion preceding Definition 8.1.3, we see that $J(\boldsymbol{\Phi}_2 \circ \boldsymbol{\Phi}_1)(\boldsymbol{t}) = 1$ and

$$\mathrm{d}(\boldsymbol{\Phi}_2 \circ \boldsymbol{\Phi}_1)(\boldsymbol{t}) = \mathrm{Id}_{\mathbb{R}^k}$$

for every $\boldsymbol{t} \in \boldsymbol{\varphi}^{-1}(D)$. Hence the chain rule applied to $\boldsymbol{\Phi}_1$ and $\boldsymbol{\Phi}_2$ gives

$$J(\boldsymbol{\Phi}_2)(\boldsymbol{\Phi}_1(\boldsymbol{t}))J(\boldsymbol{\Phi}_1)(\boldsymbol{t}) = 1 \ \ \forall \boldsymbol{t} \in \boldsymbol{\varphi}^{-1}(D),$$

i.e., for $\boldsymbol{s} = \boldsymbol{\varphi}(\boldsymbol{t})$,

$$J(\boldsymbol{\varphi}^{-1} \circ \boldsymbol{\psi})(s)J(\boldsymbol{\psi}^{-1} \circ \boldsymbol{\varphi})(\boldsymbol{t}) = 1,$$

and in particular,

$$J(\boldsymbol{\varphi}^{-1} \circ \boldsymbol{\psi})(\boldsymbol{s}) \neq 0$$

at each $\boldsymbol{s} \in \boldsymbol{\psi}^{-1}(D)$. □

Definition 8.2.2. Let M be a regular k-surface with boundary. The point $\boldsymbol{x}_0 \in M$ is said to be in the boundary of M (which we denote by ∂M) if for some coordinate system $(\mathbb{A}, \boldsymbol{\varphi})$ of M there is a point $(0, t_2^0, \ldots, t_k^0) \in \mathbb{A}$ such that

$$\boldsymbol{x}_0 = \boldsymbol{\varphi}(0, t_2^0, \ldots, t_k^0).$$

The following result guarantees that the set of boundary points is independent of the chosen coordinate systems.

Lemma 8.2.1. *Let M be a regular k-surface with boundary of class C^p. Let $\boldsymbol{x}_0 \in M$ and $(\mathbb{B}, \boldsymbol{\psi})$ a coordinate system of M such that*

$$\boldsymbol{\psi}^{-1}(\boldsymbol{x}_0) = (0, s_2^0, \ldots, s_k^0).$$

Then for each coordinate system $(\mathbb{A}, \boldsymbol{\varphi})$ with $\boldsymbol{x}_0 \in \boldsymbol{\varphi}(\mathbb{A})$ we have

$$\boldsymbol{\varphi}^{-1}(\boldsymbol{x}_0) = (0, t_2^0, \ldots, t_k^0).$$

Proof. Let

$$D := \boldsymbol{\varphi}(\mathbb{A}) \cap \boldsymbol{\psi}(\mathbb{B})$$

and put

$$\boldsymbol{s}_0 := \boldsymbol{\psi}^{-1}(\boldsymbol{x}_0), \quad \boldsymbol{t}_0 := \boldsymbol{\varphi}^{-1}(\boldsymbol{x}_0) = (t_1^0, t_2^0, \ldots, t_k^0).$$

We already know that $t_1^0 \leq 0$. Arguing by contradiction, we suppose that

$$t_1^0 < 0.$$

From Proposition 8.2.1 there exist an open set $B \subset \mathbb{R}^k$, $\boldsymbol{\psi}^{-1}(D) \subset B$, and a mapping

$$\boldsymbol{g} : B \subset \mathbb{R}^k \to \mathbb{R}^k$$

of class C^p such that

$$\boldsymbol{g}(\boldsymbol{s}) = \boldsymbol{\varphi}^{-1} \circ \boldsymbol{\psi}(\boldsymbol{s})$$

for every $\boldsymbol{s} \in \boldsymbol{\psi}^{-1}(D)$. Since the set $\boldsymbol{\varphi}^{-1}(D)$ is open in $\mathbb{H}^k$ and $t_1^0 < 0$, there exists $r > 0$ such that the open ball in $\mathbb{R}^k$ with radius r and centered at $\boldsymbol{t}_0$, denoted by $B(\boldsymbol{t}_0, r)$, is contained in $\boldsymbol{\varphi}^{-1}(D)$. Thus, the mapping

$$\boldsymbol{\psi}^{-1} \circ \boldsymbol{\varphi}$$

is well defined and of class C^p in $B(\boldsymbol{t}_0, r)$. Also

$$\boldsymbol{g} \circ (\boldsymbol{\psi}^{-1} \circ \boldsymbol{\varphi})(\boldsymbol{t}) = \boldsymbol{t}$$

for every $\boldsymbol{t} \in B(\boldsymbol{t}_0, r)$. Consequently,

$$J(\boldsymbol{\psi}^{-1} \circ \boldsymbol{\varphi})(\boldsymbol{t}) \neq 0$$

for every $\boldsymbol{t} \in B(\boldsymbol{t}_0, r)$. By the inverse function theorem (Theorem 3.1.1), the set

$$U := (\boldsymbol{\psi}^{-1} \circ \boldsymbol{\varphi})(B(\boldsymbol{t}_0, r))$$

is open in $\mathbb{R}^k$ and $(0, s_2^0, \ldots, s_k^0) = (\boldsymbol{\psi}^{-1} \circ \boldsymbol{\varphi})(\boldsymbol{t}_0) \in U$, so there exists $\delta > 0$ such that $0 < s_1 < \delta$ implies $(s_1, s_2^0, \ldots, s_k^0) \in U$, which is a contradiction, since $U \subset \mathbb{H}^k = (-\infty, 0) \times \mathbb{R}^{k-1}$. □

Remark 8.2.2. Given a subset A of $\mathbb{R}^n$, the *topological boundary* is defined to be

$$\partial^{\text{top}}(A) = \overline{A} \setminus \text{int}(A).$$

That is, the topological boundary of A consists of the points in the closure of A that do not belong to the interior of A. It is important to note that this is a different concept from the (nontopological) boundary ∂M of a regular k-surface with boundary M. For example, the unit sphere $S_{n-1} = \{(x_1, \ldots, x_n) \in \mathbb{R}^n : x_1^2 + \cdots + x_n^2 = 1\}$ is a closed set with empty interior, and hence coincides with its topological boundary, $\partial^{\text{top}}(S_{n-1}) = S_{n-1}$. However, we know from Example 3.2.4 that the sphere is a regular $(n-1)$-surface in $\mathbb{R}^n$, and so by Proposition 8.2.2, S_{n-1} has no boundary, i.e., $\partial S_{n-1} = \varnothing$. In Examples 8.3.1 and 8.3.2, the topological boundary of M coincides with all of M and is different from the boundary ∂M. There is, however, a prominent case in which the two notions coincide, and this is given in the following example.

Example 8.2.2. Let M be a compact regular n-surface with boundary in $\mathbb{R}^n$. Then $M \setminus \partial M$ is an open set in $\mathbb{R}^n$ and ∂M coincides with the topological boundary of M.

We first show that $M \setminus \partial M$ is an open set in $\mathbb{R}^n$. For every $\boldsymbol{x}_0 \in M \setminus \partial M$ let $(\mathbb{A}, \boldsymbol{\varphi})$ be a coordinate system of M,

$$\boldsymbol{\varphi} : \mathbb{A} \subset \mathbb{H}^n \to \mathbb{R}^n,$$

such that $\boldsymbol{x}_0 \in \boldsymbol{\varphi}(\mathbb{A})$. Define

$$B := \{\boldsymbol{t} \in \mathbb{A} \; ; \; t_1 < 0\},$$

which is an open subset of $\mathbb{R}^n$. Since $\boldsymbol{x}_0 \notin \partial M$, we have, according to Definition 8.2.2, that

$$\boldsymbol{x}_0 \in \boldsymbol{\varphi}(B) \subset M \setminus \partial M.$$

Moreover, the Jacobian of $\boldsymbol{\varphi}$ is nonzero at all the points of B, and by the inverse function theorem, $\boldsymbol{\varphi}(B)$ is an open set in $\mathbb{R}^n$. It follows from this that $M \setminus \partial M$ is the union of a family of open sets. Consequently,

$$M \setminus \partial M \subset M \setminus \partial^{\text{top}}(M),$$

where $\partial^{\text{top}}(M)$ is the topological boundary of M. Now we prove the reverse inclusion. For every $\boldsymbol{x}_0 \in M \setminus \partial^{\text{top}}(M)$ there exists $r > 0$ such that the open ball $B(\boldsymbol{x}_0, r)$ is contained in $M \setminus \partial^{\text{top}}(M)$. Let $\boldsymbol{R} : \mathbb{R}^n \to \mathbb{R}^n$ be an affine transformation such that

$$\mathbb{A} := \boldsymbol{R}(B(\boldsymbol{x}_0, r))$$

is contained in the interior of the half-space $\mathbb{H}^n$. Finally, define

$$\boldsymbol{\varphi} := \boldsymbol{R}^{-1} : \mathbb{A} \subset \mathbb{H}^n \to B(\boldsymbol{x}_0, r) \subset M.$$

Then $(\mathbb{A}, \boldsymbol{\varphi})$ is a coordinate system of M according to Definition 8.2.1 and $\boldsymbol{x}_0 \in \boldsymbol{\varphi}(\mathbb{A})$. Since $\mathbb{A}$ is contained in the interior of the half-space $\mathbb{H}^n$, we conclude that $\boldsymbol{x}_0 \in M \setminus \partial M$.

Proposition 8.2.2. *Let M be a regular k-surface of class C^p in $\mathbb{R}^n$. Then M is a regular k-surface with boundary of class C^p and $\partial M = \varnothing$.*

Proof. According to Definition 3.1.1, for each $\boldsymbol{x}_0 \in M$ there is coordinate system $(A, \boldsymbol{\varphi})$ of M such that $\boldsymbol{x}_0 \in \boldsymbol{\varphi}(A)$. Take $\boldsymbol{t}_0 \in A$ and $r > 0$ with $B(\boldsymbol{t}_0, r) \subset A$ and $\boldsymbol{x}_0 = \boldsymbol{\varphi}(\boldsymbol{t}_0)$. Then for the mapping

$$\boldsymbol{R} : \mathbb{R}^k \to \mathbb{R}^k, \quad \boldsymbol{R}(\boldsymbol{t}) = \boldsymbol{t} - \boldsymbol{t}_0 - r\boldsymbol{e}_1,$$

we have that

$$\mathbb{B} := \boldsymbol{R}(B(\boldsymbol{t}_0, r))$$

is contained in the interior of the half-space $\mathbb{H}^k$. Now define

$$\boldsymbol{\psi} := \boldsymbol{\varphi} \circ \boldsymbol{R}^{-1} : \mathbb{B} \subset \mathbb{H}^k \to \mathbb{R}^n.$$

Then $(\mathbb{B}, \boldsymbol{\psi})$ is a coordinate system of M according to Definition 8.2.1 and $\boldsymbol{\psi}(\mathbb{B}) = \boldsymbol{\varphi}(B(\boldsymbol{t}_0, r))$. Also, since $\boldsymbol{\psi}^{-1}(\boldsymbol{x}_0)$ is in $\mathbb{B}$, and so contained in the interior of $\mathbb{H}^k$, it turns out that the first coordinate of $\boldsymbol{\psi}^{-1}(\boldsymbol{x}_0)$ is strictly negative, and consequently,

$$\boldsymbol{x}_0 \notin \partial M.$$

□

The following proposition will be very useful in the exercises. It allows one to replace the half-space $\mathbb{H}^k$ provided by Definition 8.1.1 and used in Definition 8.2.1 of a regular surface with boundary by an arbitrary half-space.

Definition 8.2.3. A half-space in $\mathbb{R}^k$ is a set $\mathbb{S}$ of the form

$$\mathbb{S} := \{\boldsymbol{t} \in \mathbb{R}^k \ : \pi(\boldsymbol{t}) \leq a\},$$

where $\pi : \mathbb{R}^k \to \mathbb{R}$ is a nonzero linear form and $a \in \mathbb{R}$. The boundary of the half-space is defined by

$$\partial \mathbb{S} := \{\boldsymbol{t} \in \mathbb{R}^k \ : \pi(\boldsymbol{t}) = a\}.$$

An example of a half-space is $\mathbb{H}^k$, introduced in Definition 8.1.1. Another example is a set of the form

$$\mathbb{S} := \{\boldsymbol{t} \in \mathbb{R}^k \ : \ t_j \geq b\}$$

for a fixed $1 \leq j \leq k$. In fact, it suffices to consider the linear form $\pi(\boldsymbol{t}) = -t_j$ and $a = -b$. In the last example, the boundary coincides with the set of points $\boldsymbol{t} \in \mathbb{R}^k$ such that $t_j = b$.

A mapping

$$\boldsymbol{\varphi} : \mathbb{A} \subset \mathbb{S} \to M \subset \mathbb{R}^n$$

is said to be of class C^p in $\mathbb{A}$ if there exist an open set $A \subset \mathbb{R}^k$ and a mapping

$$\boldsymbol{g} : A \subset \mathbb{R}^k \to \mathbb{R}^n$$

of class C^p in A such that $A \cap \mathbb{S} = \mathbb{A}$ and $\boldsymbol{g}_{|\mathbb{A}} = \boldsymbol{\varphi}$.

Lemma 8.2.2. *For each half-space* $\mathbb{S}$ *in* $\mathbb{R}^k$ *there exists an affine mapping*

$$\boldsymbol{R} : \mathbb{R}^k \to \mathbb{R}^k$$

such that $\boldsymbol{R}(\mathbb{H}^k) = \mathbb{S}$, $\boldsymbol{R}(\partial \mathbb{H}^k) = \partial \mathbb{S}$.

Proof. Given the half-space $\mathbb{S} = \{\boldsymbol{t} \in \mathbb{R}^k \ : \pi(\boldsymbol{t}) \leq a\}$, we consider an orthonormal basis $\{\boldsymbol{v}_1, \boldsymbol{v}_2, \ldots, \boldsymbol{v}_k\}$ of $\mathbb{R}^k$ with the properties that $\pi(\boldsymbol{v}_1) > 0$ and $\{\boldsymbol{v}_2, \boldsymbol{v}_3, \ldots, \boldsymbol{v}_k\}$ is a basis of the $(k-1)$-dimensional vector subspace $\pi^{-1}(0) = \operatorname{Ker} \pi$. We take

$$\boldsymbol{L} : \mathbb{R}^k \to \mathbb{R}^k$$

to be the linear mapping defined by $\boldsymbol{L}(\boldsymbol{e}_j) = \boldsymbol{v}_j$ for $1 \leq j \leq k$. Finally, choose a vector $\boldsymbol{b} \in \mathbb{R}^k$ such that $\pi(\boldsymbol{b}) = a$ and define the affine mapping $\boldsymbol{R}$ by $\boldsymbol{R} := \boldsymbol{b} + \boldsymbol{L}$. From

$$\pi\left(\boldsymbol{R}\left(\sum_{j=1}^{k} x_j \boldsymbol{e}_j\right)\right) = a + \sum_{j=1}^{k} x_j \pi(\boldsymbol{v}_j) = a + x_1 \pi(\boldsymbol{v}_1),$$

we deduce

$$\boldsymbol{R}(\mathbb{H}^k) = \mathbb{S} \quad \text{and} \quad \boldsymbol{R}(\partial \mathbb{H}^k) = \partial \mathbb{S},$$

and the proof is finished. □

As an immediate consequence of Lemma 8.2.2, the half-space $\mathbb{S}$ is a regular k-surface with boundary of class C^∞ in $\mathbb{R}^k$, and the definition of boundary of a half-space introduced in Definition 8.2.3 is consistent with the definition of boundary of a regular surface with boundary, Definition 8.2.2.

Proposition 8.2.3. *Let $M = M_1 \cup M_2$ satisfy $M_1 \cap M_2 = \varnothing$, where M_1 is a regular k-surface of class C^p and for each point $\boldsymbol{x}_0 \in M_2$ there exist a half-space $\mathbb{S}$ in $\mathbb{R}^k$ and a mapping*

$$\boldsymbol{\varphi} : \mathbb{A} \subset \mathbb{S} \to M \subset \mathbb{R}^n$$

of class C^p such that $\boldsymbol{x}_0 \in \boldsymbol{\varphi}(\mathbb{A})$. Suppose the following hold:

(1) *There exists an open subset U of $\mathbb{R}^n$ such that $\boldsymbol{\varphi}(\mathbb{A}) = M \cap U$ and $\boldsymbol{\varphi} : \mathbb{A} \to \boldsymbol{\varphi}(\mathbb{A})$ is a homeomorphism.*
(2) *For every $\boldsymbol{t} \in \mathbb{A}$, $\mathrm{d}\boldsymbol{\varphi}(\boldsymbol{t}) : \mathbb{R}^k \to \mathbb{R}^n$ is injective.*
(3) $\boldsymbol{\varphi}^{-1}(\boldsymbol{x}_0) \in \partial\mathbb{S}$.

Then M is a regular k-surface with boundary of class C^p and $\partial M = M_2$.

Proof. For every $\boldsymbol{x}_0 \in M_1$, we can argue as in Proposition 8.2.2 in order to find a coordinate system $(\mathbb{B}, \boldsymbol{\psi})$ such that in accordance with Definition 8.2.1, $\boldsymbol{x}_0 \in \boldsymbol{\psi}(\mathbb{B})$ and the first coordinate of $\boldsymbol{\psi}^{-1}(\boldsymbol{x}_0)$ is strictly negative. Now we concentrate on giving the details only for the boundary. Fix a point $\boldsymbol{x}_0 \in M_2$ and take $\boldsymbol{\varphi} : \mathbb{A} \subset \mathbb{S} \to M$ to be a mapping of class C^p satisfying properties (1), (2), and (3) of the hypothesis. We also consider the affine mapping $\boldsymbol{R} : \mathbb{R}^k \to \mathbb{R}^k$ obtained in Lemma 8.2.2. Finally, define

$$\mathbb{B} := \boldsymbol{R}^{-1}(\mathbb{A}) \subset \mathbb{H}^k$$

and

$$\boldsymbol{\psi} : \mathbb{B} \subset \mathbb{H}^k \to M, \quad \boldsymbol{\psi} = \boldsymbol{\varphi} \circ \boldsymbol{R}.$$

It is straightforward that $(\mathbb{B}, \boldsymbol{\psi})$ is a coordinate system of M according to Definition 8.2.1 and that $\boldsymbol{\psi}^{-1}(\boldsymbol{x}_0) = (0, t_2^0, \ldots, t_k^0)$ (see the definition of boundary, Definition 8.2.2). It follows that M is a regular k-surface with boundary and $\partial M = M_2$. □

It is worth to mention that the statement of Proposition 8.2.3 is written in the most suitable way to apply to some exercises. But, if M is a subset of $\mathbb{R}^n$ with the property that for each $\boldsymbol{x}_0 \in M$ there exists $\boldsymbol{\varphi} : \mathbb{A} \subset \mathbb{S} \to M$ such that $\boldsymbol{x}_0 \in \boldsymbol{\varphi}(\mathbb{A})$ and conditions (1) and (2) of proposition 8.2.3 are satisfied, then M is a regular k-surface with boundary of class C^p and $\partial M = \left\{\boldsymbol{x} \in M : \boldsymbol{\varphi}^{-1}(\boldsymbol{x}_0) \in \partial\mathbb{S}\right\}$.

Intuitively, a regular k-surface with boundary consists of the union of a regular k-surface (without boundary) and a regular $(k-1)$-surface (that is exactly the boundary). In particular, a regular surfaces with boundary in $\mathbb{R}^3$ is the union of a regular surface and a regular curve. This is the content of the following theorem. Nevertheless, one cannot build a regular surface with boundary in $\mathbb{R}^3$ from a regular surface (without boundary) and a regular curve in $\mathbb{R}^3$, since the regular surface and the regular curve must be connected in a very special form. It is Definitions 8.2.1 and 8.2.2 that describe that exact form.

From now on we denote by $\boldsymbol{t} = (t_1, \ldots, t_k)$ an arbitrary point of $\mathbb{R}^k$ and by $\boldsymbol{s} = (s_1, \ldots, s_{k-1})$ an arbitrary point of $\mathbb{R}^{k-1}$. If $\boldsymbol{s} \in \mathbb{R}^{k-1}$, then $\boldsymbol{t} := (0, \boldsymbol{s}) \in \mathbb{R}^k$ is the point with coordinates

$$t_1 = 0 \ , \ t_2 = s_1 \ , \ \ldots, \ t_k = s_{k-1}.$$

Theorem 8.2.1. *Let M be a regular k-surface with boundary of class C^p in $\mathbb{R}^n$. Then:*

(1) ∂M is a regular $(k-1)$-surface of class C^p.
(2) $M \setminus \partial M$ is a regular k-surface of class C^p.

Proof. We will concentrate on the study of the boundary. For each $\boldsymbol{x}_0 \in \partial M$ let $(\mathbb{A}, \boldsymbol{\varphi})$ be a coordinate system of M such that $\boldsymbol{x}_0 \in \boldsymbol{\varphi}(\mathbb{A})$ and let

$$(0, t_2^0, \ldots, t_k^0) = \boldsymbol{\varphi}^{-1}(\boldsymbol{x}_0).$$

We define

$$\widehat{\mathbb{A}} := \{\boldsymbol{s} \in \mathbb{R}^{k-1} \ : \ (0, \boldsymbol{s}) \in \mathbb{A}\}.$$

In other words, if we denote by

$$\boldsymbol{j} : \mathbb{R}^{k-1} \to \mathbb{H}^k \subset \mathbb{R}^k$$

the continuous mapping $\boldsymbol{j}(\boldsymbol{s}) := (0, \boldsymbol{s})$, then

$$\widehat{\mathbb{A}} = \boldsymbol{j}^{-1}(\mathbb{A})$$

is an open set in $\mathbb{R}^{k-1}$. We also define

$$\widehat{\boldsymbol{\varphi}} : \widehat{\mathbb{A}} \subset \mathbb{R}^{k-1} \to \mathbb{R}^n \ , \quad \widehat{\boldsymbol{\varphi}}(\boldsymbol{s}) := \boldsymbol{\varphi}(0, \boldsymbol{s}).$$

If A is open in $\mathbb{R}^k$ and $\boldsymbol{g} : A \subset \mathbb{R}^k \to \mathbb{R}^n$ is a function of class C^p on A such that $A \cap \mathbb{H}^k = \mathbb{A}$ and $\boldsymbol{g} = \boldsymbol{\varphi}$ in $\mathbb{A}$ (see Definition 8.1.2), then

$$\widehat{\boldsymbol{\varphi}} = \boldsymbol{\varphi} \circ \boldsymbol{j}_{|\widehat{\mathbb{A}}} = \boldsymbol{g} \circ \boldsymbol{j}_{|\widehat{\mathbb{A}}}$$

is a mapping of class C^p in the open set $\widehat{\mathbb{A}}$ of $\mathbb{R}^{k-1}$. Moreover, if we consider $\boldsymbol{s}_0 := (t_2^0, \ldots, t_k^0) \in \widehat{\mathbb{A}} \subset \mathbb{R}^{k-1}$, then $\boldsymbol{x}_0 = \widehat{\boldsymbol{\varphi}}(\boldsymbol{s}_0)$, and by Definition 8.2.2,

$$\boldsymbol{x}_0 \in \widehat{\boldsymbol{\varphi}}(\widehat{\mathbb{A}}\,) \subset \partial M.$$

We prove that

$$(\widehat{\mathbb{A}}, \widehat{\boldsymbol{\varphi}})$$

is a coordinate system of a $(k-1)$-regular surface in the sense of Definition 3.1.1. To check condition (S1′) of Definition 3.1.1, take an open subset

$$D \subset \widehat{\mathbb{A}}$$

and observe that $(-\infty, 0] \times D$ is an open subset of $\mathbb{H}^k$ for the relative topology. Since $\boldsymbol{\varphi}$ is a homeomorphism, by condition (SB1′) for a regular surface with boundary (Definition 8.2.1), there is an open set V in $\mathbb{R}^n$ such that

$$\boldsymbol{\varphi}\left((-\infty, 0] \times D \cap \mathbb{A}\right) = M \cap V,$$

and from this we see that

$$\widehat{\boldsymbol{\varphi}}(D) \;=\; M \cap V \cap \partial M \;=\; \partial M \cap V.$$

To check the second condition for a coordinate system (that is, Definition 3.1.1(S2)), we observe that since

$$\widehat{\boldsymbol{\varphi}}(s_1, s_2, \ldots, s_{k-1}) \;= \boldsymbol{\varphi}(0, s_1, s_2, \ldots, s_{k-1}),$$

the gradient vector of the component $\widehat{\varphi}_i$ at $\boldsymbol{s}$ consists of the partial derivatives of φ_i, with respect to each of the variables except the first one, at $\boldsymbol{t} = (0, \boldsymbol{s})$. That is, the $n \times (k-1)$ Jacobian matrix

$$\widehat{\boldsymbol{\varphi}}'(\boldsymbol{s}) = \begin{pmatrix} \frac{\partial \varphi_1}{\partial t_2}(0,\boldsymbol{s}) & \frac{\partial \varphi_1}{\partial t_3}(0,\boldsymbol{s}) & \ldots & \frac{\partial \varphi_1}{\partial t_k}(0,\boldsymbol{s}) \\ \frac{\partial \varphi_2}{\partial t_2}(0,\boldsymbol{s}) & \frac{\partial \varphi_2}{\partial t_3}(0,\boldsymbol{s}) & \ldots & \frac{\partial \varphi_2}{\partial t_k}(0,\boldsymbol{s}) \\ \ldots & \ldots & \ldots & \ldots \\ \frac{\partial \varphi_n}{\partial t_2}(0,\boldsymbol{s}) & \frac{\partial \varphi_n}{\partial t_3}(0,\boldsymbol{s}) & \ldots & \frac{\partial \varphi_n}{\partial t_k}(0,\boldsymbol{s}) \end{pmatrix}$$

is obtained by eliminating the first column of the $n \times k$ Jacobian matrix

$$\boldsymbol{\varphi}'(0,\boldsymbol{s}) = \begin{pmatrix} \frac{\partial \varphi_1}{\partial t_1}(0,\boldsymbol{s}) & \frac{\partial \varphi_1}{\partial t_2}(0,\boldsymbol{s}) & \ldots & \frac{\partial \varphi_1}{\partial t_k}(0,\boldsymbol{s}) \\ \frac{\partial \varphi_2}{\partial t_1}(0,\boldsymbol{s}) & \frac{\partial \varphi_2}{\partial t_2}(0,\boldsymbol{s}) & \ldots & \frac{\partial \varphi_2}{\partial t_k}(0,\boldsymbol{s}) \\ \ldots & \ldots & \ldots & \ldots \\ \frac{\partial \varphi_n}{\partial t_1}(0,\boldsymbol{s}) & \frac{\partial \varphi_n}{\partial t_2} t & \ldots & \frac{\partial \varphi_n}{\partial t_k}(0,\boldsymbol{s}) \end{pmatrix}.$$

Since the columns of this last matrix are linearly independent, by condition (SB2) of the definition of a coordinate system for a regular surface with boundary (see Definition 8.2.1), it follows that the columns of the matrix $\widehat{\boldsymbol{\varphi}}'(\boldsymbol{s})$ are also linearly independent, and this means that condition (S2) of Definition 3.1.1 is satisfied. We have proved that ∂M is a regular $(k-1)$-surface of class C^p. □

Example 8.2.3. Let $0 < \varepsilon < 1$ be given. Then

$$M := \{(x,y,z) \in \mathbb{R}^3 \ : \ x^2+y^2+z^2 = 1 \ , \ z \geq \varepsilon\}$$

is a regular 2-surface with boundary in $\mathbb{R}^3$ and its boundary is given by

$$\partial M = \{(x,y,z) \in \mathbb{R}^3 \ : x^2+y^2 = 1-\varepsilon^2 \ , \ z = \varepsilon\}.$$

Let us decompose our surface as $M = M_1 \cup M_2$, where

$$M_1 := \{(x,y,z) \in \mathbb{R}^3 \ : \ x^2+y^2+z^2 = 1 \ , \ z > \varepsilon\}$$

and

$$M_2 := \{(x,y,z) \in \mathbb{R}^3 \ : x^2+y^2 = 1-\varepsilon^2 \ , \ z = \varepsilon\}.$$

According to Example 3.2.2, M_1 is a regular surface in $\mathbb{R}^3$. In order to apply Proposition 8.2.3, we fix a point

$$(x_0, y_0, \varepsilon) \in M_2$$

and select $\theta_0 \in \mathbb{R}$ such that $x_0 = \sqrt{1-\varepsilon^2}\cos\theta_0$, $y_0 = \sqrt{1-\varepsilon^2}\sin\theta_0$. We will parameterize the points (x,y,z) of M in a neighborhood of (x_0,y_0,ε) using the polar coordinates of the point (x,y) while bearing in mind that

$$z = \sqrt{1-x^2-y^2}.$$

That is, we consider the half-space

$$\mathbb{S} := \{(\rho,\theta) \in \mathbb{R}^2 \ : \ \rho \leq \sqrt{1-\varepsilon^2}\}$$

and the open set (for the relative topology on $\mathbb{S}$)

$$\mathbb{A} := (0, \sqrt{1-\varepsilon^2}] \ \times \ (\theta_0 - \pi, \theta_0 + \pi).$$

Now define

$$\varphi : \mathbb{A} \subset \mathbb{S} \to \mathbb{R}^3, \quad \varphi(\rho,\theta) := \left(\rho\cos\theta, \rho\sin\theta, \sqrt{1-\rho^2}\right).$$

We observe that φ is the restriction of the function

$$g : A \subset \mathbb{R}^2 \to \mathbb{R}^3, \quad g(\rho,\theta) := \left(\rho\cos\theta, \rho\sin\theta, \sqrt{1-\rho^2}\right)$$

of class C^∞ defined on the set

$$A := (0,1) \times (\theta_0 - \pi, \theta_0 + \pi).$$

To finish, we show that $\boldsymbol{\varphi}$ satisfies conditions (1), (2), and (3) of Proposition 8.2.3. To check (1), we consider the mapping

$$\boldsymbol{h} : A \subset \mathbb{R}^2 \to \mathbb{R}^2, \quad \boldsymbol{h}(\rho,\theta) := (\rho\cos\theta, \rho\sin\theta).$$

Since $\boldsymbol{h}$ is injective and its Jacobian is nonzero at each point, we deduce from the inverse function theorem that $\boldsymbol{h}(A)$ is an open set in $\mathbb{R}^2$ and $\boldsymbol{h} : A \to \boldsymbol{h}(A)$ admits an inverse of class C^∞. Then

$$\boldsymbol{\varphi}(\mathbb{A}) = (\boldsymbol{h}(A) \times \mathbb{R}) \cap M$$

is an open subset of M. We denote by $\boldsymbol{\pi} : \mathbb{R}^3 \to \mathbb{R}^2$ the projection onto the first two coordinates. It turns out that the inverse of the mapping

$$\boldsymbol{\varphi} : \mathbb{A} \to \boldsymbol{\varphi}(\mathbb{A})$$

is precisely

$$\boldsymbol{\varphi}^{-1} = \boldsymbol{h}^{-1} \circ \boldsymbol{\pi},$$

which is a continuous function. That is, $\boldsymbol{\varphi}$ is a homeomorphism onto its image. It is also possible to arrive at the same conclusion by observing that $\boldsymbol{\varphi}$ is the restriction of a coordinate system for the sphere.

Condition (2) holds is obvious since the two vectors $\frac{\partial \boldsymbol{\varphi}}{\partial \rho}$ and $\frac{\partial \boldsymbol{\varphi}}{\partial \theta}$ are linearly independent by virtue of the fact that

$$\frac{\partial \boldsymbol{\varphi}}{\partial \rho} \times \frac{\partial \boldsymbol{\varphi}}{\partial \theta}(\rho,\theta) = \left(\frac{\rho^2}{\sqrt{1-\rho^2}}\cos\theta, \frac{\rho^2}{\sqrt{1-\rho^2}}\sin\theta, \rho \right) \neq (0,0,0).$$

Concerning condition (3), it suffices to note that

$$(x_0, y_0, \varepsilon) = (\sqrt{1-\varepsilon^2}\cos\theta_0, \sqrt{1-\varepsilon^2}\sin\theta_0, \varepsilon) = \boldsymbol{\varphi}(\sqrt{1-\varepsilon^2}, \theta_0),$$

or equivalently,

$$\boldsymbol{\varphi}^{-1}(x_0, y_0, \varepsilon) = (\sqrt{1-\varepsilon^2}, \theta_0) \in \partial\mathbb{S}.$$

Remark 8.2.3. (1) The mapping $\boldsymbol{\varphi} : \mathbb{A} \subset \mathbb{S} \to \mathbb{R}^3$ of Example 8.2.3 satisfies the condition that

$$M = \boldsymbol{\varphi}(\mathbb{A}) \cup \boldsymbol{\alpha}([0, \sqrt{1-\varepsilon^2}]),$$

where $\boldsymbol{\alpha} : [0, \sqrt{1-\varepsilon^2}] \to M$ is defined by

$$\alpha(\rho) = (\rho\cos(\theta_0 - \pi), \rho\sin(\theta_0 - \pi), \sqrt{1-\rho^2}),$$

i.e., $\alpha([0, \sqrt{1-\varepsilon^2}])$ is the portion of a circle originating at point $(0,0,1)$ and ending on M_2 at point $(\sqrt{1-\varepsilon^2}\cos(\theta_0 - \pi), \sqrt{1-\varepsilon^2}\sin(\theta_0 - \pi), \varepsilon)$.

(2) Take

$$\mathbb{B} := (-\sqrt{1-\varepsilon^2}, 0] \times (\theta_0 - \pi, \theta_0 + \pi)$$

and

$$\boldsymbol{\psi} : \mathbb{B} \subset \mathbb{H}^2 \to M, \quad \boldsymbol{\psi}(s, \theta) = \boldsymbol{\varphi}(\sqrt{1-\varepsilon^2} + s, \theta).$$

Then $(\mathbb{B}, \boldsymbol{\psi})$ is a coordinate system of M in a neighborhood of (x_0, y_0, ε).

Example 8.2.4. Consider the compact set

$$M := \{(x,y,z) \in \mathbb{R}^3 \ : \ x^2 + y^2 + z^2 \leq 1\}.$$

M is a regular 3-surface with boundary of class C^∞ in $\mathbb{R}^3$. In this case, the boundary of the surface is the unit sphere (See Example 8.2.2).

Again, we form a decomposition $M = M_1 \cup M_2$, where

$$M_1 = \{(x,y,z) \in \mathbb{R}^3 \ : \ x^2 + y^2 + y^2 < 1\}$$

is an open subset of $\mathbb{R}^3$, that is, a regular 3-surface in $\mathbb{R}^3$, and

$$M_2 = \{(x,y,z) \in \mathbb{R}^3 \ : \ x^2 + y^2 + y^2 = 1\}$$

is a regular 2-surface in $\mathbb{R}^3$. Intending to apply Proposition 8.2.3, we fix an arbitrary point $(x_0, y_0, z_0) \in M_2$. Suppose that $(x_0, y_0, z_0) \notin \{(x, 0, z) : \ x \leq 0, z \in \mathbb{R}\}$. We will parameterize using the spherical coordinates (r,s,t) of a point of $\mathbb{R}^3$ (see Example 3.3.2). That is, we consider the mapping

$$\boldsymbol{g} : (0, +\infty) \times (0, \pi) \times (-\pi, \pi) \to \mathbb{R}^3$$

defined by

$$\boldsymbol{g}(r,s,t) = (r\sin s\cos t, r\sin s\sin t, r\cos s)\,.$$

We already know that $\boldsymbol{g}$ is injective that and its Jacobian, $r^2\sin s$, is nonzero. It follows from the inverse function theorem that the image of $\boldsymbol{g}$ is an open subset of $\mathbb{R}^3$ and that $\boldsymbol{g}$ has an inverse of class C^∞. In particular, $\boldsymbol{g}$ is a homeomorphism onto its image. Now consider the half-space

$$\mathbb{S} := \{(r,s,t) \in \mathbb{R}^3 \; : \; r \le 1\}$$

of $\mathbb{R}^3$, the open set

$$\mathbb{A} := (0,1] \times (0,\pi) \times (-\pi,\pi)$$

with respect to the relative topology, and the mapping

$$\boldsymbol{\varphi} : \mathbb{A} \subset \mathbb{S} \to M \subset \mathbb{R}^3$$

given by the restriction of $\boldsymbol{g}$ to the set $\mathbb{A}$. It should be clear that $\boldsymbol{\varphi}$ satisfies conditions (1) and (2) of Proposition 8.2.3. Moreover, there exist $s_0 \in (0,\pi)$ and $t_0 \in (-\pi,\pi)$ such that

$$(x_0, y_0, z_0) = \boldsymbol{\varphi}(1, s_0, t_0),$$

that is,

$$\boldsymbol{\varphi}^{-1}(x_0, y_0, z_0) \in \partial\mathbb{S}.$$

At points of $M_2 \cap \{(x,0,z) : \; x \le 0, z \in \mathbb{R}\}$ we can proceed in a similar way after appropriate modifications of the functions $\boldsymbol{g}$ and $\boldsymbol{\varphi}$, keeping in mind that the ball is invariant under rotations. We thus reach the conclusion on application of Proposition 8.2.3.

8.3 Practical Criteria

The approach used in the previous two examples of the last section is admittedly cumbersome. In this section, we provide some useful tools that simplify the task of deciding whether we have a regular surface with boundary, and further, determining the boundary. Equipping our toolbox will necessitate some complicated proofs, and the reader may safely consider them optional, but the results themselves are quite easy to apply in concrete situations and certainly worth being familiar with.

If we cut a regular surface S with a plane that is not tangent to the surface, our intuition tells us that the portion of the surface to one side of the plane is a regular surface with boundary, with the boundary being exactly the curve obtained as the intersection of S with the plane. Something analogous should occur if we replace the plane by another surface that cuts S in an appropriate way. For instance, if we perforate a sphere of radius R with the aid of a cylinder centered at an axis of the sphere and of radius r, $r < R$, then we obtain a surface with boundary whose boundary is exactly the intersection of the cylinder with the sphere (Fig. 8.3).

We are going to obtain sufficient conditions that guarantee that this intuition becomes fact. We begin by presenting the following practical criterion.

Criterion 8.3.1. *Let* $1 \leq k \leq n-1$,

$$\boldsymbol{\Phi} : U \subset \mathbb{R}^n \to \mathbb{R}^{n-k}$$

and

$$f : U \subset \mathbb{R}^n \to \mathbb{R}$$

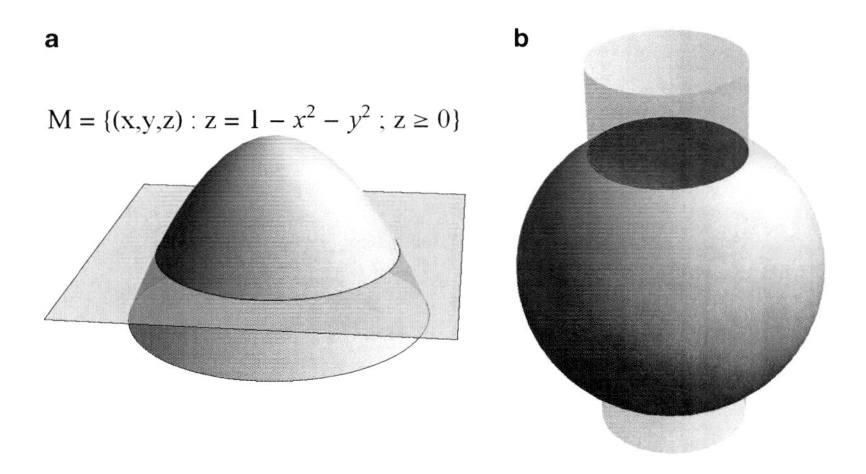

Fig. 8.3 (**a**) Paraboloid portion in the upper half-space; (**b**) cutting a sphere with a cylinder

be functions of class C^1. *Suppose that* $\boldsymbol{\Phi}'(\boldsymbol{x})$ *has maximum rank (that is, rank* $n-k$*) for every* $\boldsymbol{x} \in \boldsymbol{\Phi}^{-1}(\boldsymbol{0})$ *and also*

$$(\boldsymbol{\Phi}, f)'(\boldsymbol{x})$$

has maximum rank $(n-k+1)$ *for every* $\boldsymbol{x} \in \boldsymbol{\Phi}^{-1}(0) \cap \mathrm{f}^{-1}(0)$. *Then*

$$M := \{\boldsymbol{x} \in U \; : \; \boldsymbol{\Phi}(\boldsymbol{x}) = \boldsymbol{0} \, , \; f(\boldsymbol{x}) \leq 0\}$$

is a regular k-surface with boundary and

$$\partial M = \{\boldsymbol{x} \in U \; : \; \boldsymbol{\Phi}(\boldsymbol{x}) = \boldsymbol{0} \, , \; f(\boldsymbol{x}) = 0\}.$$

Those who need a practical criterion for dealing with surfaces with boundary can interrupt their reading at this point and proceed directly to dealing with the exercises in an effective way. The student of the mathematical theory, however, is invited to continue reading this section, where we will prove the criterion by application of the following theorem.

Theorem 8.3.2. *Let* S *be a regular k-surface,* $2 \leq k \leq n$, *contained in an open set* U *of* $\mathbb{R}^n$ *and let* $f : U \subset \mathbb{R}^n \to \mathbb{R}$ *be a function of class* C^1 *such that for every* $\boldsymbol{x} \in S \cap f^{-1}(0)$ *the gradient vector* $\nabla f(\boldsymbol{x})$ *is not orthogonal to the tangent space* $T_{\boldsymbol{x}}S$. *Then*

$$M := \{\boldsymbol{x} \in S \; : \; f(\boldsymbol{x}) \leq 0\}$$

is a regular k-surface with boundary and

$$\partial M = \{\boldsymbol{x} \in S \; : \; f(\boldsymbol{x}) = 0\}.$$

Moreover, if S *is a regular k-surface of class* C^p *and* f *is a function of class* C^p, *then* M *is a regular k-surface with boundary of class* C^p.

Proof. Write $M = M_1 \cup M_2$, where

$$M_1 := \{\boldsymbol{x} \in S \; : \; f(\boldsymbol{x}) < 0\}$$

and

$$M_2 := \{\boldsymbol{x} \in S \; : \; f(\boldsymbol{x}) = 0\}.$$

Then M_1 is a regular k-surface, since it is the intersection of S with an open set. In order to apply Proposition 8.2.3, we fix a point $\boldsymbol{x}_0$ of S with $f(\boldsymbol{x}_0) = 0$. Take a coordinate system $(B, \boldsymbol{\psi})$ of S (see Definition 3.1.1) such that $\boldsymbol{x}_0 = \boldsymbol{\psi}(\boldsymbol{t}_0)$, where $\boldsymbol{t}_0 = (t_1^0, t_2^0, \ldots, t_k^0) \in B$, and consider the function

$$\Phi : B \times \mathbb{R} \subset \mathbb{R}^k \times \mathbb{R} \to \mathbb{R} \, , \quad \Phi(\boldsymbol{t}, s) := (f \circ \boldsymbol{\psi})(\boldsymbol{t}) - s.$$

Then $\boldsymbol{\Phi}(\boldsymbol{t}_0,0)=0$. Also, since $\boldsymbol{\nabla} f(\boldsymbol{x}_0)$ is not orthogonal to $T_{\boldsymbol{x}_0}S$ and the vectors

$$\left\{\frac{\partial \boldsymbol{\psi}}{\partial t_1}(\boldsymbol{t}_0),\dots,\frac{\partial \boldsymbol{\psi}}{\partial t_k}(\boldsymbol{t}_0)\right\}$$

are a basis of the tangent space $T_{\boldsymbol{x}_0}S$ (see Theorem 3.4.1), we can assume (by a permutation of the coordinates) that

$$\left\langle \boldsymbol{\nabla} f(\boldsymbol{x}_0), \frac{\partial \boldsymbol{\psi}}{\partial t_k}(\boldsymbol{t}_0)\right\rangle \neq 0,$$

what means that, by the chain rule,

$$\frac{\partial \boldsymbol{\Phi}}{\partial t_k}(\boldsymbol{t}_0,0)=\frac{\partial (f\circ\boldsymbol{\psi})}{\partial t_k}(\boldsymbol{t}_0)=\sum_{j=1}^{n}\frac{\partial f}{\partial x_j}(\boldsymbol{x}_0)\frac{\partial \psi_j}{\partial t_k}(\boldsymbol{t}_0)\neq 0.$$

The implicit function theorem, Theorem 3.2.1 (see also [10, 4.6]), provides us with an open neighborhood W of $(\boldsymbol{t}_0,0)$ in $\mathbb{R}^{k+1}$, $W\subset B\times\mathbb{R}$, an open neighborhood A of

$$(t_1^0,\dots,t_{k-1}^0,0)$$

in $\mathbb{R}^k$ and a function $g:A\subset\mathbb{R}^k\to\mathbb{R}$ of class C^1 in A such that the following hold:

1. $g(t_1^0,\dots,t_{k-1}^0,0)=t_k^0$,
2. $\big(t_1,\dots,t_{k-1},g(t_1,\dots,t_{k-1},s),s\big)\in W\subset B\times\mathbb{R}$ whenever $(t_1,\dots,t_{k-1},s)\in A$,
3. for every $(\boldsymbol{t},s)\in W$,

$$\boldsymbol{\Phi}(\boldsymbol{t},s)=0 \text{ if and only if } (t_1,\dots,t_{k-1},s)\in A \text{ and } t_k=g(t_1,\dots,t_{k-1},s). \tag{8.5}$$

In particular, whenever $(t_1,\dots,t_{k-1},s)\in A$, we have

$$\Big(t_1,\dots,t_{k-1},g(t_1,\dots,t_{k-1},s)\Big)\in B$$

and

$$\boldsymbol{\Phi}\Big(t_1,\dots,t_{k-1},g(t_1,\dots,t_{k-1},s),s\Big)=0. \tag{8.6}$$

Now define $\mathbb{A}=A\cap\mathbb{H}^k$ and

$$\boldsymbol{\varphi}:\mathbb{A}\subset\mathbb{H}^k\to M\subset\mathbb{R}^n$$

by

$$\boldsymbol{\varphi}(t_1,\dots,t_{k-1},s):=\boldsymbol{\psi}\Big(t_1,\dots,t_{k-1},g(t_1,\dots,t_{k-1},s)\Big).$$

Letting

$$\boldsymbol{x} := \boldsymbol{\varphi}(t_1,\ldots,t_{k-1},s),$$

condition (8.6) gives $f(\boldsymbol{x}) - s = 0$. Together with the injectivity of $\boldsymbol{\psi}$, and (8.5) this implies that $\boldsymbol{\varphi}$ is injective and $\boldsymbol{\varphi}(\mathbb{A}) \subset M$. Moreover,

$$\boldsymbol{\varphi}(t_1^0,\ldots,t_{k-1}^0,0) = \boldsymbol{\psi}(t_1^0,\ldots,t_{k-1}^0,t_k^0) = \boldsymbol{x}_0.$$

It remains only to check that $(\mathbb{A},\boldsymbol{\varphi})$ satisfies the condition (1) and (2) of Proposition 8.2.3. To do this, we consider the mapping

$$\boldsymbol{\beta} : A \subset \mathbb{R}^k \to \mathbb{R}^k$$

defined by

$$\boldsymbol{\beta}(t_1,\ldots,t_{k-1},s) := \big(t_1,\ldots,t_{k-1},g(t_1,\ldots,t_{k-1},s)\big) \tag{8.7}$$

and observe that

$$\boldsymbol{\varphi} = \boldsymbol{\psi} \circ \boldsymbol{\beta}_{|\mathbb{A}}.$$

Now it suffices to check that $\boldsymbol{\beta}$ admits a local inverse of class C^1 in a neighborhood of $\boldsymbol{u}_0 := (t_1^0,\ldots,t_{k-1}^0,0) \in A$. Evaluating the partial derivative of (8.6) with respect to s, keeping in mind

$$\frac{\partial \boldsymbol{\Phi}}{\partial s}(\boldsymbol{t},s) \;=\; -1$$

at each point $(\boldsymbol{t},s) \in B \times \mathbb{R}$, we obtain

$$\frac{\partial \boldsymbol{\Phi}}{\partial t_k}\,(t_1^0,\ldots,t_{k-1}^0,g(\boldsymbol{u}_0),0)\;\cdot\;\frac{\partial g}{\partial s}(\boldsymbol{u}_0)\;-\;1\;=\;0.$$

In particular, it follows from Definition (8.7) that the Jacobian of $\boldsymbol{\beta}$ at $\boldsymbol{u}_0$ is

$$J\boldsymbol{\beta}(\boldsymbol{u}_0) \;=\; \frac{\partial g}{\partial s}(\boldsymbol{u}_0) \neq 0,$$

and an application of the inverse function theorem, Theorem 3.1.1, concludes the argument. □

We can return to the proof of Criterion 8.3.1.

Take

$$S := \boldsymbol{\Phi}^{-1}(\boldsymbol{0}),$$

which is a regular k-surface by Proposition 3.2.1. For every

$$\boldsymbol{x} \in S \cap f^{-1}(0),$$

the gradient vector $\nabla f(\boldsymbol{x})$ is not a linear combination of the vectors

$$\{\nabla \Phi_j(\boldsymbol{x})\}_{j=1}^{n-k},$$

since otherwise the rank of the matrix $(\boldsymbol{\Phi}, f)'(\boldsymbol{x})$ would be $(n-k)$. Now, by Proposition 3.4.1, the vectors

$$\{\nabla \Phi_1(\boldsymbol{x}), \nabla \Phi_2(\boldsymbol{x}), \dots, \nabla \Phi_{n-k}(\boldsymbol{x})\}$$

are a basis of the normal space to S at $\boldsymbol{x}$. Consequently, $\nabla f(\boldsymbol{x})$ is not orthogonal to the tangent space to S at $\boldsymbol{x}$. We can apply Theorem 8.3.2 to conclude that M is a regular k-surface with boundary, and moreover, the boundary consists exactly of those points of S at which the function f vanishes.

We present now another practical criterion.

Corollary 8.3.1. *Let S be a regular n-surface contained in an open set U in $\mathbb{R}^n$ and let $f : U \subset \mathbb{R}^n \to \mathbb{R}$ be a function of class C^p such that $\nabla f(x) \neq 0$ for all $x \in S \cap f^{-1}(0)$. Then*

$$M := \{x \in S : f(x) \leq 0\}$$

is a regular n-surface with boundary of class C^p and $\partial M = \{x \in S : f(x) = 0\}$.

Criterion 8.3.3. *Let M be a compact set in $\mathbb{R}^n$ such that $\partial^{\text{top}}(M)$, its topological boundary, is a regular $(n-1)$-surface of class C^p. Suppose that for each $\boldsymbol{x} \in \partial^{\text{top}}(M)$ there is $\delta > 0$ such that for every $0 < r < \delta$,*

$$B(\boldsymbol{x}, r) \setminus \partial^{\text{top}}(M)$$

has two connected components, and one component coincides with the intersection of $B(\boldsymbol{x}, r)$ with the interior of M. Then M is a regular n-surface with boundary of class C^p and $\partial M = \partial^{\text{top}}(M)$.

Proof. The hypothesis implies that the interior of M is nonempty and hence is a regular n-surface. According to Proposition 8.2.3, we need to check the conditions in Definition 8.2.1 of regular surface with boundary at each point of the topological boundary of M. Fix $\boldsymbol{x}_0 \in \partial^{\text{top}}(M)$ and take δ as guaranteed by the hypothesis. By Theorem 3.4.2, there exist $0 < r < \delta$ and

$$\Phi : B(\boldsymbol{x}_0, r) \subset \mathbb{R}^n \to \mathbb{R}$$

of class C^p in $B(\boldsymbol{x}_0, r)$ such that

$$\partial^{\text{top}}(M) \cap B(\boldsymbol{x}_0, r) = \Phi^{-1}(0)$$

and

$$\nabla \Phi(\boldsymbol{x}) \neq 0$$

for every $\boldsymbol{x} \in \boldsymbol{\Phi}^{-1}(0)$. Let U and V be the two connected components of $B(\boldsymbol{x},r) \setminus \partial^{\text{top}}(M)$ and let

$$U = B(\boldsymbol{x},r) \cap \text{Interior}(M).$$

Then the sign of $\boldsymbol{\Phi}$ is constant in U and also in V. However, the sign of $\boldsymbol{\Phi}$ is not constant in $U \cup V$, since this would imply that $0 = \boldsymbol{\Phi}(\boldsymbol{x}_0)$ would be a local extreme value of $\boldsymbol{\Phi}$, contradicting $\boldsymbol{\nabla}\boldsymbol{\Phi}(\boldsymbol{x}_0) \neq 0$. Without loss of generality (otherwise we change the sign of $\boldsymbol{\Phi}$) we can assume

$$\boldsymbol{\Phi}(\boldsymbol{x}) < 0 \text{ for every } \boldsymbol{x} \in U, \quad \boldsymbol{\Phi}(\boldsymbol{x}) > 0 \text{ for all } \boldsymbol{x} \in V.$$

Consequently,

$$B(\boldsymbol{x}_0,r) \cap M \;=\; \{\boldsymbol{x} \in B(\boldsymbol{x}_0,r) \;:\; \boldsymbol{\Phi}(\boldsymbol{x}) \leq 0\}.$$

By Corollary 8.3.1 we deduce that $B(\boldsymbol{x}_0,r) \cap M$ is a regular n-surface with boundary and also that $\boldsymbol{x}_0$ is a point of the boundary. In particular, according to Definitions 8.2.1 and 8.2.2, there is a coordinate system $(\mathbb{A},\boldsymbol{\varphi})$,

$$\boldsymbol{\varphi} : \mathbb{A} \subset \mathbb{H}^n \to M \subset \mathbb{R}^n,$$

of class C^p in $\mathbb{A}$ such that

$$\boldsymbol{\varphi}^{-1}(\boldsymbol{x}_0) = (0,t_2^0,\ldots,t_n^0).$$

□

Let M be a regular n-surface with boundary of class C^p that satisfies the hypothesis of Criterion 8.3.3. In particular, for every $\boldsymbol{x}_0 \in \partial M$, one can find $r > 0$ such that

$$B(\boldsymbol{x}_0,r) \setminus \partial^{\text{top}}(M) \tag{8.8}$$

has two connected components, U and V, where U is the intersection of $B(\boldsymbol{x}_0,r)$ with the interior of M, and V is the intersection of $B(\boldsymbol{x}_0,r)$ with the complement of M.

We are going to see that under these circumstances we can speak of normal vectors that point toward the outside or the interior of the regular surface.

Definition 8.3.1. Let $M \subset \mathbb{R}^n$ be a regular n-surface with boundary of class C^p. A normal vector $\boldsymbol{N}$ to $T_{\boldsymbol{x}_0}(\partial M)$ is said to be an *exterior normal* vector if there is $\delta > 0$ such that

$$\boldsymbol{x}_0 + t\boldsymbol{N} \in \mathbb{R}^n \setminus M,$$

whenever $0 < t < \delta$.

Proposition 8.3.1. *Let M be a regular n-surface with boundary of class C^p and $\boldsymbol{x}_0$ a point in ∂M. We assume that there exist $r > 0$ and a mapping*

$$\boldsymbol{\Phi} : B(\boldsymbol{x}_0,r) \subset \mathbb{R}^n \to \mathbb{R}$$

of class C^p such that

$$B(\boldsymbol{x}_0, r) \cap \partial M = \{\boldsymbol{x} \in B(\boldsymbol{x}_0, r) \ : \ \boldsymbol{\Phi}(\boldsymbol{x}) = 0\}$$

and

$$B(\boldsymbol{x}_0, r) \cap M = \{\boldsymbol{x} \in B(\boldsymbol{x}_0, r) \ : \ \boldsymbol{\Phi}(\boldsymbol{x}) \leq 0\}.$$

Then $\boldsymbol{N} = \nabla \boldsymbol{\Phi}(x_0)$ is an exterior normal vector to $T_{\boldsymbol{x}_0}(\partial M)$.

Proof. From Proposition 3.4.1 we already know that $\boldsymbol{N}$ is a normal vector to $T_{\boldsymbol{x}_0}(\partial M)$. Moreover,

$$0 < \langle \boldsymbol{N}, \boldsymbol{N} \rangle = D_{\boldsymbol{N}} \boldsymbol{\Phi}(\boldsymbol{x}_0) = \lim_{t \to 0} \frac{\boldsymbol{\Phi}(\boldsymbol{x}_0 + t\boldsymbol{N}) - \boldsymbol{\Phi}(\boldsymbol{x}_0)}{t}$$

$$= \lim_{t \to 0} \frac{\boldsymbol{\Phi}(\boldsymbol{x}_0 + t\boldsymbol{N})}{t},$$

which shows that if $t > 0$ is small enough, then

$$\boldsymbol{\Phi}(\boldsymbol{x}_0 + t\boldsymbol{N}) > 0,$$

and hence $\boldsymbol{x}_0 + t\boldsymbol{N} \in \mathbb{R}^n \setminus M$. □

An especially simple situation that illustrates this discussion is the following. Let $M = \mathbb{H}^n$ be the half-space of Definition 8.1.1 in the case $k = n$. For $\boldsymbol{\Phi}(\boldsymbol{x}) = x_1$ we have

$$M = \{\boldsymbol{x} \in \mathbb{R}^n \ : \ \boldsymbol{\Phi}(\boldsymbol{x}) \leq 0\}.$$

In this case, the boundary of M coincides with the hyperplane $x_1 = 0$, and the vector

$$\boldsymbol{\nabla} \boldsymbol{\Phi}(\boldsymbol{x}_0) = \boldsymbol{e}_1$$

points toward the exterior of the half-space.

Example 8.3.1. Let

$$M := \{(x, y, z) \in \mathbb{R}^3 \ : \ x^2 + y^2 + z^2 = 1, \ z \geq 0\}$$

be the upper hemisphere of the unit sphere. Then M is a regular 2-surface with boundary in $\mathbb{R}^3$, and the boundary is the unit circle in the xy-plane,

$$\partial M = \{(x, y, z) \in \mathbb{R}^3 \ : x^2 + y^2 = 1, \ z = 0\}.$$

To check this, consider the mappings given by

$$f(x, y, z) = -z, \quad \boldsymbol{\Phi}(x, y, z) = x^2 + y^2 + z^2 - 1.$$

Then

$$\nabla \Phi(x,y,z) = (2x,2y,2z) \neq (0,0,0)$$

in $\mathbb{R}^3 \setminus \{(0,0,0)\}$, and in particular, $\Phi'(x,y,z)$ has rank 1 at every point on the sphere. Moreover,

$$(\Phi, f)'(x,y,z) = \begin{pmatrix} 2x & 2y & 2z \\ 0 & 0 & -1 \end{pmatrix}$$

has rank 2 except at points of the z-axis, but since the z-axis does not intersect the unit circle in the xy-plane, an appeal to Criterion 8.3.1 concludes our argument.

Example 8.3.2. The truncated paraboloid

$$M := \{(x,y,z) \in \mathbb{R}^3 \; : \; z = x^2 + y^2, \; z \leq 1\}$$

is a regular 2-surface with boundary whose boundary is the unit circle in the plane $z = 1$,

$$\partial M = \{(x,y,z) \in \mathbb{R}^3 \; : \; x^2 + y^2 = z = 1\}.$$

Again we apply Criterion 8.3.1, this time with the mappings

$$\Phi(x,y,z) = x^2 + y^2 - z \, , \; f(x,y,z) = z - 1.$$

Easily we see that

$$\nabla \Phi(x,y,z) = (2x, 2y, -1) \neq (0,0,0),$$

which means that $\Phi'(x,y,z)$ has rank 1 at every point of $\mathbb{R}^3$. Also,

$$(\Phi, f)'(x,y,z) = \begin{pmatrix} 2x & 2y & -1 \\ 0 & 0 & 1 \end{pmatrix}$$

has rank 2 except at points of the z-axis. Since the z-axis does not intersect the unit circle in the plane $z = 1$, Criterion 8.3.1 provides our conclusion.

Let us indicate briefly how one can analyze this example, to the same end, by use of Proposition 8.2.3 rather than Criterion 8.3.1. We decompose $M = M_1 \cup M_2$, where

$$M_1 = \{(x,y,z) \in \mathbb{R}^3 \; : \; z = x^2 + y^2 \; , \; z < 1\},$$

being the intersection of the paraboloid with an open set, is a regular 2-surface in $\mathbb{R}^3$, and

$$M_2 = \{(x,y,z) \in \mathbb{R}^3 \; : \; x^2 + y^2 = z = 1\}.$$

Now we fix an arbitrary point $(x_0, y_0, z_0) \in M_2$. The level curves obtained on intersecting M with horizontal planes are circles, and this suggests the use of polar coordinates. Thus, we put

$$x = \rho \cos \theta,\ y = \rho \sin \theta,\ z = x^2 + y^2 = \rho^2.$$

Since $z \leq 1$, it follows that $\rho \leq 1$. We consider the half-plane

$$\mathbb{S} = \{(\rho, \theta) \in \mathbb{R}^2\ :\ \rho \leq 1\}$$

and select $\theta_0 \in \mathbb{R}$ such that

$$x_0 = \cos \theta_0,\ y_0 = \sin \theta_0.$$

Finally, we define

$$\mathbb{A} := (0, 1] \times (\theta_0 - \pi, \theta_0 + \pi)$$

and

$$\varphi : \mathbb{A} \subset \mathbb{S} \to M \subset \mathbb{R}^3, \quad \varphi(\rho, \theta) = (\rho \cos \theta, \rho \sin \theta, \rho^2).$$

The verification of conditions (1) and (2) of Proposition 8.2.3 are left as an exercise, but it is similar to the corresponding verification in Example 8.2.3. To finish, we note that

$$(x_0, y_0, z_0) = \varphi(1, \theta_0) \in \varphi\left(\mathbb{A} \cap \partial\mathbb{S}\right).$$

Example 8.3.3. Already we have seen in Example 8.2.4 that the closed unit ball

$$M := \{(x,y,z) \in \mathbb{R}^3\ :\ x^2 + y^2 + z^2 \leq 1\}$$

is a regular 3-surface with boundary in $\mathbb{R}^3$ and that the boundary is the unit sphere.

This conclusion can alternatively be arrived by application of Theorem 8.3.2 in the limiting case of $k = n = 3$ and $S = \mathbb{R}^3$ upon considering the function

$$f : \mathbb{R}^3 \to \mathbb{R}\ ;\ f(x,y,z) = x^2 + y^2 + z^2 - 1.$$

The normal space to S at an arbitrary point is simply $(0,0,0)$, and so, in order to apply Theorem 8.3.2, we have only to check that $\boldsymbol{\nabla} f(x,y,z) \neq (0,0,0)$ at every point at which the function f vanishes, and this much is clear.

The next example shows again the difference between the boundary (which, for emphasis, we might call the *geometric* boundary) of a surface and the topological boundary of that surface considered as a set.

Example 8.3.4. The annulus $M := \{(x,y) \in \mathbb{R}^2\ : 1 < x^2 + y^2 \leq 4\}$ is a regular 2-surface with boundary. The boundary is the circle with radius 2 centered at the origin.

Again we apply Theorem 8.3.2 in a limiting case. Take

$$S := \{(x,y) \in \mathbb{R}^2\ : 1 < x^2 + y^2\},$$

which is a regular 2-surface in $\mathbb{R}^2$ because it is an open set. Now define

$$f : \mathbb{R}^2 \to \mathbb{R}, \quad f(x,y) = x^2 + y^2 - 4.$$

Since $\nabla f(x,y) = (2x, 2y) \neq (0,0)$ whenever $x^2 + y^2 = 4$, Theorem 8.3.2 provides our assertion.

On the other hand the topological boundary of M is

$$\overset{top}{\partial} M = \{(x,y) \in \mathbb{R}^2 : x^2 + y^2 = 1\} \cup \{(x,y) \in \mathbb{R}^2 : x^2 + y^2 = 4\}.$$

8.4 Orientation of Surfaces with Boundary

In Chap. 5, we studied the orientation of regular surfaces without boundary, and so if M is a regular k-surface with boundary, we already know what is meant by an orientation of the regular k-surface $M \setminus \partial M$ and an orientation of the regular $(k-1)$-surface ∂M. Of course, one cannot assume that either of these is orientable, and a priori, there is no reason to believe that the orientability of one is related to the orientability of the other. However, the aim of this section is to prove that if $M \setminus \partial M$ is orientable, then so is ∂M. Moreover, of the two possible orientations of ∂M (we suppose ∂M is connected), one of them is canonically determined by the chosen orientation in $M \setminus \partial M$. As one might expect, the connection between these two orientations is closely related to the notion of *orientation of a surface with boundary*.

Definition 8.4.1. Let M be a regular k-surface with boundary of class C^p in $\mathbb{R}^n$. Then M is said to be *orientable* if there exists an atlas $\{(\mathbb{A}_i, \varphi_i) : i \in I\}$ such that for each pair of coordinate systems $(\mathbb{A}_1, \varphi_1)$ and $(\mathbb{A}_2, \varphi_2)$ with $\varphi_1(\mathbb{A}_1) \cap \varphi_2(\mathbb{A}_2) \neq \varnothing$ one has $J(\varphi_1^{-1} \circ \varphi_2) > 0$ on its domain of definition. Also we will say that the atlas defines an orientation in M.

Let us define

$$\mathbb{R}^k_- := \{y \in \mathbb{R}^k \; : \; y_1 < 0\}, \quad \mathbb{R}^k_0 := \{y \in \mathbb{R}^k \; : \; y_1 = 0\},$$

and note that $\mathbb{H}^k = \mathbb{R}^k_- \cup \mathbb{R}^k_0$. Also, for each subset $\mathbb{A} \subset \mathbb{H}^k$ we put

$$\mathbb{A}_- := \mathbb{A} \cap \mathbb{R}^k_-.$$

If $(\mathbb{A}, \varphi)$ is a coordinate system of M, then $(\mathbb{A}_-, \varphi_{|\mathbb{A}_-})$ is a coordinate system of the regular k-surface $M \setminus \partial M$.

Some authors say that a regular k-surface M with boundary is orientable if the 1-regular surface $M \setminus \partial M$ is orientable. Next proposition shows that their and our definitions are equivalent.

Proposition 8.4.1. *Let M be a regular k-surface with boundary in $\mathbb{R}^n$. Then M is orientable if and only if $M \setminus \partial M$ is.*

Proof. Indeed, if the orientation of M is determined by the atlas

$$\{(\mathbb{A}_i, \varphi_i) : i \in I\},$$

then in $M \setminus \partial M$, we consider the orientation defined by the atlas

$$\{((\mathbb{A}_i)_-, \varphi_i) : i \in I\}.$$

Conversely, suppose that $M \setminus \partial M$ is an orientable regular k-surface and let

$$\mathscr{A} = \{(\mathbb{A}_i, \boldsymbol{\varphi}_i) : i \in I\}$$

be an arbitrary atlas of M, where each $(\mathbb{A}_i)_-$ is a connected set. We suppose that we have chosen an orientation in $M \setminus \partial M$, and we set about obtaining a new atlas of M that defines an orientation according to Definition 8.4.1. We proceed as follows. First consider the linear mapping $\boldsymbol{L} : \mathbb{R}^k \to \mathbb{R}^k$ defined by

$$\boldsymbol{L}(\boldsymbol{t}) = (t_1, \dots, t_{k-2}, t_k, t_{k-1}).$$

Now for every $i \in I$, we analyze whether the coordinate system $((\mathbb{A}_i)_-, \boldsymbol{\varphi}_i)$ of $M \setminus \partial M$ is compatible with the chosen orientation of $M \setminus \partial M$. If this is the case, we define

$$(\mathbb{B}_i, \boldsymbol{\psi}_i) := (\mathbb{A}_i, \boldsymbol{\varphi}_i),$$

but if not, we let

$$(\mathbb{B}_i, \boldsymbol{\psi}_i) := (\boldsymbol{L}^{-1}(\mathbb{A}_i), \boldsymbol{\varphi}_i \circ \boldsymbol{L}).$$

According to Lemma 5.2.3, the coordinate system $((\mathbb{B}_i)_-, \boldsymbol{\psi}_i)$ is compatible with the orientation of $M \setminus \partial M$. To finish, we show that the atlas

$$\mathscr{B} := \{(\mathbb{B}_i, \boldsymbol{\psi}_i) \ : \ i \in I\}$$

defines an orientation of M. Suppose, therefore, that $(\mathbb{B}_1, \boldsymbol{\psi}_1)$ and $(\mathbb{B}_2, \boldsymbol{\psi}_2)$ are two coordinate systems in the atlas such that

$$D := \boldsymbol{\psi}_1(\mathbb{B}_1) \cap \boldsymbol{\psi}_2(\mathbb{B}_2) \neq \varnothing.$$

Since the two coordinate systems

$$((\mathbb{B}_1)_-, \boldsymbol{\psi}_1) \text{ and } ((\mathbb{B}_2)_-, \boldsymbol{\psi}_2)$$

of $M \setminus \partial M$ are compatible with the orientation of $M \setminus \partial M$ and

$$\boldsymbol{\psi}_1((\mathbb{B}_1)_-) \cap \boldsymbol{\psi}_2((\mathbb{B}_2)_-) = D \setminus \partial M,$$

we deduce from Theorem 5.2.1 that

$$J(\boldsymbol{\psi}_1^{-1} \circ \boldsymbol{\psi}_2) > 0$$

on the set $\boldsymbol{\psi}_2^{-1}(D \setminus \partial M)$. Note that it is implicit in the proof of Theorem 8.2.1 that $D \setminus \partial M \neq \varnothing$. Since $\Psi_2^{-1}(D)$ is contained in the closure of $\psi_2^{-1}(D \setminus \partial M)$, by continuity,

$$J(\boldsymbol{\psi}_1^{-1} \circ \boldsymbol{\psi}_2) \geq 0$$

on $\boldsymbol{\psi}_2^{-1}(D)$. Moreover, according to Proposition 8.2.1, the mapping

$$\boldsymbol{\psi}_1^{-1} \circ \boldsymbol{\psi}_2 : \boldsymbol{\psi}_2^{-1}(D) \subset \mathbb{H}^k \to \mathbb{R}^k$$

has nonzero Jacobian at each point. Finally we deduce

$$J(\boldsymbol{\psi}_1^{-1} \circ \boldsymbol{\psi}_2) > 0$$

in the set $\boldsymbol{\psi}_2^{-1}(D)$. □

From now on, if M is an orientable regular surface with boundary and its orientation is defined by the atlas $\{(\mathbb{A}_i, \boldsymbol{\varphi}_i) : i \in I\}$, then we will consider the orientation of $M \setminus \partial M$ defined by the atlas

$$\{((\mathbb{A}_i)_-, \boldsymbol{\varphi}_i) : i \in I\}$$

and will refer to this as the *orientation induced by the orientation of* M. In the next theorem, we will show how the orientation of M also induces an orientation of the $(k-1)$-dimensional regular surface ∂M.

Theorem 8.4.1. *Let M be a regular k-surface with boundary in $\mathbb{R}^n$ that is orientable and of class C^p. Then ∂M is an orientable regular $(k-1)$-surface of class C^p.*

Proof. According to Theorem 8.2.1, ∂M is a regular $(k-1)$-surface of class C^p. Given $\boldsymbol{x}_0 \in \partial M$, let $(\mathbb{A}, \boldsymbol{\varphi})$ be a coordinate system of M such that $\boldsymbol{x}_0 = \boldsymbol{\varphi}(\boldsymbol{t}_0)$, where $\boldsymbol{t}_0 = (0, t_2^0, \ldots, t_k^0) \in \mathbb{A}$. As in Theorem 8.2.1, define

$$\widehat{\mathbb{A}} := \{\boldsymbol{s} \in \mathbb{R}^{k-1} \ : \ (0, \boldsymbol{s}) \in \mathbb{A}\},$$

which is an open set in $\mathbb{R}^{k-1}$, and

$$\widehat{\boldsymbol{\varphi}} : \widehat{\mathbb{A}} \subset \mathbb{R}^{k-1} \to \mathbb{R}^n, \quad \widehat{\boldsymbol{\varphi}}(\boldsymbol{s}) := \boldsymbol{\varphi}(0, \boldsymbol{s}), \tag{8.9}$$

which is a function of class C^p. As we saw in the proof of Theorem 8.2.1, $(\widehat{\mathbb{A}}, \widehat{\boldsymbol{\varphi}})$ is a coordinate system of ∂M. Suppose that we have chosen an orientation in M. We are going to show that if $\{(\mathbb{A}_\alpha, \boldsymbol{\varphi}_\alpha)\}$ is an atlas of M compatible with that orientation, where $\mathbb{A}_\alpha$ is connected, then $\{(\widehat{\mathbb{A}}_\alpha, \widehat{\boldsymbol{\varphi}}_\alpha)\}$ induces an orientation of ∂M. Here, $\widehat{\boldsymbol{\varphi}}_\alpha$ is related to $\boldsymbol{\varphi}_\alpha$ as in (8.9), that is,

$$\widehat{\boldsymbol{\varphi}}_\alpha(\boldsymbol{s}) := \boldsymbol{\varphi}_\alpha(0, \boldsymbol{s}).$$

Take $(\mathbb{A}_1, \boldsymbol{\varphi}_1)$ and $(\mathbb{A}_2, \boldsymbol{\varphi}_2)$ to be two arbitrary coordinate systems in this atlas of M. We need to check that $\widehat{\boldsymbol{\varphi}}_2^{-1} \circ \widehat{\boldsymbol{\varphi}}_1$ has positive Jacobian in its domain of definition, which we denote by V, and assume that $V \neq \varnothing$. To this end, let

$$\boldsymbol{g} := (g_1, g_2, \ldots, g_k)$$

be a C^p extension of $\varphi_2^{-1}\circ\varphi_1$ to an open subset of $\mathbb{R}^k$. From the identity

$$(\varphi_2^{-1}\circ\varphi_1)(0,s) \;=\; \big(g_1(0,s),\ldots,g_k(0,s)\big)$$

and Lemma 8.2.1, it follows that $g_1(0,s)=0$ for every $s\in V$. Thus

$$\widehat{\varphi}_2\big(g_2(0,s),\ldots,g_k(0,s)\big) \;=\; \varphi_2\big(g(0,s)\big)=\varphi_1(0,s)=\widehat{\varphi}_1(s).$$

We conclude that

$$(\widehat{\varphi}_2^{-1}\circ\widehat{\varphi}_1)(s)=\big(g_2(0,s),\ldots,g_k(0,s)\big).$$

Now, bearing in mind that $g_1(0,s)$ is identically zero on V, we have

$$\frac{\partial g_1}{\partial t_j}(0,s)=0$$

for each $2\le j\le k$ and every $s\in V$, and so obtain the following relation between the Jacobian of $\varphi_2^{-1}\circ\varphi_1$ and that of $\widehat{\varphi}_2^{-1}\circ\widehat{\varphi}_1$:

$$0 \;<\; J(\varphi_2^{-1}\circ\varphi_1)(0,s) = \left|\begin{matrix} \frac{\partial g_1}{\partial t_1} & 0 & \cdots & 0 \\ \frac{\partial g_2}{\partial t_1} & \frac{\partial g_2}{\partial t_2} & \cdots & \frac{\partial g_2}{\partial t_k} \\ \cdots & \cdots & \cdots & \cdots \\ \frac{\partial g_k}{\partial t_1} & \frac{\partial g_k}{\partial t_2} & \cdots & \frac{\partial g_k}{\partial t_k} \end{matrix}\right|_{(0,s)}$$

$$= \frac{\partial g_1}{\partial t_1}(0,s)\cdot J(\widehat{\varphi}_2^{-1}\circ\widehat{\varphi}_1)(s) \tag{8.10}$$

for every $s\in V$. On the other hand, since $g_1(0,s)=0$ and $g_1(h,s)\le 0$ for every $h<0$ such that $|h|$ is small enough, we have

$$\frac{\partial g_1}{\partial t_1}(0,s) \;=\; \lim_{h\to 0^-}\frac{g_1(h,s)-g_1(0,s)}{h}\ge 0,$$

and conclude that

$$J(\widehat{\varphi}_2^{-1}\circ\widehat{\varphi}_1)(s)>0,$$

for every $s\in V$, as we wanted to prove. □

Definition 8.4.2. Let M be an oriented regular k-surface with boundary in $\mathbb{R}^n$ and $\{(\mathbb{A}_\alpha,\varphi_\alpha)\}$ an atlas of M compatible with its orientation. The *orientation induced on the boundary* ∂M is the orientation defined by the atlas $\{(\widehat{\mathbb{A}}_\alpha,\widehat{\varphi}_\alpha)\}$, $\widehat{\varphi}_\alpha(s)=\varphi_\alpha(0,s)$ introduced in Theorem 8.4.1.

Definition 8.4.3. Let $\mathscr{H}$ be a hyperplane in $\mathbb{R}^n$ and $\boldsymbol{N}$ an orthogonal vector to $\mathscr{H}$. The orientation induced by $\boldsymbol{N}$ on $\mathscr{H}$ is the orientation

$$\mathscr{O} := [\boldsymbol{v}_1, \ldots, \boldsymbol{v}_{n-1}],$$

where $\{\boldsymbol{v}_1, \ldots, \boldsymbol{v}_{n-1}\}$ is a basis of $\mathscr{H}$ such that

$$\{\boldsymbol{N}, \boldsymbol{v}_1, \ldots, \boldsymbol{v}_{n-1}\}$$

is a positively oriented basis of $\mathbb{R}^n$.

From Proposition 7.2.2 we can see that, $\boldsymbol{N}$ is a positive multiple of the cross product $\boldsymbol{v}_1 \times \cdots \times \boldsymbol{v}_{n-1}$.

Example 8.4.1. We consider the half-space

$$\mathbb{S} := \{\boldsymbol{t} \in \mathbb{R}^n \ : \pi(\boldsymbol{t}) \leq a\}$$

(see Definition 8.2.3). Suppose this is oriented so that $(\mathbb{S} \setminus \partial\mathbb{S}, \mathrm{id})$ is a coordinate system of $\mathbb{S} \setminus \partial\mathbb{S}$ compatible with the orientation, where id denotes the identity mapping

$$\mathrm{id} : \mathbb{S} \setminus \partial\mathbb{S} \to \mathbb{S} \setminus \partial\mathbb{S}.$$

Let $\boldsymbol{N}$ be the (unique) unit vector that is orthogonal to the hyperplane $\mathscr{H} := \operatorname{Ker} \pi$ and satisfies $\pi(\boldsymbol{N}) > 0$. Then the orientation induced by $\boldsymbol{N}$ in $\mathscr{H}$ coincides with the orientation induced in the boundary by the orientation of $\mathbb{S}$.

We observe that the tangent space to $\partial\mathbb{S}$ at a point is exactly the $(n-1)$-dimensional vector subspace $\operatorname{Ker} \pi$ (see, for instance, Proposition 3.4.1). Put $\boldsymbol{v}_1 := \boldsymbol{N}$ and consider an orthonormal basis $\{\boldsymbol{v}_2, \boldsymbol{v}_3, \ldots, \boldsymbol{v}_n\}$ of $\operatorname{Ker} \pi$ such that

$$\{\boldsymbol{v}_1, \boldsymbol{v}_2, \ldots, \boldsymbol{v}_n\}$$

is a positively oriented basis of $\mathbb{R}^n$. As in Lemma 8.2.2, take $\boldsymbol{L} : \mathbb{R}^n \to \mathbb{R}^n$ to be the linear mapping defined by $\boldsymbol{L}(\boldsymbol{e}_j) = \boldsymbol{v}_j$, $1 \leq j \leq n$. Finally, let $\boldsymbol{b} \in \mathbb{R}^n$ be a vector such that $\pi(\boldsymbol{b}) = a$ and define $\boldsymbol{R}$ as the affine mapping $\boldsymbol{R} := \boldsymbol{b} + \boldsymbol{L}$. Then

$$\boldsymbol{R} : \mathbb{H}^n \to \mathbb{S}$$

is a coordinate system of $\mathbb{S}$. The chosen orientation in $\mathbb{S}$ is the one for which, for every $\boldsymbol{x} \in U := \mathbb{S} \setminus \partial\mathbb{S}$, the orientation in the tangent space (isomorphic to $\mathbb{R}^n$) $T_{\boldsymbol{x}}U$ is

$$\theta_{\boldsymbol{x}} = [\boldsymbol{e}_1, \boldsymbol{e}_2, \ldots, \boldsymbol{e}_n].$$

Since for each $\boldsymbol{t} \in \mathbb{R}^n$ and $t_1 < 0$, we have

$$\frac{\partial \boldsymbol{R}}{\partial t_j}(\boldsymbol{t}) = \boldsymbol{L}(\boldsymbol{e}_j) = \boldsymbol{v}_j$$

and $\{\boldsymbol{v}_1, \boldsymbol{v}_2, \ldots, \boldsymbol{v}_n\}$ is a positively oriented basis of $\mathbb{R}^n$, we conclude that $(\mathbb{H}^n, \boldsymbol{R})$ is a coordinate system compatible with the orientation of $\mathbb{S}$. According to Definition 8.4.2, $(\mathbb{R}^{n-1}, \boldsymbol{\psi})$ is a coordinate system of $\partial\mathbb{S}$ compatible with its orientation, where

$$\boldsymbol{\psi} : \mathbb{R}^{n-1} \to \partial\mathbb{S}, \quad \boldsymbol{\psi}(\boldsymbol{s}) = \boldsymbol{R}(0, \boldsymbol{s}).$$

From

$$\frac{\partial \boldsymbol{\psi}}{\partial s_i}(\boldsymbol{s}) = \frac{\partial \boldsymbol{R}}{\partial t_{i+1}}(0, \boldsymbol{s}) = \boldsymbol{v}_{i+1}$$

for every $1 \leq i \leq n-1$, and every $\boldsymbol{s}$ in $\mathbb{R}^{n-1}$ we deduce that the orientation in Ker π is

$$\mathscr{O} = [\boldsymbol{v}_2, \ldots, \boldsymbol{v}_n].$$

It follows from Definition 8.4.3 that this is the orientation induced by the normal vector $\boldsymbol{N}$.

Example 8.4.2. Let $0 < \varepsilon < 1$ and

$$M := \{(x, y, z) \in \mathbb{R}^3 \; : \; x^2 + y^2 + z^2 = 1 \; , \; z \geq \varepsilon\}$$

be given. We wish to analyze the orientation induced on $M \setminus \partial M$ and on ∂M by the coordinate system $(\mathbb{B}, \boldsymbol{\psi})$ of Remark 8.2.3 (see also Example 8.2.3).

Recall that $\mathbb{B} := (-r, 0] \times (\theta_0 - \pi, \theta_0 + \pi)$, where $r^2 = 1 - \varepsilon^2$, and

$$\boldsymbol{\psi} : \mathbb{B} \subset \mathbb{H}^2 \to M$$

is given by

$$\boldsymbol{\psi}(s, \theta) := \left((r+s)\cos\theta, (r+s)\sin\theta, \sqrt{1-(r+s)^2} \right).$$

For every $-r < s < 0$, we have

$$\left(\frac{\partial \boldsymbol{\psi}}{\partial s} \times \frac{\partial \boldsymbol{\psi}}{\partial \theta}\right)(s, \theta) = \begin{vmatrix} \boldsymbol{e}_1 & \boldsymbol{e}_2 & \boldsymbol{e}_3 \\ \cos\theta & \sin\theta & \frac{-(r+s)}{\sqrt{1-(r+s)^2}} \\ -(r+s)\sin\theta & (r+s)\cos\theta & 0 \end{vmatrix}.$$

Since

$$\left\langle \left(\frac{\partial \boldsymbol{\psi}}{\partial s} \times \frac{\partial \boldsymbol{\psi}}{\partial \theta}\right)(s, \theta), \boldsymbol{e}_3 \right\rangle = \begin{vmatrix} \cos\theta & \sin\theta \\ -(r+s)\sin\theta & (r+s)\cos\theta \end{vmatrix} = r + s > 0,$$

the normal vector defined by $(\mathbb{B}, \boldsymbol{\psi})$ at the points of $M \setminus \partial M$ point to the *exterior* of the sphere. In other words, the orientation induced by the coordinate system $(\mathbb{B}, \boldsymbol{\psi})$ on $M \setminus \partial M$ coincides with the orientation induced by a vector field of unit normal

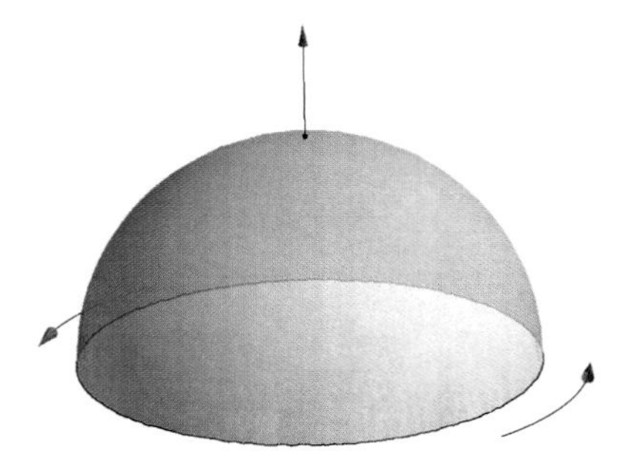

Fig. 8.4 Surface of Example 8.4.2

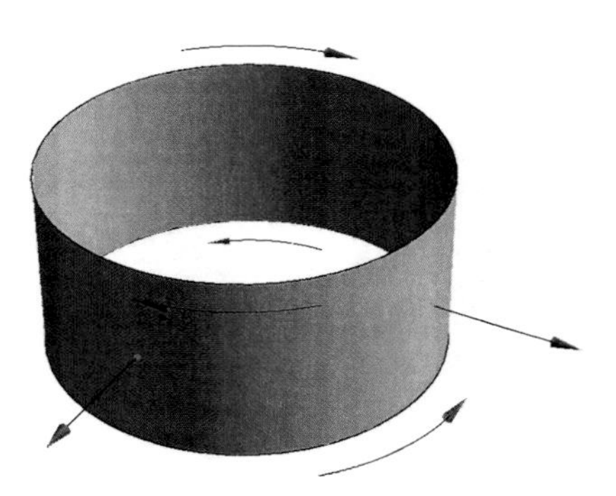

Fig. 8.5 Cylinder of Example 8.4.3

vectors pointing to the exterior of the sphere (see Definition 5.2.3). Moreover, according to Definition 8.4.2, a coordinate system of the boundary of M compatible with the orientation is given by

$$\big((\theta_0-\pi,\theta_0+\pi),\boldsymbol{\gamma}\big),$$

where

$$\boldsymbol{\gamma}:(\theta_0-\pi,\theta_0+\pi)\to\partial M,\quad \boldsymbol{\gamma}(\theta)=\boldsymbol{\psi}(0,\theta).$$

From

$$\boldsymbol{\gamma}(\theta)=(\sqrt{1-\varepsilon^2}\cos\theta,\sqrt{1-\varepsilon^2}\sin\theta,\varepsilon),$$

we see that the orientation induced on ∂M consists of *traveling counterclockwise about the circle of radius* $\sqrt{1-\varepsilon^2}$ *in the plane* $z=\varepsilon$ (Fig. 8.4).

Example 8.4.3. Consider the cylinder

$$M:=\{(x,y,z)\in\mathbb{R}^3\ :\ x^2+y^2=1\ ;\ 0\le z\le 1\},$$

and the orientation on it defined by the vector field $\boldsymbol{F}(x,y,z):=(x,y,0)$. We determine the orientation this induces on the boundary of M (Fig. 8.5).

Form the decomposition $M = S_1 \cup S_2 \cup S_3$, where

$$S_1 := \{(x,y,z) \in \mathbb{R}^3 \ : \ x^2+y^2 = 1 \ ; \ 0 < z < 1\}$$

is a regular surface and S_2, S_3 are two circles in horizontal planes:

$$S_2 := \{(x,y,0) \in \mathbb{R}^3 \ : \ x^2+y^2 = 1\}, \quad S_3 := \{(x,y,1) \in \mathbb{R}^3 \ : \ x^2+y^2 = 1\}.$$

For every $(x,y,z) \in S_1$, the vector $\boldsymbol{F}(x,y,z)$ has norm one and is orthogonal to S_1 at (x,y,z) (see, for instance, Proposition 3.4.1). Therefore, the vector field $\boldsymbol{F}$ defines an orientation of S_1, according Definition 5.2.3. We now analyze the orientation of the boundary.

1. Fix a point $(x_0, y_0, 0) \in S_2$ and find a coordinate neighborhood of this point in M that is compatible with the orientation of S_1. Indeed, observe that the points of M can be described as

$$x = \cos s, \quad y = \sin s, \quad z = -t,$$

which suggests consideration of the mapping

$$\boldsymbol{\varphi}(t,s) = (\cos s, \sin s, -t).$$

The negative sign in the third variable is due to the fact that the domain of the coordinate system should be an open set in the left half-plane and the z-coordinate of an arbitrary point of M satisfies $0 \le z \le 1$. More precisely, take θ_0 such that $x_0 = \cos\theta_0$, $y_0 = \sin\theta_0$ and define

$$\mathbb{A} := \{(t,s) \ : -1 < t \le 0 \ ; \theta_0 - \pi < s < \theta_0 + \pi\}$$

and

$$\boldsymbol{\varphi} : \mathbb{A} \subset \mathbb{H}^2 \to \mathbb{R}^3, \quad \boldsymbol{\varphi}(t,s) = (\cos s, \sin s, -t).$$

Then

$$\left(\frac{\partial \boldsymbol{\varphi}}{\partial t} \times \frac{\partial \boldsymbol{\varphi}}{\partial s}\right)(t,s) \ = \ (\cos s, \ \sin s, \ 0) \ = \ \boldsymbol{F}(\boldsymbol{\varphi}(t,s)),$$

whenever $t < 0$. This proves that $(\mathbb{A}, \boldsymbol{\varphi})$ is a coordinate system of M compatible with the orientation of S_1. According to Definition 8.4.2, the mapping

$$\boldsymbol{\gamma} : (\theta_0 - \pi, \theta_0 + \pi) \to \mathbb{R}^3 \ ; \boldsymbol{\gamma}(s) = (\cos s, \ \sin s, \ 0)$$

defines a coordinate system of $S_2 \subset \partial M$ compatible with the orientation induced in ∂M. This means that ∂M is oriented in such a way that the circle S_2 is traversed *counterclockwise*.

2. Fix a point $(x_0, y_0, 1) \in S_3$ and find a coordinate neighborhood of this point in M compatible with the orientation in S_1. This time, we observe that the points of M can be described by

$$x = \cos s, \quad y = \sin s, \quad z = 1 + t,$$

and so we consider the mapping

$$\varphi(t,s) = (\cos s, \sin s, 1+t).$$

We write the third coordinate in the form $1+t$, so that $t = 0$ corresponds to the points in S_3, and the negative values of t correspond to $M \setminus S_3$. Proceeding as before, we take θ_0 such that $x_0 = \cos\theta_0$, $y_0 = \sin\theta_0$ and define

$$\mathbb{A} := (-1,0] \times (\theta_0 - \pi, \theta_0 + \pi)$$

and

$$\varphi : \mathbb{A} \subset \mathbb{H}^2 \to \mathbb{R}^3, \quad \varphi(t,s) = (\cos s, \sin s, 1+t).$$

Then

$$\left(\frac{\partial \varphi}{\partial t} \times \frac{\partial \varphi}{\partial s}\right)(t,s) = (-\cos s, \; -\sin s, \; 0) = -\boldsymbol{F}(\varphi(t,s)),$$

which means that $(\mathbb{A}, \varphi)$ reverses the orientation of S_1. Hence, we replace the previous coordinate system by $(\mathbb{B}, \psi)$, where

$$\mathbb{B} := \{(t,s) \; : (t,-s) \in \mathbb{A}\}$$

and

$$\psi(t,s) := \varphi(t,-s).$$

This new coordinate system is compatible with the orientation of S_1. According to Definition 8.4.2, the mapping

$$\gamma(s) = \psi(0,s) = \varphi(0,-s)$$

defines a coordinate system in $S_3 \subset \partial M$ compatible with the orientation induced in ∂M. Since

$$\gamma(s) = (\cos s, \; -\sin s, \; 1),$$

we see that ∂M is oriented in such a way that S_3 is traversed *clockwise*.

The orientation of ∂M in the previous example is consistent with the following general practical rule, which will be the focus of our attention in the next section.

Criterion 8.4.2. *Suppose M is an oriented regular surface with boundary. If we traverse ∂M according to its orientation while we keep our head pointing in the direction $\boldsymbol{N}$ of the vector field that gives the orientation of $M \setminus \partial M$, then the surface M remains to our left.*

8.5 Orientation in Classical Vector Calculus

It is important at this point for us to form a bridge between two different concepts of orientation. The first is from *vector analysis*, which sees orientation as an equivalence of bases in $\mathbb{R}^n$, while the second is from *vector calculus* in $\mathbb{R}^2$ and $\mathbb{R}^3$, where orientation is based on clear geometric intuitions.

Let M be a compact subset of $\mathbb{R}^n$ whose topological boundary, $\partial^{\text{top}}(M)$, is a regular $(n-1)$-surface of class C^1 and suppose that for every $\boldsymbol{x} \in \partial^{\text{top}}(M)$, there exists $\delta > 0$ such that for each $0 < r < \delta$,

$$B(\boldsymbol{x}, r) \setminus \partial^{\text{top}}(M) \tag{8.11}$$

has two connected components, one of which coincides with the intersection of $B(\boldsymbol{x}, r)$ with the interior of M. Then, as we proved in Criterion 8.3.3, M is a regular n-surface with boundary of class C^1 and $\partial M = \partial^{\text{top}}(M)$.

We have already seen that choosing an orientation of M is equivalent to choosing an orientation of the open set $U := M \setminus \partial M$ (see Proposition 8.4.1).

From now on, we will always assume that M is oriented in such a way that the identity mapping

$$\text{id} : M \setminus \partial M \to M \setminus \partial M$$

defines a coordinate system $(M \setminus \partial M, \text{id})$ of $M \setminus \partial M$ that is compatible with its orientation. This means that if $(A, \boldsymbol{\varphi})$, with

$$\boldsymbol{\varphi} : A \subset \mathbb{R}^n \to \mathbb{R}^n,$$

is a coordinate system of $M \setminus \partial M$ compatible with the orientation, then for every $\boldsymbol{t} \in A$, the basis of $\mathbb{R}^n$

$$\left\{ \frac{\partial \boldsymbol{\varphi}}{\partial t_1}(\boldsymbol{t}), \ldots, \frac{\partial \boldsymbol{\varphi}}{\partial t_n}(\boldsymbol{t}) \right\}$$

is positively oriented, that is, its orientation coincides with that of

$$\{\boldsymbol{e}_1, \ldots, \boldsymbol{e}_n\},$$

or equivalently, the Jacobian of $\boldsymbol{\varphi}$ is positive (see Proposition 5.2.1 and the comment prior to Proposition 5.1.1). Our objective in the following subsections is to analyze the orientation that is subsequently induced on the boundary. We restrict our attention to the cases $\mathbb{R}^2$ and $\mathbb{R}^3$.

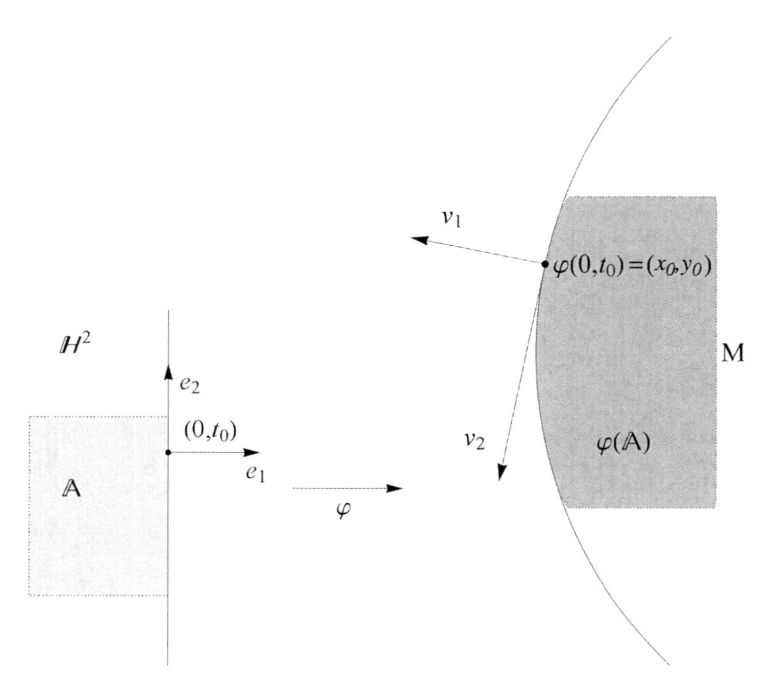

Fig. 8.6 Orientation of the boundary of a compact set with boundary in $\mathbb{R}^2$

8.5.1 Compact Sets with Boundary in $\mathbb{R}^2$

We first consider the case $n = 2$ (Fig. 8.6). Let (x_0, y_0) be a point in the boundary of M and $(\mathbb{A}, \boldsymbol{\varphi})$,

$$\boldsymbol{\varphi} : \mathbb{A} \subset \mathbb{H}^2 \to M \subset \mathbb{R}^2,$$

a coordinate system compatible with the orientation of M so that

$$(x_0, y_0) = \boldsymbol{\varphi}(0, t_0).$$

Then

$$\widehat{\mathbb{A}} := \{t \in \mathbb{R} : \ (0,t) \in \mathbb{A}\}$$

is an open neighborhood of t_0 in $\mathbb{R}$. Let (a,b) be an open interval such that $t_0 \in (a,b) \subset \widehat{\mathbb{A}}$. It turns out that

$$\boldsymbol{\gamma} : (a,b) \to \mathbb{R}^2 \ ; \ \boldsymbol{\gamma}(t) := \boldsymbol{\varphi}(0,t)$$

is a path of class C^1 whose trajectory is contained in the boundary of M (see Definition 8.2.2) and passes through (x_0, y_0). Moreover, $((a,b), \boldsymbol{\gamma})$ is a coordinate system of ∂M compatible with the induced orientation on the boundary, according to Theorem 8.4.1 and Definition 8.4.2. The tangent vector to $\boldsymbol{\gamma}$ at $(x_0, y_0) = \boldsymbol{\gamma}(t_0)$ is

$$\boldsymbol{v}_2 := \boldsymbol{\gamma}'(t_0) = \frac{\partial \boldsymbol{\varphi}}{\partial t}(0, t_0).$$

On the other hand, for every $s < 0$ such that $(s, t_0) \in \mathbb{A}$ we have that

$$\left\{ \frac{\partial \boldsymbol{\varphi}}{\partial s}(s, t_0), \ \frac{\partial \boldsymbol{\varphi}}{\partial t}(s, t_0) \right\}$$

is a positively oriented basis of $\mathbb{R}^2$, and consequently,

$$\det \left[\frac{\partial \boldsymbol{\varphi}}{\partial s}(s, t_0), \frac{\partial \boldsymbol{\varphi}}{\partial t}(s, t_0) \right] > 0.$$

Take

$$\boldsymbol{v}_1 := \frac{\partial \boldsymbol{\varphi}}{\partial s}(0, t_0).$$

Then from the continuity of the partial derivatives of $\boldsymbol{\varphi}$, we obtain

$$\det[\boldsymbol{v}_1, \boldsymbol{v}_2] \geq 0.$$

By condition (SB2) of the definition of coordinate system, Definition 8.2.1, this determinant is nonzero, and so it follows that

$$\det[\boldsymbol{v}_1, \boldsymbol{v}_2] > 0.$$

That is, $\{\boldsymbol{v}_1, \boldsymbol{v}_2\}$ is a positively oriented basis of $\mathbb{R}^2$. This means the following:

If we decompose the plane into two half-planes separated by the tangent line to ∂M at the point (x_0, y_0) and move along the tangent line in the direction provided by the vector $\boldsymbol{v}_2$, then $-\boldsymbol{v}_1$ (supported at point (x_0, y_0)) points toward the left half-plane. See the comments after Proposition 5.1.1.

We see now, by means of a heuristic argument, that the vector $-\boldsymbol{v}_1$ points toward the set M. Indeed, for $s > 0$ and $|s|$ small enough, we have $\boldsymbol{\varphi}(-s, t_0) \in M$, and also, according to the definition of the differential of a function,

$$\boldsymbol{\varphi}(-s, t_0) \approx \boldsymbol{\varphi}(0, t_0) + \frac{\partial \boldsymbol{\varphi}}{\partial s}(0, t_0)(-s) + \frac{\partial \boldsymbol{\varphi}}{\partial t}(0, t_0) \cdot 0 = (x_0, y_0) - s\boldsymbol{v}_1.$$

The previous considerations allow us to obtain the following criterion.

Criterion 8.5.1. *If we move along ∂M according to the orientation induced by the orientation of M, then the set M always remains to our left (Fig. 8.7).*

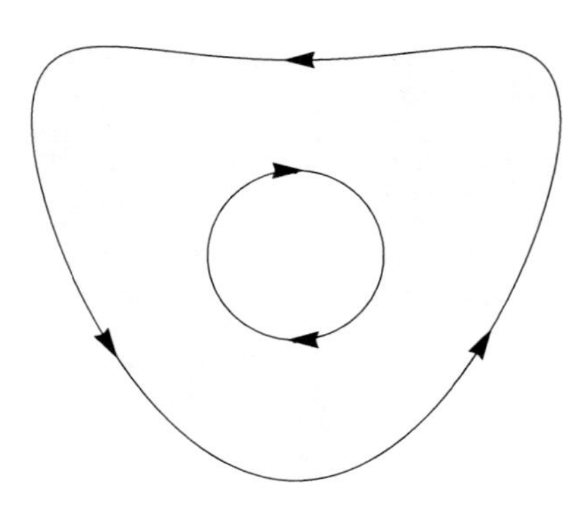

Fig. 8.7 Illustration of Criterion 8.5.1

8.5.2 Compact Sets with Boundary in $\mathbb{R}^3$

Consider M a compact set whose topological boundary is a regular surface of class C^1 in $\mathbb{R}^3$ and with the property that for every $(x_0, y_0, z_0) \in \partial M$, there exists $\delta > 0$ such that

$$B((x_0, y_0, z_0), r) \setminus \partial^{\mathrm{top}}(M) \tag{8.12}$$

for each $0 < r < \delta$ has two connected components, U and V, where U is the intersection of $B((x_0, y_0, z_0), r)$ with the interior of M and V is the intersection of $B((x_0, y_0, z_0), r)$ with the complement of M. According to Criterion 8.3.3, M is a regular 3-surface with boundary of class C^1 and $\partial M = \partial^{\mathrm{top}}(M)$.

Let $(\mathbb{A}, \boldsymbol{\varphi})$ be a coordinate system compatible with the orientation of M,

$$\boldsymbol{\varphi} : \mathbb{A} \subset \mathbb{H}^3 \to M,$$

so that

$$\boldsymbol{\varphi}(\mathbb{A}) \subset B((x_0, y_0, z_0), r) \cap M$$

and

$$(x_0, y_0, z_0) = \boldsymbol{\varphi}(0, t_2^0, t_3^0).$$

Then according to Theorem 8.4.1 and Definition 8.4.2,

$$\boldsymbol{\psi}(s, t) := \boldsymbol{\varphi}(0, s, t)$$

defines a coordinate system $(\widehat{\mathbb{A}}, \boldsymbol{\psi})$ of ∂M in a neighborhood of (x_0, y_0, z_0) that is compatible with the orientation induced by the orientation of M. Now define

$$\boldsymbol{v}_1 := \frac{\partial \boldsymbol{\varphi}}{\partial t_1}(0, t_2^0, t_3^0)\ ,\ \boldsymbol{v}_2 := \frac{\partial \boldsymbol{\varphi}}{\partial t_2}(0, t_2^0, t_3^0)\ ,\ \boldsymbol{v}_3 := \frac{\partial \boldsymbol{\varphi}}{\partial t_3}(0, t_2^0, t_3^0).$$

Then $\{\boldsymbol{v}_2, \boldsymbol{v}_3\}$ is a positively oriented basis of the tangent space to ∂M at (x_0, y_0, z_0) and $\{\boldsymbol{v}_1, \boldsymbol{v}_2, \boldsymbol{v}_3\}$ is a positively oriented basis of $\mathbb{R}^3$. This means that

$$\det \begin{bmatrix} \boldsymbol{v}_1 \\ \boldsymbol{v}_2 \\ \boldsymbol{v}_3 \end{bmatrix} > 0,$$

and hence $\langle \boldsymbol{v}_1, \boldsymbol{v}_2 \times \boldsymbol{v}_3 \rangle > 0$. That is, if $\theta \in (-\pi, \pi]$ is the angle between the vectors $\boldsymbol{v}_1$ and $\boldsymbol{v}_2 \times \boldsymbol{v}_3$, it turns out that $\cos\theta > 0$, or equivalently, $\theta \in (-\frac{\pi}{2}, \frac{\pi}{2})$. Since the vector $\boldsymbol{v}_2 \times \boldsymbol{v}_3$ is orthogonal to the tangent plane, we deduce that the two vectors $\boldsymbol{v}_1$ and $\boldsymbol{v}_2 \times \boldsymbol{v}_3$ (supported at the point (x_0, y_0, z_0)) are on the same side of the tangent plane to ∂M at (x_0, y_0, z_0).

Suppose now that

$$\boldsymbol{\Phi} : B((x_0, y_0, z_0), r) \to \mathbb{R}$$

is a function of class C^1 such that

$$\boldsymbol{\Phi}^{-1}(0) = B((x_0, y_0, z_0), r) \cap \partial M$$

and

$$\{\boldsymbol{x} \in B((x_0, y_0, z_0), r) : \boldsymbol{\Phi}(x, y, z) < 0\} = U$$

(see the proof of Criterion 8.3.3). We define

$$f(s) := \boldsymbol{\Phi}(\boldsymbol{\varphi}(s, t_2^0, t_3^0)),$$

so that

$$f(0) = \boldsymbol{\Phi}(x_0, y_0, z_0) = 0,$$

while

$$f(s) < 0,$$

whenever $s < 0$ has modulus small enough. Consequently, by the chain rule,

$$\begin{aligned} \langle \boldsymbol{\nabla} \boldsymbol{\Phi}(x_0, y_0, z_0), \boldsymbol{v}_1 \rangle &= \left\langle \boldsymbol{\nabla} \boldsymbol{\Phi}(\boldsymbol{\varphi}(0, t_2^0, t_3^0)), \tfrac{\partial \boldsymbol{\varphi}}{\partial t_1}(0, t_2^0, t_3^0) \right\rangle \\ &= f'(0) = \lim_{s \to 0^-} \frac{f(s) - f(0)}{s} \geq 0. \end{aligned}$$

Since $\boldsymbol{\nabla} \boldsymbol{\Phi}(x_0, y_0, z_0)$ is orthogonal to the tangent plane to ∂M at (x_0, y_0, z_0) (see Proposition 3.4.1) and $\boldsymbol{v}_1$ is not in that tangent plane, we obtain

$$\langle \boldsymbol{\nabla} \boldsymbol{\Phi}(x_0, y_0, z_0), \boldsymbol{v}_1 \rangle > 0.$$

This proves, with reference to the discussion after Definition 8.3.1, that

$$(x_0, y_0, z_0) - t\boldsymbol{v}_1 \in U \subset M,$$

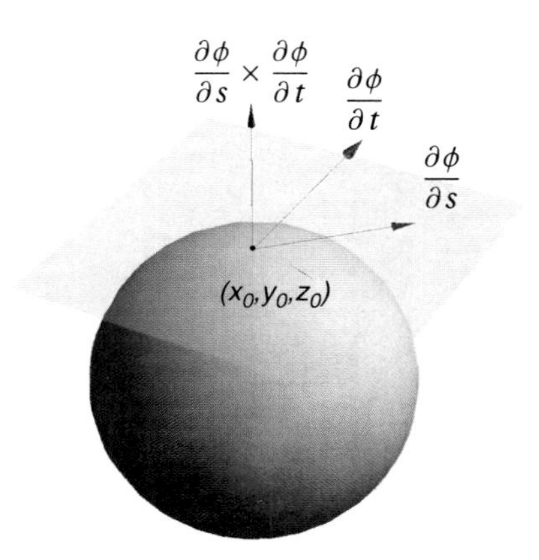

Fig. 8.8 Illustration of Criterion 8.5.2

whenever $t > 0$ is small enough. That is, the vector $-\boldsymbol{v}_1$ points toward the set M and the vector $\boldsymbol{v}_2 \times \boldsymbol{v}_3 = \left(\frac{\partial \psi}{\partial s} \times \frac{\partial \psi}{\partial t}\right)(t_2^0, t_3^0)$, which is normal to the tangent plane and is on the same side of the tangent plane as $\boldsymbol{v}_1$, points to the exterior of M. Of course we get the same conclusion for any coordinate system (B, ϕ) of ∂M that is compatible with the orientation induced on ∂M by that of M.

From the above considerations we reach another practical criterion. We denote by (s,t) the coordinates of an arbitrary point of $\mathbb{R}^2$.

Criterion 8.5.2. *Let M be a compact set in $\mathbb{R}^3$ that is a regular 3-surface with boundary of class C^1. If $\boldsymbol{\phi}$ is a parameterization of the surface ∂M compatible with the orientation, then $\frac{\partial \boldsymbol{\phi}}{\partial s} \times \frac{\partial \boldsymbol{\phi}}{\partial t}$ is a normal vector to that surface that points toward the exterior of M (Fig. 8.8).*

8.5.3 Regular 2-Surfaces with Boundary in $\mathbb{R}^3$

To finish, we analyze the case that M is an oriented regular surface with boundary in $\mathbb{R}^3$. Let

$$\boldsymbol{F} : M \setminus \partial M \to \mathbb{R}^3$$

be the continuous vector field of unit normal vectors that define the orientation of $M \setminus \partial M$ (see Theorem 5.2.2 and Definition 5.2.3). From the expression obtained in Theorem 5.2.2 for the vector field $\boldsymbol{F}$ in terms of a coordinate system, it is obvious that $\boldsymbol{F}$ can be extended to a continuous vector field $\overline{\boldsymbol{F}} : M \to \mathbb{R}^3$. Indeed, let $\{(\mathbb{A}_j, \boldsymbol{\varphi}_j) : j \in J\}$ be an atlas of M compatible with the orientation, and for every $j \in J$, put $M_j := \boldsymbol{\varphi}_j(\mathbb{A}_j)$ and

$$\boldsymbol{G}_j(s,t) := \frac{\frac{\partial \boldsymbol{\varphi}_j}{\partial s}(s,t) \times \frac{\partial \boldsymbol{\varphi}_j}{\partial t}(s,t)}{\left\| \frac{\partial \boldsymbol{\varphi}_j}{\partial s}(s,t) \times \frac{\partial \boldsymbol{\varphi}_j}{\partial t}(s,t) \right\|},$$

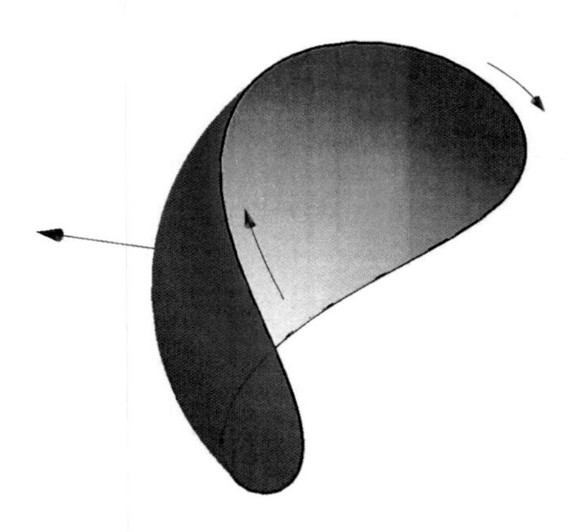
Fig. 8.9 Illustration of Criterion 8.5.3

which is a continuous function in $\mathbb{A}_j$, since $\boldsymbol{\varphi}$ is of class C^1 in $\mathbb{A}_j$. As in Theorem 5.2.2, it suffices to define $\overline{\boldsymbol{F}} : M \to \mathbb{R}^3$ so that

$$\overline{\boldsymbol{F}}(\boldsymbol{\varphi}_j(s,t)) := \boldsymbol{G}_j(s,t) \quad \text{for } (s,t) \in \mathbb{A}_j.$$

From now on, we will simply denote the vector field $\overline{\boldsymbol{F}}$ by

$$\boldsymbol{F} : M \to \mathbb{R}^3.$$

We will try to convince the reader of the following general principle, of doubtless practical interest:

Criterion 8.5.3. *If we move along ∂M according to the orientation induced by that of M in such a way that our head points toward the direction given by the vector field $\boldsymbol{F}$, then the surface M remains to our left (Fig. 8.9).*

Indeed, suppose that $(\mathbb{A}, \boldsymbol{\varphi})$,

$$\boldsymbol{\varphi} : \mathbb{A} \subset \mathbb{H}^2 \to \mathbb{R}^3,$$

is a coordinate system of M compatible with the orientation and

$$(x_0, y_0, z_0) = \boldsymbol{\varphi}(0, t_0).$$

The vector

$$\frac{\partial \boldsymbol{\varphi}}{\partial s}(0,t_0) \times \frac{\partial \boldsymbol{\varphi}}{\partial t}(0,t_0)$$

is a positive multiple of the unit vector $\boldsymbol{F}(\boldsymbol{\varphi}(0,t_0))$.

Now observe that the set

$$\widehat{\mathbb{A}} := \{t \in \mathbb{R} : (0,t) \in \mathbb{A}\}$$

is an open neighborhood of t_0 in $\mathbb{R}$; hence there is an open interval (a,b) such that $t_0 \in (a,b) \subset \widehat{\mathbb{A}}$. It turns out that

$$\boldsymbol{\gamma} : (a,b) \to \mathbb{R}^3, \quad \boldsymbol{\gamma}(t) := \boldsymbol{\varphi}(0,t)$$

is a path of class C^1 whose trajectory is contained in the boundary of M (See Definition 8.2.2) and passes through (x_0,y_0,z_0). Moreover, $\boldsymbol{\gamma}$ defines a coordinate system of ∂M compatible with the orientation induced in the boundary, according to Theorem 8.4.1 and Definition 8.4.2. The tangent vector to $\boldsymbol{\gamma}$ at $(x_0,y_0,z_0) = \boldsymbol{\gamma}(t_0)$ is

$$\boldsymbol{v}_2 := \boldsymbol{\gamma}'(t_0) = \frac{\partial \boldsymbol{\varphi}}{\partial t}(0,t_0).$$

If we put

$$\boldsymbol{v}_1 := \frac{\partial \boldsymbol{\varphi}}{\partial s}(0,t_0),$$

then as indicated before,

$$\boldsymbol{v}_3 := \boldsymbol{v}_1 \times \boldsymbol{v}_2$$

is a positive multiple of $\boldsymbol{F}(x_0,y_0,z_0)$. Now observe that the plane through the point (x_0,y_0,z_0) determined by the vectors $\boldsymbol{v}_2$ and $\boldsymbol{v}_3$ decomposes $\mathbb{R}^3$ into two half-spaces. If we stand at (x_0,y_0,z_0), with our head pointing in the direction $\boldsymbol{v}_3$ and face the direction $\boldsymbol{v}_2$ (that is, according to the orientation induced in ∂M), then the open half-space to our left consists exactly of those points $(x,y,z) \in \mathbb{R}^3$ with the property that

$$\langle (x,y,z) - (x_0,y_0,z_0), \boldsymbol{v} \rangle > 0,$$

where $\boldsymbol{v} = \boldsymbol{v}_3 \times \boldsymbol{v}_2$. This means that the angle θ between the vectors $(x,y,z) - (x_0,y_0,z_0)$ and $\boldsymbol{v}$ varies between $-\frac{\pi}{2}$ and $\frac{\pi}{2}$.

Taking, if necessary, a suitable open subset, we can assume that $\mathbb{A}$ is a convex set. If $(s,t_0) \in \mathbb{A}$ and $s < 0$, then we can apply the mean value theorem to the function

$$f : [s,0] \to \mathbb{R} \ ; \ f(r) = \left\langle \boldsymbol{\varphi}(r,t_0), \boldsymbol{v} \right\rangle$$

to deduce that there exists ξ, $s < \xi < 0$, such that

$$\left\langle \boldsymbol{\varphi}(s,t_0) - \boldsymbol{\varphi}(0,t_0), \boldsymbol{v} \right\rangle = s \left\langle \frac{\partial \boldsymbol{\varphi}}{\partial s}(\xi,t_0), \boldsymbol{v} \right\rangle.$$

Since

$$\left\langle \frac{\partial \boldsymbol{\varphi}}{\partial s}(0,t_0), \boldsymbol{v} \right\rangle = \left\langle \boldsymbol{v}_1, \boldsymbol{v}_3 \times \boldsymbol{v}_2 \right\rangle$$

$$= -\det \begin{pmatrix} \boldsymbol{v}_3 \\ \boldsymbol{v}_1 \\ \boldsymbol{v}_2 \end{pmatrix}$$

$$= -||\boldsymbol{v}_3||^2 < 0,$$

and the partial derivatives are continuous functions, we conclude that

$$\left\langle \boldsymbol{\varphi}(s,t_0) - \boldsymbol{\varphi}(0,t_0), \boldsymbol{v} \right\rangle > 0,$$

whenever $s < 0$ has modulus small enough. In other words, the point $(x,y,z) = \boldsymbol{\varphi}(s,t_0) \in M$ is in the half-space located to our left.

8.6 The n-dimensional cube

An important case where a physicist will actually apply Stokes's theorem is when studying the flux through a cube. The cube, unfortunately, is not a regular surface. In this section, we will show how we can, from a formal point of view, overcome that difficulty. We will see that if we remove the edges and vertices of the cube $[0,1]^n$ then we get a regular n-surface with boundary in $\mathbb{R}^n$. We will also discuss what is the orientation induced on the boundary by the canonical orientation of $\mathbb{R}^n$.

For $n \geq 3$, put $I := (0,1)^n$ and suppose that it is oriented so that the identity mapping defines a coordinate system compatible with the orientation. For every $1 \leq j \leq n$ let

$$I_{j,0} := \{\boldsymbol{x} \in \mathbb{R}^n \ : x_j = 0\,; 0 < x_i < 1 \text{ if } i \neq j\}$$

and

$$I_{j,1} := \{\boldsymbol{x} \in \mathbb{R}^n \ : x_j = 1\,; 0 < x_i < 1 \text{ if } i \neq j\}.$$

The set

$$M := I \cup \bigcup_{j=1}^{n} (I_{j,0} \cup I_{j,1})$$

is called, for obvious reason, the *cube without edges*. Now we fix $1 \leq j \leq n$ and study $I_{j,1}$. Take

$$\mathbb{A} := (-1,0] \times (0,1)^{n-1} \subset \mathbb{H}^n$$

and define

$$\boldsymbol{\varphi} : \mathbb{A} \to \mathbb{R}^n \ ; \ \boldsymbol{\varphi}(t_1,\ldots,t_j,\ldots,t_n) := (t_j,\ldots,1+t_1,\ldots,t_n)$$

for $2 \leq j \leq n$ and $\boldsymbol{\varphi}(t_1,t_2,\ldots,t_n) = (1+t_1,t_2,\ldots,t_n)$ for $j = 1$. Then

$$(\mathbb{A},\boldsymbol{\varphi})$$

is a coordinate system for a regular surface with boundary and

$$\boldsymbol{\varphi}(\{0\} \times (0,1)^{n-1}) = I_{j,1}.$$

Observe that, while it may not seem natural to exchange first and j-th variables in the definition of $\boldsymbol{\varphi}$, one must proceed in this way in order to obtain a coordinate system since the domain should be a subset of the half-space $\mathbb{H}^n$. Also,

$$\boldsymbol{\varphi}((-1,0) \times (0,1)^{n-1}) = I.$$

As the Jacobian of φ is

$$J(\boldsymbol{\varphi})(\boldsymbol{t}) = \begin{cases} -1, & j \geq 2 \\ 1, & j = 1 \end{cases}$$

for every $\boldsymbol{t} \in \mathbb{A}$, it follows that the mapping $\boldsymbol{\psi} : (0,1)^{n-1} \to \mathbb{R}^n$

$$\boldsymbol{\psi}(s_1, \ldots, s_{n-1}) := \boldsymbol{\varphi}(0, s_1, \ldots, s_{n-1})$$

$$= \begin{cases} (s_{j-1}, s_1, \ldots, s_{j-2}, 1, s_j, \ldots, s_{n-1}), & 2 \leq j \leq n \\ (1, s_1, \ldots, s_{n-1}), & j = 1 \end{cases}$$

keeps the positive orientation of $I_{j,1}$ for $j = 1$ and changes the orientation for $j \geq 2$. The normal vector to this face, which (according to Definition 8.4.3) corresponds to the orientation induced on ∂M by that of M (Theorem 8.4.1 and Definition 8.4.2) will be

$$\left(\frac{\partial \boldsymbol{\psi}}{\partial s_1}(\boldsymbol{s}) \times \ldots \times \frac{\partial \boldsymbol{\psi}}{\partial s_{n-1}}(\boldsymbol{s})\right) = \boldsymbol{e}_2 \times \ldots \times \boldsymbol{e}_{n-1} = \boldsymbol{e}_1$$

for $j = 1$ and

$$-\left(\frac{\partial \boldsymbol{\psi}}{\partial s_1}(\boldsymbol{s}) \times \ldots \times \frac{\partial \boldsymbol{\psi}}{\partial s_{n-1}}(\boldsymbol{s})\right) = -(\boldsymbol{e}_2 \times \ldots \times \boldsymbol{e}_1 \times \boldsymbol{e}_{j+1} \times \ldots \times \boldsymbol{e}_n) = \boldsymbol{e}_j$$

for $j \geq 2$. (In this cross product $\boldsymbol{e}_1$ is located at position $(j-1)$ while $\boldsymbol{e}_j$ is absent.) That is, the face $I_{j,1}$ is oriented according the normal vector $\boldsymbol{e}_j$ pointing towards the *exterior* of M for every $1 \leq j \leq n$. In the case $2 \leq j \leq n$, by Lemma 5.2.3, the compatible orientation of $I_{j,1}$ is given by $(\boldsymbol{L}^{-1}(\mathbb{A}), \boldsymbol{\varphi} \circ \boldsymbol{L})$, where $\boldsymbol{L} : \mathbb{R}^n \to \mathbb{R}^n$ is given by $\boldsymbol{L}(u_1, \ldots, u_n) = (u_1, \ldots, u_n, u_{n-1})$.

Now consider $I_{j,0}$. As before, put

$$\mathbb{A} := (-1, 0] \times (0,1)^{n-1} \subset \mathbb{H}^n$$

and define

$$\boldsymbol{\varphi} : \mathbb{A} \to \mathbb{R}^n \ ; \ \boldsymbol{\varphi}(t_1, t_2, \ldots, t_j, \ldots, t_n) := (t_j, t_2, \ldots, -t_1, \ldots, t_n)$$

for $j \geq 2$ and $\boldsymbol{\varphi}(t_1, t_2, \ldots, t_n) = (-t_1, t_2, \ldots, t_n)$ for $j = 1$. Then

$$(\mathbb{A}, \boldsymbol{\varphi})$$

is a coordinate system of a regular surface with boundary and

$$\boldsymbol{\varphi}(\{0\} \times (0,1)^{n-1}) = I_{j,0}.$$

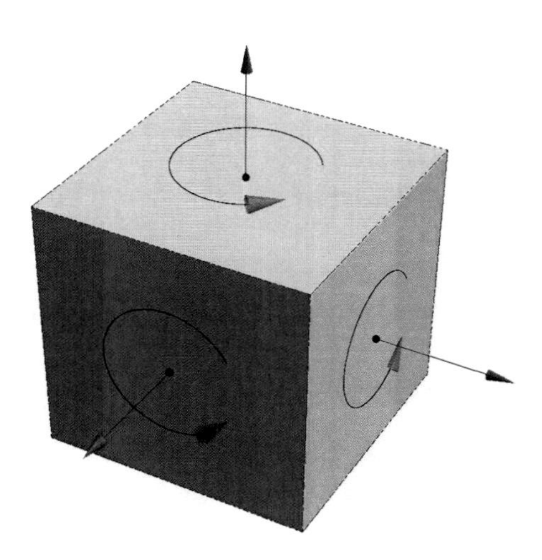

Fig. 8.10 Orientation of the boundary of the cube

From

$$J(\boldsymbol{\varphi})(\boldsymbol{t}) = \begin{cases} 1, & j \geq 2 \\ -1, & j = 1 \end{cases}$$

for every $\boldsymbol{t} \in \mathbb{A}$, we get that

$$\boldsymbol{\psi}(s_1, \ldots, s_{n-1}) := \boldsymbol{\varphi}(0, s_1, \ldots, s_{n-1})$$

$$= \begin{cases} (s_{j-1}, s_1, \ldots, s_{j-2}, 0, s_j, \ldots, s_{n-1}), & j \geq 2 \\ (0, s_1, \ldots, s_{n-1}), & j = 1 \end{cases}$$

(with 0 at the j-th coordinate) is a positively oriented coordinate system of $I_{j,0}$ except for $j = 1$. The normal vector to this face consistent with that orientation is the cross product of the $(n-1)$ vectors

$$\{\boldsymbol{e}_2, \boldsymbol{e}_3, \ldots, \boldsymbol{e}_1, \boldsymbol{e}_{j+1}, \ldots, \boldsymbol{e}_n\}$$

for $2 \leq j \leq n$ and $-\boldsymbol{e}_2 \times \ldots \times \boldsymbol{e}_n = -\boldsymbol{e}_1$ for $j = 1$. That is, $I_{j,0}$ is oriented according the vector normal $-\boldsymbol{e}_j$, which points towards the *exterior* of M (Fig. 8.10) for every $1 \leq j \leq n$.

Note that the conclusion is consistent with Criterion 8.5.2.

To make things easier for the student we include coordinate systems of the 6 faces of the unit cube in $\mathbb{R}^3$ which are compatible with the orientation. Let $A = (0,1) \times (0,1)$ be given and define $\boldsymbol{\psi}_{j,a} : A \subset \mathbb{R}^2 \to \mathbb{R}^3$ as

$$\boldsymbol{\psi}_{1,1}(u,v) = (1,u,v), \ \boldsymbol{\psi}_{2,1}(u,v) = (v,1,u), \ \boldsymbol{\psi}_{3,1}(u,v) = (u,v,1)$$

and also

$$\boldsymbol{\psi}_{1,0}(u,v) = (0,v,u),\ \boldsymbol{\psi}_{2,0}(u,v) = (u,0,v),\ \boldsymbol{\psi}_{3,0}(u,v) = (v,u,0).$$

Then $(A, \boldsymbol{\psi}_{j,a})$ is a coordinate system of $I_{j,a}$ which is compatible with the orientation for every $j = 1,2,3$ and $a = 0,1$. In fact, it suffices to observe that

$$\frac{\partial \boldsymbol{\psi}_{j,a}}{\partial u} \times \frac{\partial \boldsymbol{\psi}_{j,a}}{\partial v}(u,v)$$

coincides with the the normal vector to $I_{j,a}$ pointing toward the exterior of the unit cube.

Finally we consider the case of the unit square in the plane. That square, $[0,1] \times [0,1]$, is a simple region, where a precise counter-clockwise parametrization of its boundary is very easily given. Hence the version of Green Theorem stated in theorem 2.6.4 is more than enough for the applications. But a reader could be interested in producing a seamless connection with surfaces with boundary and with the Divergence theorem obtained from the general Stokes's theorem in the next chapter. For that kind of reader we produce explicitly atlases for the square without the vertexes $U_2 = [0,1] \times [0,1] \setminus \{(0,0),(0,1),(1,0),(1,1)\}$ and ∂U_2, that are positively oriented, i.e., compatible with the identity as coordinate system in the interior of U_2.

The atlas on U_2 is $\{\varphi_1, \varphi_2, \varphi_3, \varphi_4\}$ that generates on ∂U_2 the atlas $\{\widehat{\varphi}_1, \widehat{\varphi}_2, \widehat{\varphi}_3, \widehat{\varphi}_4\}$,

$$\varphi_j : (-1,0] \times (0,1) \to U_2, \quad \widehat{\varphi}_j : (0,1) \to (0,1), \text{ for } j = 1,2,3,4,$$

defined as

$$\begin{aligned}
\varphi_1(s,t) &= (s+1,t) &&, \quad \widehat{\varphi}_1(t) = (1,t) \\
\varphi_2(s,t) &= (1-t,s+1) &&, \quad \widehat{\varphi}_2(t) = (1-t,1) \\
\varphi_3(s,t) &= (-s,1-t) &&, \quad \widehat{\varphi}_3(t) = (0,1-t) \\
\varphi_4(s,t) &= (t,-s) &&, \quad \widehat{\varphi}_4(t) = (t,0)
\end{aligned}$$

Observe that $J\varphi_j(s,t) = 1$ for every (s,t) in $(-1,0] \times (0,1)$ and $j = 1,2,3,4$.

The next result will be useful in solving exercises. It allows one to parameterize regular surfaces with boundary in $\mathbb{R}^3$ by taking the domain of parameters to be a regular surface with boundary in $\mathbb{R}^2$ (usually a closed disk or a rectangle without vertices).

Criterion 8.6.1. *Let S be a regular k-surface with boundary in $\mathbb{R}^k$ oriented so that the identity mapping on $S \setminus \partial S$ defines a coordinate system $(S \setminus \partial S, \mathrm{id})$ compatible with its orientation. Let M be an oriented regular k-surface with boundary in $\mathbb{R}^n$ and $\boldsymbol{f} : S \to M$ a homeomorphism admitting a C^1 extension to an open neighborhood of S such that $\boldsymbol{f}'$ has maximal rank on S. Then*

(1) If $(\mathbb{A}, \boldsymbol{\varphi})$ is a coordinate system of S then $(\mathbb{A}, \boldsymbol{f} \circ \boldsymbol{\varphi})$ is a coordinate system of M.
(2) $\boldsymbol{f}(\partial S) = \partial M$, $\boldsymbol{f}(S \setminus \partial S) = M \setminus \partial M$ and $(S \setminus \partial S, \boldsymbol{f})$ is a coordinate system of $M \setminus \partial M$.
(3) If $(B, \boldsymbol{\psi})$ is a coordinate system of ∂S then $(B, \boldsymbol{f} \circ \boldsymbol{\psi})$ is a coordinate system of ∂M.
(4) If the coordinate system $(S \setminus \partial S, \boldsymbol{f})$ is compatible with the orientation of M and $(B, \boldsymbol{\psi})$ is a coordinate system of ∂S compatible with its orientation then $(B, \boldsymbol{f} \circ \boldsymbol{\psi})$ is a coordinate system of ∂M compatible with its orientation.

Proof.

(1) Let $\mathbb{B}$ be an open subset of $\mathbb{A}$ for the relative topology. By condition (SB1') of Definition 8.2.1, there exists an open set $U \subset \mathbb{R}^k$ such that $\boldsymbol{\varphi}(\mathbb{B}) = S \cap U$. Since $\boldsymbol{f}$ is an homeomorphism there is an open subset $V \subset \mathbb{R}^n$ such that $\boldsymbol{f}(S \cap U) = M \cap V$. Hence $\boldsymbol{f} \circ \boldsymbol{\varphi}(\mathbb{B}) = M \cap V$. Also $(\boldsymbol{f} \circ \boldsymbol{\varphi})'(\boldsymbol{t}) = \boldsymbol{f}'(\boldsymbol{\varphi}(\boldsymbol{t})) \circ \boldsymbol{\varphi}'(\boldsymbol{t})$ has rank k for every $\boldsymbol{t} \in \mathbb{A}$. According to Definition 8.2.1, $(\mathbb{A}, \boldsymbol{f} \circ \boldsymbol{\varphi})$ is a coordinate system of M.

(2) Let $\{(\mathbb{A}_\alpha, \boldsymbol{\varphi}_\alpha)\}_{\alpha \in I}$, $\boldsymbol{\varphi}_\alpha : \mathbb{A}_\alpha \subset \mathbb{H}^k \to S$, an atlas of S. Then $\{(\mathbb{A}_\alpha, \boldsymbol{f} \circ \boldsymbol{\varphi}_\alpha)\}_{\alpha \in I}$ is an atlas of M. In particular,

$$\boldsymbol{f}(\partial S) = \cup_{\alpha \in I} \boldsymbol{f}\left(\boldsymbol{\varphi}_\alpha(\mathbb{A}_\alpha \cap \{0\} \times \mathbb{R}^{k-1})\right) = \partial M.$$

Hence also $\boldsymbol{f}(S \setminus \partial S) = M \setminus \partial M$. On the other hand, $S \setminus \partial S$ is an open set in $\mathbb{R}^k$ (Example 8.2.2) and from Corollary 3.3.1 we deduce that $(S \setminus \partial S, \boldsymbol{f})$ is a coordinate system of $M \setminus \partial M$.

(3) Since $\boldsymbol{f} \circ \boldsymbol{\psi} : B \subset \mathbb{R}^{k-1} \to \partial M$ is injective and C^1 and also $(\boldsymbol{f} \circ \boldsymbol{\psi})'(\boldsymbol{s}) = \mathbf{f}'(\boldsymbol{\psi}(\boldsymbol{s})) \circ \boldsymbol{\psi}'(\boldsymbol{s})$ has rank $(k-1)$ at each point $\boldsymbol{s} \in B$ it follows from Corollary 3.3.1 that $(\boldsymbol{f} \circ \boldsymbol{\psi}, B)$ is a coordinate system of ∂M.

(4) Fix $\boldsymbol{t}_0 \in \boldsymbol{\psi}(B) \subset \partial S$ and select

$$\boldsymbol{\varphi} : \mathbb{A} \subset \mathbb{H}^k \subset \mathbb{R}^k \to S$$

compatible with the orientation of S and such that $\boldsymbol{t}_0 = \boldsymbol{\varphi}(\boldsymbol{s}_0)$. We observe that the coordinate system $(\mathbb{A}, \boldsymbol{f} \circ \boldsymbol{\varphi})$ is also compatible with the orientation of M. Indeed, it suffices to note that

$$\boldsymbol{f}^{-1} \circ (\boldsymbol{f} \circ \boldsymbol{\varphi}) = \boldsymbol{\varphi}$$

has positive Jacobian at each point (see for instance Proposition 5.2.1). Consequently, it turns out that $(\widehat{\mathbb{A}}, \boldsymbol{\gamma})$,

$$\boldsymbol{\gamma} : \widehat{\mathbb{A}} \to \mathbb{R}^n\ ; \boldsymbol{\gamma}(\boldsymbol{t}) = (\boldsymbol{f} \circ \boldsymbol{\varphi})(0, \boldsymbol{t}),$$

is compatible with the orientation of ∂M. Now put

$$\boldsymbol{\xi} : \widehat{\mathbb{A}} \to \mathbb{R}^k\ ; \boldsymbol{\xi}(\boldsymbol{t}) := \boldsymbol{\varphi}(0, \boldsymbol{t}).$$

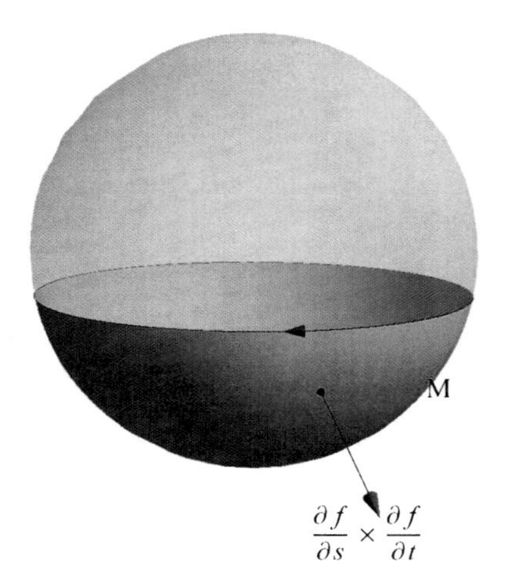

Fig. 8.11 The portion of sphere of Example 8.6.1

Then $(\mathbb{A}, \boldsymbol{\xi})$ is a coordinate system of ∂S which is compatible with its orientation. By hypothesis, $\boldsymbol{\psi}^{-1} \circ \boldsymbol{\xi}$ has positive Jacobian on its domain (Proposition 5.2.1). From the fact that

$$((\boldsymbol{f} \circ \boldsymbol{\psi})^{-1} \circ \boldsymbol{\gamma})(t) = (\boldsymbol{\psi}^{-1} \circ \boldsymbol{f}^{-1} \circ \boldsymbol{f} \circ \boldsymbol{\varphi})(0,t) = (\boldsymbol{\psi}^{-1} \circ \boldsymbol{\xi})(t)$$

has positive Jacobian we conclude, on applying again Proposition 5.2.1, that $(B, \boldsymbol{f} \circ \boldsymbol{\psi})$ is a coordinate system of ∂M which is compatible with its orientation.

□

Example 8.6.1. Fix $0 < \varepsilon < 1$ and let M be the portion of sphere (Fig. 8.11)

$$M := \{(x,y,z) \in \mathbb{R}^3 \ : \ x^2 + y^2 + z^2 = 1 \ ; \ z \leq -\varepsilon\}.$$

We suppose this is oriented according to the normal vector pointing to the exterior of the sphere (see Definition 5.2.3). Take

$$D := \{(s,t) \in \mathbb{R}^2 \ : s^2 + t^2 \leq 1 - \varepsilon^2\}$$

and consider the mapping

$$\boldsymbol{f} : D \to M \ ; \boldsymbol{f}(s,t) := (t, s, -\sqrt{1 - s^2 - t^2}).$$

Then

$$\frac{\partial \boldsymbol{f}}{\partial s}(s,t) \times \frac{\partial \boldsymbol{f}}{\partial t}(s,t) = \left(\frac{t}{\sqrt{1 - s^2 - t^2}}, \frac{s}{\sqrt{1 - s^2 - t^2}}, -1 \right)$$

is a positive multiple of the vector $\boldsymbol{f}(s,t)$.

Setting $r^2 = 1 - \varepsilon^2$, we deduce from Criterion 8.6.1 that

$$\boldsymbol{\gamma} : (\theta_0 - \pi, \theta_0 + \pi) \to \mathbb{R}^3 \; ; \; \boldsymbol{\gamma}(t) = \boldsymbol{f}(r\cos t, \; r\sin t)$$

is a positively oriented coordinate system of ∂M. Since

$$\boldsymbol{\gamma}(t) = (r\sin t, \; r\cos t, -\varepsilon),$$

we conclude that the boundary of M (that is, the circle in the plane $z = -\varepsilon$) is oriented clockwise.

8.7 Exercises

Exercise 8.7.1. Show that the portion of the cylinder $\{(x,y,z) \in \mathbb{R}^3 : x^2 + y^2 = a^2, 0 \leq z < 1\}$ is a regular surface with boundary.

Exercise 8.7.2. Let M be the portion of the plane $x + y + z = 1$ whose projection onto the xy-plane is the closed disk centered at the origin and with radius 1.

(1) Show that M is a regular surface with boundary.
(2) If M is oriented according to the normal vector $(1,1,1)$ and we consider the induced orientation in ∂M, find

$$\int_{\partial M} x \, \mathrm{d}y + y \, \mathrm{d}z.$$

Exercise 8.7.3. Let M be the lower hemisphere of the unit sphere, i.e., the surface specified by

$$z = -\sqrt{1 - x^2 - y^2}.$$

Suppose the sphere is oriented according to the exterior normal vector. Evaluate

$$\int_{\partial M} -y \, \mathrm{d}x + x \, \mathrm{d}y.$$

Chapter 9
The General Stokes's Theorem

Let ω be a differential form of degree $k-1$ and class C^1 in a neighborhood of a compact regular k-surface with boundary M of class C^2. The general Stokes's theorem gives a relationship between the integral of ω over the boundary of M and the integral of the exterior differential $d\omega$ over M. It can be viewed as a generalization of Green's theorem to higher dimensions, and it plays a role not unlike that of the fundamental theorem of calculus in an elementary course of analysis. Particular cases of the general Stokes's theorem that are of great importance are the divergence theorem, which relates a triple integral with a surface integral and what we know as the classical Stokes's theorem, which relates a surface integral with a line integral.

A. Galbis and M. Maestre, *Vector Analysis Versus Vector Calculus*, Universitext,
DOI 10.1007/978-1-4614-2200-6_9,

9.1 Integration on Surfaces with Boundary

The definition of integral of a continuous differential form of degree k over an oriented regular k-surface with boundary is completely analogous to the definition studied in Chap. 7.

Definition 9.1.1. Let M be an oriented regular k-surface with boundary and $(\mathbb{A}, \boldsymbol{\varphi})$ a coordinate system of M compatible with the orientation. Let $\boldsymbol{\omega} : M \subset \mathbb{R}^n \to \Lambda^k(\mathbb{R}^n)$ be a continuous differential form of degree k on M. For each measurable subset $\Delta \subset \mathbb{A}$ we define

$$\int_{\boldsymbol{\varphi}(\Delta)} \boldsymbol{\omega} := \int_{\Delta} \boldsymbol{\omega}(\boldsymbol{\varphi}(\boldsymbol{t})) \left(\frac{\partial \boldsymbol{\varphi}}{\partial t_1}(\boldsymbol{t}), \ldots, \frac{\partial \boldsymbol{\varphi}}{\partial t_k}(\boldsymbol{t}) \right) \mathrm{d}\boldsymbol{t},$$

whenever the integral exists.

Observe that if $\mathbb{A}$ is an open set for the relative topology of $\mathbb{H}^k$, then by definition, $\mathbb{A}$ is the intersection of an open subset of $\mathbb{R}^n$ with the closed set $\mathbb{H}^k$ defined and studied in Sect. 8.1. Hence $\mathbb{A}$ is a measurable set in $\mathbb{R}^n$.

Let $A \subset \mathbb{R}^k$ be an open set such that $\mathbb{A} = A \cap \mathbb{H}^k$, and $\boldsymbol{g} : A \subset \mathbb{R}^k \to \mathbb{R}^n$ a function of class C^p such that

$$\boldsymbol{g}_{|\mathbb{A}} = \boldsymbol{\varphi}.$$

Then for every $\boldsymbol{t} \in \mathbb{A}$, we define the *pullback of* $\boldsymbol{\omega}$*, a differential form of degree* k*, under the mapping* φ as

$$(\boldsymbol{\varphi}^* \boldsymbol{\omega})(\boldsymbol{t}) := (\boldsymbol{g}^* \boldsymbol{\omega})(\boldsymbol{t}),$$

that is, the restriction of the pullback of $\boldsymbol{\omega}$ under the mapping g to the set $\mathbb{A}$. According to Definition 8.1.4 and the comments following Definition 8.1.2, we see that the pullback is independent of the choice of A and $\boldsymbol{g}$. In fact,

$$\begin{aligned}(\boldsymbol{\varphi}^* \boldsymbol{\omega})(\boldsymbol{t})(\boldsymbol{v}_1, \ldots, \boldsymbol{v}_k) &= \boldsymbol{\omega}(\boldsymbol{g}(\boldsymbol{t}))(\mathrm{d}\boldsymbol{g}(\boldsymbol{t})(\boldsymbol{v}_1), \ldots, \mathrm{d}\boldsymbol{g}(\boldsymbol{t})(\boldsymbol{v}_k)) \\ &= \boldsymbol{\omega}(\boldsymbol{\varphi}(\boldsymbol{t}))(\mathrm{d}\boldsymbol{\varphi}(\boldsymbol{t})(\boldsymbol{v}_1), \ldots, \mathrm{d}\boldsymbol{\varphi}(\boldsymbol{t})(\boldsymbol{v}_k)).\end{aligned}$$

The reader should be aware that every time we apply the properties of pullback proven in Chap. 6, we actually apply them to $\boldsymbol{g}^* \boldsymbol{\omega}$ rather than to $\boldsymbol{\varphi}^* \boldsymbol{\omega}$. That is, we simply use $\boldsymbol{g}^* \boldsymbol{\omega}$ applied to points of $\mathbb{A}$. However, the presentation of proofs will be less cumbersome if we keep the notation $\boldsymbol{\varphi}^* \boldsymbol{\omega}$ instead of $\boldsymbol{g}^* \boldsymbol{\omega}$. For example,

$$(\boldsymbol{\varphi}^* \boldsymbol{\omega})(\boldsymbol{t}) = f(\boldsymbol{t}) \cdot \mathrm{d}t_1 \wedge \cdots \wedge \mathrm{d}t_k,$$

where f is a continuous function on $\mathbb{A}$ (see Theorem 6.4.1).

Proceeding exactly as in Chap. 7 (see Theorem 7.1.1), we have

$$\int_{\boldsymbol{\varphi}(\Delta)} \boldsymbol{\omega} = \int_{\Delta} f(\boldsymbol{t})\,\mathrm{d}\boldsymbol{t}. \tag{9.1}$$

Moreover, as in Proposition 7.1.1, one can check that the definition of this integral does not depend on the chosen coordinate system, as long as we restrict our choice to coordinate systems that define the same orientation.

If M is an oriented regular k-surface with boundary having a finite atlas

$$\{(\mathbb{A}_j, \boldsymbol{\varphi}_j)\}_{j=1}^{m}$$

compatible with the orientation, then there is a partition of M into subsets $\{Y_j\}$ such that $Y_j = \boldsymbol{\varphi}_j(\Delta_j)$, where Δ_j is a measurable subset of $\mathbb{A}_j$. The proof is as in Lemma 4.2.2.

Under these conditions, given a continuous differential form

$$\boldsymbol{\omega} : M \to \Lambda^k(\mathbb{R}^n)$$

of degree k, the following definition makes sense, provided all Lebesgue integrals involved are defined and finite.

Definition 9.1.2. $\displaystyle\int_M \boldsymbol{\omega} := \sum_{j=1}^{m} \int_{\boldsymbol{\varphi}_j(\Delta_j)} \boldsymbol{\omega}.$

We omit the proof of the following result, since it is almost identical to its analogue for surfaces without boundary obtained in Chap. 7 (see Theorem 7.3.1).

This definition does not depend on the chosen partition. This claim can be obtained as in the proof of proposition 7.1.1.

Theorem 9.1.1. *Suppose that $M \subset \mathbb{R}^n$ is an oriented and compact regular k-surface with boundary and let $\boldsymbol{\omega} : M \to \Lambda^k(\mathbb{R}^n)$ be a continuous differential form of degree k. Then*

$$\int_M \boldsymbol{\omega}$$

is well defined.

Our next lemma is well known in Lebesgue integration theory, but we provide its proof, since it is going to play an important role in subsequent proofs and also in the solution of several exercises.

Lemma 9.1.1. *Let $\boldsymbol{f} : U \subset \mathbb{R}^k \to \mathbb{R}^n$, $1 \leq k \leq n$, be a mapping of class C^1 on the open set U. If $A \subset U$ is a null set in $\mathbb{R}^k$, then $\boldsymbol{f}(A)$ is a null set in $\mathbb{R}^n$.*

Proof. Consider $\boldsymbol{t}_0 \in U$ and $r > 0$ such that the closed ball $\overline{B(\boldsymbol{t}_0, r)}$ centered at $\boldsymbol{t}_0$ and of radius r is contained in U. Let

$$H := \max\left\{ \left| \frac{\partial f_i}{\partial t_j}(\boldsymbol{t}) \right| : i = 1, \ldots, n,\ j = 1, \ldots, k,\ \boldsymbol{t} \in \overline{B(\boldsymbol{t}_0, r)} \right\}$$

and let $Q = \Pi_{j=1}^{k}[a_j, b_j]$ be an k-cube (i.e., $b_j - a_j = l$, $j = 1, \ldots, k$) contained in $\overline{B(\boldsymbol{t}_0, r)}$ such that $l < 1$. If $\boldsymbol{t} \in Q$, the segment $[\boldsymbol{a}, \boldsymbol{t}]$ is contained in $\overline{B(\boldsymbol{t}_0, r)}$. Thus the path

$$\gamma : [0,1] \to \mathbb{R}^n, \quad \boldsymbol{\gamma}(s) = \boldsymbol{f}((1-s)\boldsymbol{a} + s\boldsymbol{t}) = (f_i((1-s)\boldsymbol{a} + s\boldsymbol{t}))_{i=1}^{n},$$

is well defined and of class C^1 on $[0,1]$. If we apply the mean value theorem to each $\gamma_i(s) = f_i((1-s)\boldsymbol{a} + s\boldsymbol{t})$, we see that there exists $0 < s_i < 1$ such that

$$f_i(\boldsymbol{t}) - f_i(\boldsymbol{a}) = \gamma_i(1) - \gamma_i(0) = \gamma_i'(s_i) = \sum_{j=1}^{k} \frac{\partial f_i}{\partial t_j}((1-s_i)\boldsymbol{a} + s_i\boldsymbol{t})(t_j - a_j),$$

and we have

$$|f_i(\boldsymbol{t}) - f_i(\boldsymbol{a})| \leq kHl.$$

We have obtained that

$$\boldsymbol{f}(Q) \subset \Pi_{i=1}^{n}[f_i(\boldsymbol{a}) - kHl, f_i(\boldsymbol{a}) + kHl]$$

and

$$m(\Pi_{i=1}^{n}[f_i(\boldsymbol{a}) - kHl, f_i(\boldsymbol{a}) + kHl]) = (2kHl)^n \leq (2kH)^n l^k = (2kH)^n m(Q), \quad (9.2)$$

where the last inequality follows from $l < 1$ and $n \geq k$.

Now we can show that if $A \subset \overline{B(\boldsymbol{t}_0, \frac{r}{2})}$ is null, then $\boldsymbol{f}(A)$ is null. Indeed, given $0 < \varepsilon < 1$ we take a sequence of k-cubes (I_p) contained in $\overline{B(\boldsymbol{t}_0, r)}$ such that $A \subset \cup_{p=1}^{\infty} I_p$ and

$$\sum_{p=1}^{\infty} m(I_p) < \frac{\varepsilon}{(2kH)^n}.$$

By (9.2) there exists a sequence of n-cubes (J_p) such that $\boldsymbol{f}(I_p) \subset J_p$ and $m(J_p) \leq (2kH)^n m(I_p)$ for all p. Hence (J_p) covers $\boldsymbol{f}(A)$ and

$$\sum_{p=1}^{\infty} m(J_p) \leq (2kH)^n \sum_{p=1}^{\infty} m(I_p) < \varepsilon.$$

Finally, the open set U can be covered by a sequence B_q of closed balls, $B_q = \overline{B(\boldsymbol{t}, \frac{r}{2})}$, with the property that $\overline{B(\boldsymbol{t}, r)}$ is contained in U (e.g., take balls centered at points of U with rational coordinates and with radii rational and small enough), and we can write

$$A = \cup_{q=1}^{\infty} A \cap B_q.$$

Then $\boldsymbol{f}(A)$ is a countable union of the null sets $\boldsymbol{f}(A \cap B_q)$, which implies that $\boldsymbol{f}(A)$ is null also. □

The next proposition shows that in order to integrate over an oriented regular k-surface with boundary, it is necessary to integrate over only the *proper* regular k-surface, that is, one can neglect the boundary in calculating the integral.

Proposition 9.1.1. *Let M be a subset of $\mathbb{R}^n$ that is an oriented regular k-surface with boundary having a finite atlas and let $\boldsymbol{\omega} : M \to \Lambda^k(\mathbb{R}^n)$ be a continuous differential form of degree k. Then*

$$\int_M \boldsymbol{\omega} = \int_{M\setminus\partial M} \boldsymbol{\omega}.$$

In fact, ∂M is a null set of $\mathbb{R}^n$.

Proof. Let

$$\{(\mathbb{A}_j, \boldsymbol{\varphi}_j)\}_{j=1}^m$$

be a finite atlas of M (compatible with the orientation) and $\{Y_j\}$ a partition of M such that $Y_j = \boldsymbol{\varphi}_j(\Delta_j)$, where Δ_j is a measurable subset of $\mathbb{A}_j$. Recall from Proposition 8.4.1 that

$$\{((\mathbb{A}_j)_-, \boldsymbol{\varphi}_j)\}_{j=1}^m$$

is a finite atlas of $M\setminus\partial M$ and the sets $\{Z_j := \boldsymbol{\varphi}_j(\Delta_j \cap \mathbb{H}^k_-)\}$ form a partition of $M\setminus\partial M$. Put

$$(\boldsymbol{\varphi}_j^*\,\boldsymbol{\omega})(\boldsymbol{t}) = f_j(\boldsymbol{t})\cdot \mathrm{d}y_1\wedge\cdots\wedge\mathrm{d}y_k.$$

Since the hyperplanes in $\mathbb{R}^k$ have zero Lebesgue measure and the set Δ_j is contained in the half-space $\mathbb{H}^k$ (see Definition 8.1.1), we conclude that the characteristic functions of the sets Δ_j and $\Delta_j\cap\mathbb{H}^k_-$ coincide almost everywhere. Hence

$$\begin{aligned}\int_M \boldsymbol{\omega} &= \sum_{j=1}^m \int_{\boldsymbol{\varphi}_j(\Delta_j)} \boldsymbol{\omega} = \sum_{j=1}^m \int_{\Delta_j} f_j(\boldsymbol{t})\,\mathrm{d}\boldsymbol{t} = \sum_{j=1}^m \int_{\mathbb{R}^k} f_j(\boldsymbol{t})\chi_{\Delta_j}(\boldsymbol{t})\,\mathrm{d}\boldsymbol{t}\\ &= \sum_{j=1}^m \int_{\mathbb{R}^k} f_j(\boldsymbol{t})\chi_{\Delta_j\cap\mathbb{H}^k_-}(\boldsymbol{t})\,\mathrm{d}\boldsymbol{t} = \sum_{j=1}^m \int_{\Delta_j\cap\mathbb{H}^k_-} f_j(\boldsymbol{t})\,\mathrm{d}\boldsymbol{t} = \sum_{j=1}^m \int_{Z_j} \boldsymbol{\omega}\\ &= \int_{M\setminus\partial M} \boldsymbol{\omega}.\end{aligned}$$

Since

$$\partial M = \cup_{j=1}^m \boldsymbol{\varphi}_j(\{0\}\times\hat{\mathbb{A}}_j),$$

by Lemma 9.1.1, ∂M is a finite union of null sets; hence it is a null set in $\mathbb{R}^n$. □

9.2 Partitions of Unity

Let M be a compact regular k-surface with boundary in $\mathbb{R}^n$ and $\boldsymbol{\omega}$ a differential form of degree k defined on a neighborhood of M. We aim to decompose $\boldsymbol{\omega}$ as a finite sum of differential forms, each one having support in a coordinate neighborhood. This will allow us to simplify the proof of Stokes's theorem. In this section, we provide the requisite tools for forming such a decomposition.

Consider the function $f_0 : \mathbb{R} \to \mathbb{R}$ of class C^∞ defined by

$$f_0(t) := \begin{cases} \mathrm{e}^{-\frac{1}{t}}, & t > 0, \\ 0, & t \leq 0. \end{cases}$$

As in Lemma 2.1.1, one can prove that f_0 has derivatives of every order at every point and that

$$f_0^{(n)}(t) = 0$$

for every $n \in \mathbb{N}$ and $t \leq 0$. It follows that

$$g_0(t) := f_0(1+t)f_0(1-t)$$

is a function of class C^∞ on the real line that vanishes whenever $|t| \geq 1$ and is strictly positive for $t \in (-1,1)$ (Fig. 9.1).

Proposition 9.2.1. *Let $\boldsymbol{x}_0 \in \mathbb{R}^n$ and $r > 0$ be given. There is a function $h : \mathbb{R}^n \to \mathbb{R}$ of class C^∞ such that $h(\boldsymbol{x}) > 0$ if $\boldsymbol{x} \in B(\boldsymbol{x}_0, r)$ and $h(\boldsymbol{x}) = 0$ if $\boldsymbol{x} \notin B(\boldsymbol{x}_0, r)$.*

Proof. Put $h(\boldsymbol{x}) := g_0\left(\frac{\|\boldsymbol{x}-\boldsymbol{x}_0\|^2}{r^2}\right)$. Since

$$\varphi(\boldsymbol{x}) := \| \boldsymbol{x} - \boldsymbol{x}_0 \|^2 = \sum_{j=1}^{n} (x_j - x_{0j})^2$$

is a function of class C^∞ on $\mathbb{R}^n$, we deduce that h is the composition of two C^∞ functions. □

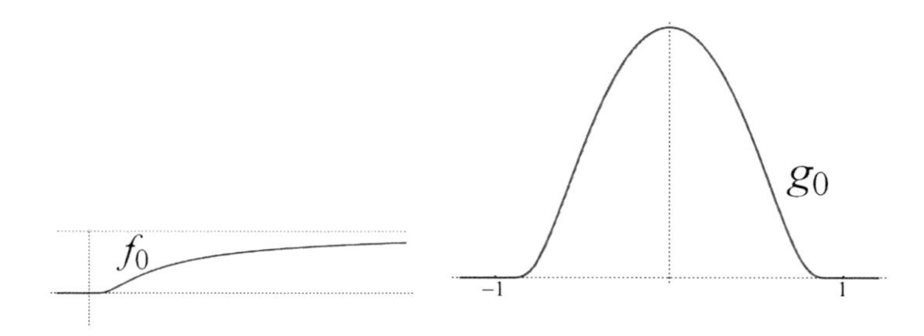

Fig. 9.1 The functions f_0 and g_0

Definition 9.2.1. The *support of a function* $f : D \subset \mathbb{R}^n \to \mathbb{R}$ is the closure in D of the set of points where f does not vanish, that is,

$$\operatorname{supp} f := \overline{\{\boldsymbol{x} \in D \,:\, f(\boldsymbol{x}) \neq 0\}}^{D} = \overline{\{\boldsymbol{x} \in D \,:\, f(\boldsymbol{x}) \neq 0\}}^{\mathbb{R}^n} \cap D.$$

In particular, if $D = \mathbb{R}^n$, then $\operatorname{supp} f$ is the closure of the set $f^{-1}(\mathbb{R} \setminus \{0\})$.

The complement in D of the support of f is the largest open set, relative to D, in which the function is identically zero.

Lemma 9.2.1. (a) *Let $V \subset \mathbb{R}^n$ be an open set, $K \subset V$ a closed subset, and $h \in C^p(V)$ a function of class C^p on V, $1 \leq p \leq \infty$, such that $h(\boldsymbol{x}) = 0$ if $\boldsymbol{x} \in V \setminus K$. Then*

$$f(\boldsymbol{x}) := \begin{cases} h(\boldsymbol{x}), & \boldsymbol{x} \in V, \\ 0, & \boldsymbol{x} \in \mathbb{R}^n \setminus K, \end{cases}$$

is well defined and is of class C^p on $\mathbb{R}^n$.

(b) *Let $V, W \subset \mathbb{R}^n$ be open sets such that $\overline{W} \subset V$ and suppose that $g, h : \mathbb{R}^n \to \mathbb{R}$ are functions of class C^p, $1 \leq p \leq \infty$, such that $h(\boldsymbol{x}) = 0$ whenever $\boldsymbol{x} \in \mathbb{R}^n \setminus W$ and $h(\boldsymbol{x}) + g(\boldsymbol{x}) > 0$ whenever $\boldsymbol{x} \in V$. Then, the function*

$$f(\boldsymbol{x}) := \begin{cases} \dfrac{h(\boldsymbol{x})}{h(\boldsymbol{x}) + g(\boldsymbol{x})}, & \boldsymbol{x} \in V, \\ 0, & \boldsymbol{x} \in \mathbb{R}^n \setminus W, \end{cases}$$

is well defined and of class C^p on $\mathbb{R}^n$.

Proof. (a) Since $h(\boldsymbol{x}) = 0$ on the open set $V \cap (\mathbb{R}^n \setminus K)$ and $V \cup (\mathbb{R}^n \setminus K) = \mathbb{R}^n$, we see that $f : \mathbb{R}^n \to \mathbb{R}$ is well defined. The function is of class C^p on V because it coincides with h, and it is also of class C^p on the open set $\mathbb{R}^n \setminus K$, since it is identically zero there. From $V \cup (\mathbb{R}^n \setminus K) = \mathbb{R}^n$ we conclude that f is a function of class C^p on $\mathbb{R}^n$.

(b) Since $h(\boldsymbol{x}) = 0$ on $V \cap (\mathbb{R}^n \setminus W)$ and $V \cup (\mathbb{R}^n \setminus W) = \mathbb{R}^n$, we have that $f : \mathbb{R}^n \to \mathbb{R}$ is well defined. Also, f is of class C^p on the open set V because h and g are functions of class C^p whose sum does not vanish on V. In addition, f is of class C^p on the open set $\mathbb{R}^n \setminus \overline{W}$, since it is zero in that open set. Since the two open sets V and $\mathbb{R}^n \setminus \overline{W}$ together cover $\mathbb{R}^n$, we deduce that f is a function of class C^p on $\mathbb{R}^n$.

□

Theorem 9.2.1. *Let $K \subset \mathbb{R}^n$ be a compact set and $\{U_1, \ldots, U_m\}$ an open cover of K. Then one can find an open neighborhood U of K and functions $\pi_j : \mathbb{R}^n \to [0,1]$ $(1 \leq j \leq m)$ such that*

(1) each π_j is a function of class C^∞ on $\mathbb{R}^n$,
(2) the support of π_j is contained in U_j,
(3) $\sum_{j=1}^{m} \pi_j(\boldsymbol{x}) = 1$ for every $\boldsymbol{x} \in U$.

Proof. For every $\boldsymbol{x} \in K$ there exist $r_{\boldsymbol{x}} > 0$ and $1 \leq j \leq m$ such that $\overline{B(\boldsymbol{x}, r_{\boldsymbol{x}})} \subset U_j$. By compactness, we can find a finite number of points $\boldsymbol{x}_1, \ldots, \boldsymbol{x}_s \in K$ with the property that

$$K \subset \bigcup_{i=1}^{s} B(\boldsymbol{x}_i, r_{\boldsymbol{x}_i})$$

and every open set U_j $(1 \leq j \leq m)$ contains at least one ball $B(\boldsymbol{x}_i, r_{\boldsymbol{x}_i})$. Now put

$$W := \bigcup_{i=1}^{s} B(\boldsymbol{x}_i, r_{\boldsymbol{x}_i})$$

and for each $1 \leq i \leq s$ take a function

$$h_i : \mathbb{R}^n \to [0, +\infty)$$

of class C^∞ on $\mathbb{R}^n$ such that

$$h_i(\boldsymbol{x}) > 0 \text{ if } \boldsymbol{x} \in B(\boldsymbol{x}_i, r_{\boldsymbol{x}_i})$$

and

$$h_i(\boldsymbol{x}) = 0 \text{ if } \boldsymbol{x} \notin \overline{B(\boldsymbol{x}_i, r_{\boldsymbol{x}_i})}.$$

Then

$$\sum_{i=1}^{s} h_i(\boldsymbol{x}) > 0$$

for every $\boldsymbol{x} \in W$. Since W is a bounded set, its (topological) boundary is compact. Moreover, the points of K are contained in the interior of W, because W is open and $K \subset W$. Thus, the boundary of W is a compact set disjoint from K. By compactness, there are points $\boldsymbol{y}_1, \ldots, \boldsymbol{y}_l \in \partial^{\text{top}}(W)$ and radii $s_1, \ldots s_l > 0$ such that

$$\partial^{\text{top}}(W) \subset \bigcup_{k=1}^{l} B(\boldsymbol{y}_k, s_k)$$

and

$$\overline{B(\boldsymbol{y}_j, s_j)} \cap K = \varnothing.$$

For every $1 \leq k \leq l$ take a function

$$g_k : \mathbb{R}^n \to [0, +\infty)$$

of class C^∞ on $\mathbb{R}^n$ with

$$g_k(\boldsymbol{x}) > 0 \text{ if } \boldsymbol{x} \in B(\boldsymbol{y}_k, s_k)$$

and

$$g_k(\boldsymbol{x}) = 0 \text{ if } \boldsymbol{x} \notin B(\boldsymbol{y}_k, s_k).$$

Now define

$$V := W \cup \left(\bigcup_{k=1}^{l} B(\mathbf{y}_k, s_k) \right)$$

and observe that

$$\overline{W} = W \cup \partial^{\text{top}}(W) \subset V.$$

Define, for every $1 \leq i \leq s$,

$$f_i : \mathbb{R}^n \to [0,1]$$

by

$$f_i(\mathbf{x}) := \begin{cases} \dfrac{h_i(\mathbf{x})}{\sum_{j=1}^{s} h_j(\mathbf{x}) + \sum_{k=1}^{l} g_k(\mathbf{x})}, & \mathbf{x} \in V \\ 0, & \mathbf{x} \notin W. \end{cases}$$

From

$$\sum_{i=1}^{s} h_i(\mathbf{x}) + \sum_{k=1}^{l} g_k(\mathbf{x}) > 0$$

whenever $\mathbf{x} \in V$ and Lemma 9.2.1 we conclude that the function f_i is well defined and of class C^∞ on all of $\mathbb{R}^n$. Now take an open neighborhood U of K such that $U \subset W$ and

$$U \cap \left(\bigcup_{k=1}^{l} \overline{B(\mathbf{y}_k, s_k)} \right) = \varnothing.$$

Then

$$\sum_{i=1}^{s} h_i(\mathbf{x}) > 0 \quad \text{and} \quad \sum_{k=1}^{l} g_k(\mathbf{x}) = 0,$$

whenever $\mathbf{x} \in U$. Consequently,

$$\sum_{i=1}^{s} f_i(\mathbf{x}) = 1$$

for every $\mathbf{x} \in U$. Moreover, the support of f_i is contained in $\overline{B(\mathbf{x}_i, r_{\mathbf{x}_i})}$, which is contained in some open set U_j. To finish, it suffices to select a natural number $1 \leq r(i) \leq m$ for each $1 \leq i \leq s$ such that the support of f_i is contained in $U_{r(i)}$ and $\{r(1), \ldots, r(s)\} = \{1, \ldots, m\}$. Now for every $1 \leq j \leq m$, put

$$I_j := \{i \, : \, r(i) = j\},$$

which is a nonempty set, and define

$$\pi_j := \sum_{i \in I_j} f_i.$$

By construction, π_j has the desired properties. □

The previous theorem says that the functions $\{\pi_j\}$ form a *partition of unity* over a neighborhood of K and subordinated to the cover $\{U_j\}$.

Definition 9.2.2. Let M be a regular k-surface with boundary and

$$\boldsymbol{\omega} : M \to \Lambda^k(\mathbb{R}^n)$$

a continuous differential form of degree k. The *support* of $\boldsymbol{\omega}$ *relative* to M is

$$\operatorname{supp}_M \boldsymbol{\omega} := \overline{\{\boldsymbol{x} \in M \, : \boldsymbol{\omega}(\boldsymbol{x}) \neq \boldsymbol{0}\}}^M.$$

This slightly modified notion of support is necessary if one wants to prove Stokes's theorem for abstract differentiable manifolds (Theorem 9.3.1).

As a result of Theorem 9.2.1 on the existence of partitions of unity, we have the following corollary, which we will use in the proof of the general Stokes's theorem.

Corollary 9.2.1. *Let M be a compact regular k-surface with boundary and*

$$\boldsymbol{\omega} : U \to \Lambda^r(\mathbb{R}^n)$$

a differential form of degree r and class C^1 on an open neighborhood U of M. Then there is an open neighborhood V of M, $V \subset U$, for which we can form a decomposition

$$\boldsymbol{\omega} = \sum_{j=1}^{m} \boldsymbol{\omega}_j$$

such that

$$\boldsymbol{\omega}_j : V \to \Lambda^r(\mathbb{R}^n)$$

is a differential form of degree r and class C^1 whose support (relative to M) is contained in some set $\boldsymbol{\varphi}_j(\Delta_j)$, where Δ_j is contained in a compact subset of $\mathbb{A}_j$ and $(\mathbb{A}_j, \boldsymbol{\varphi}_j)$ is a coordinate system of M. Moreover, the support of $\mathrm{d}\boldsymbol{\omega}_j$ (relative to M) is contained in $\boldsymbol{\varphi}_j(\Delta_j)$.

Proof. By compactness, one can find a finite family $\{(\mathbb{A}_j, \boldsymbol{\varphi}_j)\}_{j=1}^m$ of coordinate systems and bounded open sets $\mathbb{B}_j$ (for the relative topology of $\mathbb{H}^k$) such that

$$\mathbb{B}_j \subset \overline{\mathbb{B}_j} \subset \mathbb{A}_j \subset \mathbb{H}^k$$

and

$$\{(\mathbb{B}_j, \boldsymbol{\varphi}_j)\}_{j=1}^m$$

form an atlas of M. Let U_j be an open set in $\mathbb{R}^n$, contained in U, such that

$$U_j \cap M = \boldsymbol{\varphi}_j(\mathbb{B}_j).$$

Then $\{U_j\}_{j=1}^m$ is an open cover of the compact set M. By Theorem 9.2.1, there exist an open neighborhood V of M, $V \subset U$, and functions $\pi_j : \mathbb{R}^n \to [0,1]$ $(1 \leq j \leq m)$ of class C^∞ with the property that the support of π_j is contained in U_j and

$$\sum_{j=1}^{m} \pi_j(\boldsymbol{x}) = 1$$

for every $\boldsymbol{x} \in V$. Finally, put

$$\boldsymbol{\omega}_j := \pi_j \cdot \boldsymbol{\omega} : V \to \Lambda^r(\mathbb{R}^n),$$

so that

$$\boldsymbol{\omega} = \sum_{j=1}^{m} \boldsymbol{\omega}_j.$$

Moreover,

$$\operatorname{supp}_M \boldsymbol{\omega}_j \subset \overline{\{\boldsymbol{x} \in \mathbb{R}^n \ : \ \pi_j(\boldsymbol{x}) \neq 0\}} \cap M \subset U_j \cap M = \boldsymbol{\varphi}_j(\mathbb{B}_j).$$

But since the support in $\mathbb{R}^n$ of π_j is contained in the open set U_j, if $\mathrm{d}\boldsymbol{\omega}_j(x) \neq 0$, then x must to belong to U_j. Hence

$$\operatorname{supp}_M \mathrm{d}\boldsymbol{\omega}_j \subset \overline{\{\boldsymbol{x} \in \mathbb{R}^n \ : \ \pi_j(\boldsymbol{x}) \neq 0\}} \cap M \subset U_j \cap M = \boldsymbol{\varphi}_j(\mathbb{B}_j).$$

□

We remark that if additionally M is assumed to be oriented then, by applying Lemma 5.2.2. if necessary, the coordinate system obtained in Corollary 9.2.1 can be chosen compatible with the orientation.

9.3 Stokes's Theorem

In this section, we will prove our general formulation of Stokes's theorem for compact regular k-surfaces with boundary in $\mathbb{R}^n$. Subsequently, in Sect. 9.4, we will discuss the classical theorems of vector analysis (Green's theorem, divergence theorem, and the classical Stokes's theorem), which can be seen as particular cases of the general result Theorem 9.3.1. We finish with a discussion on the possibility of extending the divergence theorem to some regions of $\mathbb{R}^3$ with topological boundary the union of a regular surface with some *vertices* and *edges*.

As in Sect. 8.2, we will denote by $\boldsymbol{x} = (x_1,\dots,x_n)$ the points of $\mathbb{R}^n$, by $\boldsymbol{t} = (t_1,\dots,t_k)$ the points of $\mathbb{R}^k$, and by $\boldsymbol{s} = (s_1,\dots,s_{k-1})$ the points of $\mathbb{R}^{k-1}$. If $\boldsymbol{s} \in \mathbb{R}^{k-1}$, then $\boldsymbol{t} := (0,\boldsymbol{s}) \in \mathbb{R}^k$ is the point with coordinates

$$t_1 = 0\ ,\ t_2 = s_1\ ,\ \dots,\ t_k = s_{k-1}.$$

Also, according to the notation used in the proof of Theorems 8.2.1 and 8.4.1, for a given coordinate system $(\mathbb{A},\boldsymbol{\varphi})$ of a regular k-surface with boundary M we put

$$\widehat{\mathbb{A}} := \{\boldsymbol{s} = (s_1,\dots,s_{k-1}) \in \mathbb{R}^{k-1}\ :\ (0,\boldsymbol{s}) \in \mathbb{A}\},$$

which is an open set in $\mathbb{R}^{k-1}$, and

$$\widehat{\boldsymbol{\varphi}} : \widehat{\mathbb{A}} \subset \mathbb{R}^{k-1} \to \mathbb{R}^n\ ,\quad \widehat{\boldsymbol{\varphi}}(\boldsymbol{s}) := \boldsymbol{\varphi}(0,\boldsymbol{s}).$$

Then $(\widehat{\mathbb{A}},\widehat{\boldsymbol{\varphi}})$ is a coordinate system of ∂M (see Theorem 8.2.1). In the following lemmas we make use again of the notation introduced in Definition 6.1.3.

Remark 9.3.1. Let $\boldsymbol{\omega}$ be a differential form of degree $k-1$ on an open neighborhood U of M, a regular k-surface with boundary in $\mathbb{R}^n$ of class C^p, and let $(\mathbb{A},\boldsymbol{\varphi})$ be a coordinate system of M. We claim that we can always take $\boldsymbol{g} : W \to U$ of class C^p on W an open subset of $\mathbb{R}^k$ such that $W \cap \mathbb{H}^k = \mathbb{A}$ and $\boldsymbol{g}(\boldsymbol{t}) = \boldsymbol{\varphi}(\boldsymbol{t})$ for every $\boldsymbol{t} \in \mathbb{A}$. Indeed, consider $A \subset \mathbb{R}^k$ open in $\mathbb{R}^k$ such that $A \cap \mathbb{H}^k = \mathbb{A}$ and $\boldsymbol{g} : A \to \mathbb{R}^n$ a mapping of class C^p on A such that $\boldsymbol{g}(\boldsymbol{t}) = \varphi(t)$ whenever $\boldsymbol{t} \in \mathbb{A}$. By continuity, $\boldsymbol{g}^{-1}(U)$ is an open subset of $\mathbb{R}^k$ such that $\mathbb{A} \subset \boldsymbol{g}^{-1}(U) \subset A$. Thus, if we define $W := \boldsymbol{g}^{-1}(U)$, we are going to have $\boldsymbol{g} : W \to U$ of class C^p on W, and clearly $W \cap \mathbb{H}^k = \mathbb{A}$.

In the next lemma and also in Stokes's theorem (Theorem 9.3.1), every time that we apply properties of the pullback $\boldsymbol{\varphi}^*\boldsymbol{\omega}$ or $\boldsymbol{\varphi}^*\mathrm{d}\boldsymbol{\omega}$, we are actually applying them to $\boldsymbol{g}^*\boldsymbol{\omega}$ or $\boldsymbol{g}^*\mathrm{d}\boldsymbol{\omega}$ on the open set W just described, and only later do we take the restriction to the corresponding subset of $\mathbb{A}$.

Lemma 9.3.1. *Let $\boldsymbol{\omega}$ be a differential form of degree $k-1$ and class C^1 on an open neighborhood of a regular k-surface with boundary M in $\mathbb{R}^n$. Let $(\mathbb{A},\boldsymbol{\varphi})$ be a coordinate system of M. Writing the differential form $\boldsymbol{\varphi}^*\boldsymbol{\omega}$ of degree $k-1$ with respect to the basis $\{\mathrm{d}t_1 \wedge\cdots\wedge \widehat{\mathrm{d}t_j} \wedge\cdots\wedge \mathrm{d}t_k : j = 1,\dots,k\}$ at each point $(t_1,\dots,t_k) \in \mathbb{A}$ as*

$$\boldsymbol{\varphi}^*\boldsymbol{\omega}(t_1,\dots,t_k) = \sum_{j=1}^{k} f_j(t_1,\dots,t_k)\cdot \mathrm{d}t_1 \wedge\cdots\wedge \widehat{\mathrm{d}t_j} \wedge\cdots\wedge \mathrm{d}t_k,$$

we have that

$$(\widehat{\boldsymbol{\varphi}}^* \boldsymbol{\omega})(\boldsymbol{s}) = f_1(0, s_1, \ldots, s_{k-1}) \cdot \mathrm{d}s_1 \wedge \cdots \wedge \mathrm{d}s_{k-1},$$

for all $\boldsymbol{s} = (s_1, \ldots, s_{k-1}) \in \widehat{\mathbb{A}}$.

Proof. Note that $\widehat{\boldsymbol{\varphi}} = \boldsymbol{\varphi} \circ \boldsymbol{J}$, where $\boldsymbol{J}$ is the inclusion

$$\boldsymbol{J} : \widehat{\mathbb{A}} \to \mathbb{A}, \quad \boldsymbol{J}(\boldsymbol{s}) := (0, \boldsymbol{s}).$$

Making the *change of variables*

$$t_1 = 0, t_2 = s_1, \ldots, t_k = s_{k-1},$$

we deduce that

$$\boldsymbol{J}^*(\mathrm{d}t_j) = \begin{cases} 0, & j = 1, \\ \mathrm{d}s_{j-1}, & j > 1, \end{cases}$$

and hence (see Proposition 6.4.1(1) and Remark 9.3.1)

$$\widehat{\boldsymbol{\varphi}}^* \boldsymbol{\omega} = \boldsymbol{J}^*(\boldsymbol{\varphi}^* \boldsymbol{\omega})$$

has the representation

$$\begin{aligned}(\widehat{\boldsymbol{\varphi}}^* \boldsymbol{\omega})(\boldsymbol{s}) &= f_1(0, s_1, \ldots, s_{k-1}) \cdot \boldsymbol{J}^*(\mathrm{d}t_2) \wedge \cdots \wedge \boldsymbol{J}^*(\mathrm{d}t_k) \\ &\quad + \sum_{j=2}^{k} f_j(0, s_1, \ldots, s_{k-1}) \cdot \boldsymbol{J}^*(\mathrm{d}t_1) \wedge \cdots \wedge \widehat{\boldsymbol{J}^*(\mathrm{d}t_j)} \wedge \cdots \wedge \boldsymbol{J}^*(\mathrm{d}t_k) \\ &= f_1(0, s_1, \ldots, s_{k-1}) \cdot \mathrm{d}s_1 \wedge \cdots \wedge \mathrm{d}s_{k-1}.\end{aligned}$$

□

Lemma 9.3.2. *The exterior differential of the differential form* $\boldsymbol{\nu}$ *of degree* $(k-1)$ *and class* C^1 *defined on an open subset of* $\mathbb{R}^k$ *by*

$$\sum_{j=1}^{k} f_j(t_1, \ldots, t_k) \cdot \mathrm{d}t_1 \wedge \cdots \wedge \widehat{\mathrm{d}t_j} \wedge \cdots \wedge \mathrm{d}t_k$$

is given by

$$\mathrm{d}\boldsymbol{\nu} = \left(\sum_{j=1}^{k} (-1)^{j-1} \frac{\partial f_j}{\partial t_j} \right) \cdot \mathrm{d}t_1 \wedge \cdots \wedge \mathrm{d}t_k.$$

Proof. According to Definition 6.3.1,

$$\mathrm{d}\boldsymbol{\nu}(t_1, \ldots, t_k) = \sum_{j=1}^{k} \mathrm{d}f_j(t_1, \ldots, t_k) \wedge \mathrm{d}t_1 \wedge \cdots \wedge \widehat{\mathrm{d}t_j} \wedge \cdots \wedge \mathrm{d}t_k.$$

Moreover, by Example 2.3.1,

$$\mathrm{d}f_j = \sum_{l=1}^{k} \frac{\partial f_j}{\partial t_l}\,\mathrm{d}t_l.$$

On the other hand, by Propositions 6.1.1 and 6.2.1 we have

$$\mathrm{d}t_l \wedge \mathrm{d}t_1 \wedge \cdots \wedge \widehat{\mathrm{d}t_j} \wedge \cdots \wedge \mathrm{d}t_k = \begin{cases} 0 & \text{if } l \neq j, \\ (-1)^{j-1}\mathrm{d}t_1 \wedge \cdots \wedge \mathrm{d}t_k & \text{if } l = j. \end{cases}$$

The conclusion easily follows. □

The next lemma will be useful in the proof of Theorem 9.3.1 (Case II).

Lemma 9.3.3. *Let $\mathbb{A}$ be an open set in $\mathbb{H}^k$, $L \subset \mathbb{A}$ a compact subset, and $f : \mathbb{A} \subset \mathbb{H}^k \to \mathbb{R}$ a function of class C^1 on $\mathbb{A}$ such that $f(\boldsymbol{t}) = 0$ for every $\boldsymbol{t} \in \mathbb{A} \setminus L$. Then there is a function $h : \mathbb{R}^k \to \mathbb{R}$ of class C^1 on $\mathbb{R}^k$ with compact support such that $h_{|\mathbb{A}} = f$ and such that $h(\boldsymbol{t}) = 0$ and $\frac{\partial h}{\partial t_i}(\boldsymbol{t}) = 0$, for every $\boldsymbol{t} \in \mathbb{H}^k \setminus L$ and $i = 1, \ldots, k$.*

Proof. Let $g : A \subset \mathbb{R}^k \to \mathbb{R}$ be a mapping of class C^1 on the open set $A \subset \mathbb{R}^k$ such that $A \cap \mathbb{H}^k = \mathbb{A}$ and $g_{|\mathbb{A}} = f$. For every $\boldsymbol{t} \in L$ there exists $r_{\boldsymbol{t}} > 0$ with $\overline{B(\boldsymbol{t}, r_{\boldsymbol{t}})} \subset A$. Since L is compact, we can select $\boldsymbol{t}_1, \ldots \boldsymbol{t}_m \in L$ such that $\{B(\boldsymbol{t}_1, r_{\boldsymbol{t}_1}), \ldots B(\boldsymbol{t}_m, r_{\boldsymbol{t}_m})\}$ is an open covering of L. According to Theorem 9.2.1, we can find an open neighborhood Z of L,

$$Z \subset W := \bigcup_{j=1}^{m} B(\boldsymbol{t}_j, r_{\boldsymbol{t}_j}) \subset \overline{W} \subset A,$$

and functions of class C^∞, $\pi_j : \mathbb{R}^k \to [0,1]$ $(1 \leq j \leq m)$, such that the support of π_j is contained in $B(\boldsymbol{t}_j, r_{\boldsymbol{t}_j})$, and $\sum_{j=1}^{m} \pi_j(\boldsymbol{x}) = 1$ for every $\boldsymbol{x} \in Z$. We now define

$$\tilde{h} := \left(\sum_{j=1}^{m} \pi_j \right) g : A \subset \mathbb{R}^k \to \mathbb{R}.$$

Then $\tilde{h}$ is a function of class C^1 on A whose support is contained in the compact set $\overline{W} \subset A$ and with the property that $\tilde{h}(\boldsymbol{t}) = g(\boldsymbol{t})$ for all $\boldsymbol{t} \in Z$. Moreover, by hypothesis, $g(\boldsymbol{t}) = f(\boldsymbol{t}) = 0$ for all $\boldsymbol{t} \in A \cap \mathbb{H}^k \setminus L = \mathbb{A} \setminus L$. Thus $\tilde{h}(\boldsymbol{t}) = 0$ for all $\boldsymbol{t} \in A \cap \mathbb{H}^k \setminus L$.

From Lemma 9.2.1(a), the function

$$h(\boldsymbol{t}) := \begin{cases} \tilde{h}(\boldsymbol{t}), & \boldsymbol{t} \in A, \\ 0, & \boldsymbol{t} \in \mathbb{R}^k \setminus \overline{W}, \end{cases}$$

is well defined and is of class C^1 on $\mathbb{R}^k$. In addition, $h_{|\mathbb{A}} = f$, and $h(\boldsymbol{t}) = 0$ for every $\boldsymbol{t} \in \mathbb{H}^k \setminus L$.

With respect to the partial derivatives of h, one must take a little extra care. Since L is a compact set, it is closed, and hence $\mathbb{H}^k \setminus L$ is an open set with respect to the relative topology of $\mathbb{H}^k$. An analogous discussion to that given in (8.1)–(8.3) shows that to calculate any partial derivative of h at any point $\boldsymbol{t}$ belonging to $\mathbb{H}^k \setminus L$, we need to take into account the values of h only on that set. More precisely,

$$\frac{\partial h}{\partial t_i}(\boldsymbol{t}) = \lim_{\substack{s\to 0\\ \boldsymbol{t}+s\boldsymbol{e}_i\in\mathbb{H}^k\setminus L}} \frac{h(\boldsymbol{t}+s\boldsymbol{e}_i)-h(\boldsymbol{t})}{s}$$

$$= \lim_{\substack{s\to 0\\ \boldsymbol{t}+s\boldsymbol{e}_i\in\mathbb{H}^k\setminus L}} \frac{0}{s} = 0,$$

for every $\boldsymbol{t} \in \mathbb{H}^k \setminus L$, and every $i = 1, \ldots, k$.

□

The reader will observe in passing that we have shown, under the hypothesis of the above proposition, that $\frac{\partial f}{\partial t_i}(\boldsymbol{t}) = 0$ for every $\boldsymbol{t} \in \mathbb{A} \setminus L$ and $i = 1, \ldots, k$.

In our main result, which now follows, we assume that ∂M has the orientation induced by the orientation of M, as obtained in Theorem 8.4.1.

Theorem 9.3.1 (Stokes's theorem). *Let $M \subset \mathbb{R}^n$ be a compact orientable regular k-surface with boundary of class C^2 and let $\boldsymbol{\omega}$ be a differential form of degree $k-1$ and class C^1 on an open neighborhood U of M. Then*

$$\int_{\partial M} \boldsymbol{\omega} = \int_M \mathrm{d}\boldsymbol{\omega}.$$

Proof. Our proof is based on that of Do Carmo [6]. Let $K := \operatorname{supp}_M \boldsymbol{\omega}$, which, as a closed subset of the compact set M, is itself a compact subset of $\mathbb{R}^n$. To begin, let us suppose that there is a coordinate system $(\mathbb{A}, \boldsymbol{\varphi})$ of M of class C^2 such that

$$K = \operatorname{supp}_M \boldsymbol{\omega} \subset \boldsymbol{\varphi}(\mathbb{A}) \quad \text{and} \quad \operatorname{supp}_M \mathrm{d}\boldsymbol{\omega} \subset \boldsymbol{\varphi}(\mathbb{A}).$$

Since $\boldsymbol{\varphi}$ is of class C^2, we see that $\boldsymbol{\varphi}^*\boldsymbol{\omega}$ is a differential form of degree $k-1$ and class C^1 on an open subset of $\mathbb{H}^k$ (see for instance Corollary 6.4.2), and we can represent it as

$$\boldsymbol{\varphi}^*\boldsymbol{\omega}(t_1, \ldots, t_k) = \sum_{j=1}^{k} f_j(t_1, \ldots, t_k) \cdot \mathrm{d}t_1 \wedge \cdots \wedge \widehat{\mathrm{d}t_j} \wedge \cdots \wedge \mathrm{d}t_k,$$

for all $\boldsymbol{t} = (t_1, \ldots, t_k) \in \mathbb{A}$, where f_j is a function of class C^1 on $\mathbb{A} \subset \mathbb{H}^k$. We note that $\boldsymbol{\varphi}^{-1}(K)$ is a compact subset of $\mathbb{A}$ (because $\boldsymbol{\varphi}$ is a homeomorphism) with the property that if $\boldsymbol{t} \notin \boldsymbol{\varphi}^{-1}(K)$, then $(\boldsymbol{\varphi}^*\boldsymbol{\omega})(\boldsymbol{t}) = \boldsymbol{0}$. That is,

$$\boldsymbol{t} \notin \boldsymbol{\varphi}^{-1}(K) \Rightarrow f_j(\boldsymbol{t}) = 0,$$

for every $1 \leq j \leq k$.

By Lemma 9.3.2, the differential is given by

$$\mathrm{d}(\boldsymbol{\varphi}^*\boldsymbol{\omega})(\boldsymbol{t}) = \left(\sum_{j=1}^{k}(-1)^{j-1}\frac{\partial f_j}{\partial t_j}(\boldsymbol{t})\right)\cdot \mathrm{d}t_1 \wedge \cdots \wedge \mathrm{d}t_k$$

for all $\boldsymbol{t} = (t_1,\ldots,t_k) \in \mathbb{A}$. On the other hand, since $\boldsymbol{\varphi}$ is of class C^2 on $\mathbb{A}$ and $\boldsymbol{\omega}$ is of class C^1 on U, by Remark 9.3.1 and Proposition 6.4.2, we have $\boldsymbol{\varphi}^*(\mathrm{d}\boldsymbol{\omega})(\boldsymbol{t}) = \mathrm{d}(\boldsymbol{\varphi}^*\boldsymbol{\omega})(\boldsymbol{t})$ for all $\boldsymbol{t} \in \mathbb{A}$. Consequently,

$$\boldsymbol{\varphi}^*(\mathrm{d}\boldsymbol{\omega})(\boldsymbol{t}) = \left(\sum_{j=1}^{k}(-1)^{j-1}\frac{\partial f_j}{\partial t_j}(\boldsymbol{t})\right)\cdot \mathrm{d}t_1 \wedge \cdots \wedge \mathrm{d}t_k. \tag{9.3}$$

If we apply Lemma 9.3.3, we can extend each f_j to a function $\tilde{f}_j : \mathbb{R}^k \to \mathbb{R}$ of class C^1 on all of $\mathbb{R}^k$ such that $\tilde{f}_j(\boldsymbol{t}) = 0$ and $\frac{\partial \tilde{f}_j}{\partial t_i}(\boldsymbol{t}) = 0$, for every $\boldsymbol{t} \in \mathbb{H}^k \setminus \boldsymbol{\varphi}^{-1}(K)$ and $i, j = 1,\ldots,k$. In particular,

$$\frac{\partial f_j}{\partial t_i}(\boldsymbol{t}) = 0 \tag{9.4}$$

for every $\boldsymbol{t} \in \mathbb{A} \setminus \boldsymbol{\varphi}^{-1}(K)$ and $i, j = 1,\ldots,k$. To ease notation, from now on we will identify $\tilde{f}_j$ and f_j. By hypothesis, $\mathrm{supp}_M\, \mathrm{d}\boldsymbol{\omega} \subset \boldsymbol{\varphi}(\mathbb{A})$. Thus

$$\int_M \mathrm{d}\boldsymbol{\omega} = \int_{(\mathbb{A},\boldsymbol{\varphi})} \mathrm{d}\boldsymbol{\omega},$$

and by (9.1), we have

$$\int_M \mathrm{d}\boldsymbol{\omega} = \int_{\mathbb{A}} h(t_1,\ldots,t_k)\, \mathrm{d}(t_1,\ldots,t_k),$$

where $\boldsymbol{\varphi}^*(\mathrm{d}\boldsymbol{\omega})(t_1,\ldots,t_k) = h(t_1,\ldots,t_k)\mathrm{d}t_1 \wedge \cdots \wedge \mathrm{d}t_k$. From (9.3), we get

$$\int_M \mathrm{d}\boldsymbol{\omega} = \int_{\mathbb{A}} \left(\sum_{j=1}^{k}(-1)^{j-1}\frac{\partial f_j}{\partial t_j}\right)(\boldsymbol{t})\, \mathrm{d}(t_1,\ldots,t_k).$$

But by (9.4) and the fact that all the above partial derivatives of (the extensions of) f_j vanish on $\mathbb{H}^k \setminus \boldsymbol{\varphi}^{-1}(K)$, we deduce

$$\int_M \mathrm{d}\boldsymbol{\omega} = \int_{Q} \left(\sum_{j=1}^{k}(-1)^{j-1}\frac{\partial f_j}{\partial t_j}\right)(\boldsymbol{t})\, \mathrm{d}(t_1,\ldots,t_k), \tag{9.5}$$

for any measurable set Q such that $\boldsymbol{\varphi}^{-1}(K) \subset Q \subset \mathbb{H}^k$.

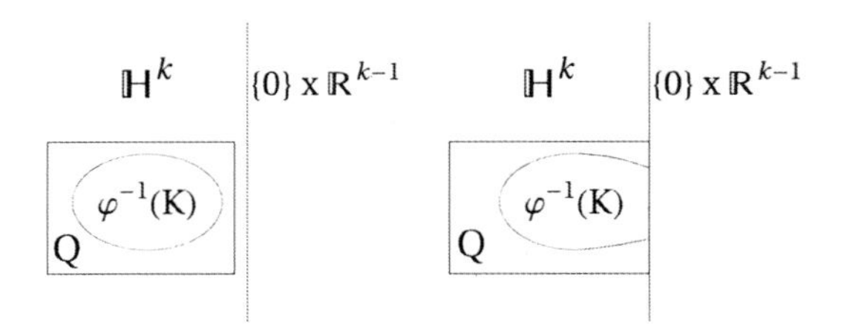

Fig. 9.2 Proof of Theorem 9.3.1

We will distinguish two cases (Fig. 9.2).

Case I: $K \cap \partial M = \varnothing$.

In this case, $\boldsymbol{\omega}$ vanishes on ∂M, and so

$$\int_{\partial M} \boldsymbol{\omega} = 0.$$

We will show that $\int_M \mathrm{d}\boldsymbol{\omega}$ also vanishes. Indeed, by hypothesis, $\boldsymbol{\varphi}^{-1}(K)$ is disjoint from $\{0\} \times \mathbb{R}^{k-1}$, and consequently, there is a k-rectangle $Q \subset \mathbb{H}^k$ defined by

$$Q := \{\boldsymbol{t} \in \mathbb{R}^k \ : t_j^0 \leq t_j \leq t_j^1 \ ; \ 1 \leq j \leq k\}$$

with $t_1^1 < 0$ and such that $\boldsymbol{\varphi}^{-1}(K)$ is contained in the interior of Q. We can apply Fubini's theorem in (9.5) to obtain

$$\begin{aligned}\int_Q \frac{\partial f_j}{\partial t_j}(\boldsymbol{t}) \ \mathrm{d}(t_1,\ldots,t_k) &= \int_{Q_j} \left(\int_{t_j^0}^{t_j^1} \frac{\partial f_j}{\partial t_j}(\boldsymbol{t}) \ \mathrm{d}t_j \right) \mathrm{d}\widehat{\boldsymbol{t}_j} \\ &= \int_{Q_j} (f_j(t_1,\ldots,t_{j-1},t_j^1,t_{j+1},\ldots,t_k) \\ &\qquad -f_j(t_1,\ldots,t_{j-1},t_j^0,t_{j+1},\ldots,t_k)) \ \mathrm{d}\widehat{\boldsymbol{t}_j},\end{aligned}$$

where $\mathrm{d}\widehat{\boldsymbol{t}_j} = \mathrm{d}(t_1,\ldots,t_{j-1},t_{j+1},\ldots,t_n)$ and

$$Q_j := \prod_{i \neq j} [t_i^0, t_i^1] \subset \mathbb{R}^{k-1}.$$

Since the points

$$(t_1,\ldots,t_{j-1},t_j^0,t_{j+1},\ldots,t_k)$$

and

$$(t_1,\ldots,t_{j-1},t_j^1,t_{j+1},\ldots,t_k)$$

are located in the boundary of Q, we deduce that the function f_j vanishes at each of them, and thus the previous integral is zero. It follows that

$$\int_M \mathrm{d}\boldsymbol{\omega} = 0.$$

Case II: $K \cap \partial M \neq \varnothing$.

Define Q as in Case I but this time with $t_1^1 = 0$ and $\boldsymbol{\varphi}_{-1}(k)$ is contained in the interior of Q in $\mathbb{H}^k$ i.e., in the union of interior of Q in $\mathbb{R}^k$ and $(\{0\} \times \mathbb{R}^{k-1}) \cap Q$. Note that Q is still contained in $\mathbb{H}^k$ but one of its faces is in $\{0\} \times \mathbb{R}^{k-1}$. For each $j \geq 2$, the two points

$$(t_1, \ldots, t_{j-1}, t_j^0, t_{j+1}, \ldots, t_k)$$

and

$$(t_1, \ldots, t_{j-1}, t_j^1, t_{j+1}, \ldots, t_k)$$

are outside the support of f_j whenever $t_1 < 0$. This means that

$$f_j(t_1, \ldots, t_{j-1}, t_j^0, t_{j+1}, \ldots, t_k) = f_j(t_1, \ldots, t_{j-1}, t_j^1, t_{j+1}, \ldots, t_k) = 0,$$

for all $\widehat{\boldsymbol{t}_j}$ in $Q_j \subset \mathbb{R}^{k-1}$ except may be on the set $Q_j \cap (\{0\} \times \mathbb{R}^{k-2})$, which is a null set in $\mathbb{R}^{k-1}$. Hence

$$\begin{aligned}
\int_Q \frac{\partial f_j}{\partial t_j}(\boldsymbol{t})\, \mathrm{d}(t_1, \ldots, t_k) &= \int_{Q_j} \left(\int_{t_j^0}^{t_j^1} \frac{\partial f_j}{\partial t_j}(\boldsymbol{t})\, \mathrm{d}t_j \right) \mathrm{d}\hat{\boldsymbol{t}_j} \\
&= \int_{Q_j} f_j(t_1, \ldots, t_{j-1}, t_j^1, t_{j+1}, \ldots, t_k)\, \mathrm{d}\hat{\boldsymbol{t}_j} \\
&\quad - \int_{Q_j} f_j(t_1, \ldots, t_{j-1}, t_j^0, t_{j+1}, \ldots, t_k)\, \mathrm{d}\hat{\boldsymbol{t}_j} \\
&= 0.
\end{aligned}$$

By (9.5) it follows that

$$\int_M \mathrm{d}\boldsymbol{\omega} = \int_Q \frac{\partial f_1}{\partial t_1}(\boldsymbol{t})\, \mathrm{d}(t_1, \ldots, t_k).$$

If we apply Fubini's theorem and use the fact that

$$f_1(t_1^0, t_2, \ldots, t_k) = 0,$$

then we can conclude, as above, that

$$\int_M \mathrm{d}\boldsymbol{\omega} = \int_{Q_1} f_1(0, t_2, \ldots, t_k)\, \mathrm{d}(t_2, \ldots, t_k).$$

It remains to calculate $\int_{\partial M} \boldsymbol{\omega}$. For this, observe that the support of $\boldsymbol{\omega}$ in ∂M is contained in the neighborhood coordinate $\widehat{\boldsymbol{\varphi}}(\widehat{\mathbb{A}})$. Moreover, from the definition of orientation of the boundary (see Theorem 8.4.1), we have that $(\widehat{\mathbb{A}}, \widehat{\boldsymbol{\varphi}})$ is a coordinate system of ∂M compatible with its orientation. Also, by Lemma 9.3.1,

$$(\widehat{\varphi}^*\omega)(s) = f_1(0, s_1, \ldots, s_{k-1}) \cdot \mathrm{d}s_1 \wedge \cdots \wedge \mathrm{d}s_{k-1}$$

for $s \in \widehat{\mathbb{A}}$. Hence

$$\int_{\partial M} \omega = \int_{\widehat{\mathbb{A}}} f_1(0, s_1, \ldots, s_{k-1})\, \mathrm{d}(s_1, \ldots, s_{k-1}),$$

and we conclude with the identity

$$\int_{\partial M} \omega = \int_M \mathrm{d}\omega.$$

In the general case (that is, when our initial supposition does not apply), Corollary 9.2.1 implies the existence of an open neighborhood V of M and differential forms $\omega_1, \ldots \omega_p$ of degree $k-1$ on V with $\omega = \sum_{j=1}^p \omega_j$ on V and coordinate systems $(\mathbb{A}_j, \varphi_j)$ such that $\mathrm{supp}_M \omega_j \subset \varphi_j(\mathbb{A}_j)$ and also $\mathrm{supp}_M \mathrm{d}\omega_j \subset \varphi_j(\mathbb{A}_j)$ for every $j = 1, \ldots, p$. It follows that

$$\int_{\partial M} \omega = \sum_{j=1}^{p} \int_{\partial M} \omega_j = \sum_{j=1}^{p} \int_M \mathrm{d}\omega_j = \int_M \mathrm{d}\omega.$$

□

We note that $\mathrm{d}\omega = \sum_{j=1}^p \mathrm{d}(\omega_j)$ is a consequence of the fact that $\omega = \sum_{j=1}^p \omega_j$ in the open set V that contains M.

9.4 The Classical Theorems of Vector Analysis

We are going to present connections between the general Stokes's theorem and the classical theorems of vector analysis. We first study the divergence theorem, taking an approach similar to that of Edwards [9].

Let U be an open set in $\mathbb{R}^n$ and

$$\boldsymbol{F} : U \subset \mathbb{R}^n \to \mathbb{R}^n, \quad \boldsymbol{F} = (f_1, \dots, f_n),$$

a vector field of class C^1 on U. Recall (see Definition 1.2.10) that the divergence of $\boldsymbol{F}$ is the scalar function defined on U by

$$\mathrm{Div}\, \boldsymbol{F} : U \to \mathbb{R}\,, \quad \mathrm{Div}\, \boldsymbol{F}(\boldsymbol{x}) := \sum_{j=1}^{n} \frac{\partial f_j}{\partial x_j}(\boldsymbol{x}).$$

In the theorems that follow, M is a compact regular n-surface with boundary in $\mathbb{R}^n$. Then the boundary of M coincides with its topological boundary, and in particular, $V := M \setminus \partial M$ is an open subset of $\mathbb{R}^n$. According to Proposition 8.4.1, in order to define an orientation on M, it suffices to define an orientation on the regular n-surface without boundary $M \setminus \partial M$. As in Sect. 8.5, from now on *we will suppose that M is oriented in such a way that the identity mapping*

$$\mathrm{id} : V \to M \setminus \partial M$$

defines a coordinate system (V, id) of $M \setminus \partial M$ that is compatible with the orientation.

In the next result we use the notation introduced in Definition 6.1.3.

Theorem 9.4.1 (Divergence theorem, first version). *Let M be a compact n-regular surface with boundary of class C^2 in $\mathbb{R}^n$ and let $\boldsymbol{F} = (f_1, \dots, f_n)$ be a vector field of class C^1 on an open neighborhood of M. Then*

$$\int_M \mathrm{Div}\, \boldsymbol{F}(\boldsymbol{x})\, \mathrm{d}\boldsymbol{x} = \int_{\partial M} \sum_{j=1}^{n} (-1)^{j-1} f_j\, \mathrm{d}x_1 \wedge \cdots \wedge \widehat{\mathrm{d}x_j} \wedge \cdots \wedge \mathrm{d}x_n.$$

Proof. We take

$$\omega := \sum_{j=1}^{n} (-1)^{j-1} f_j\, \mathrm{d}x_1 \wedge \cdots \wedge \widehat{\mathrm{d}x_j} \wedge \cdots \wedge \mathrm{d}x_n.$$

From Lemma 9.3.2 we have

$$\mathrm{d}\omega = \left(\sum_{j=1}^{n} \frac{\partial f_j}{\partial x_j} \right) \mathrm{d}x_1 \wedge \cdots \wedge \mathrm{d}x_n,$$

while by Stokes's theorem (Theorem 9.3.1) and Proposition 9.1.1,

$$\begin{aligned}\int_{\partial M} \omega &= \int_M \left(\sum_{j=1}^n \frac{\partial f_j}{\partial x_j} \right) \mathrm{d}x_1 \wedge \cdots \wedge \mathrm{d}x_n \\ &= \int_{M \setminus \partial M} \left(\sum_{j=1}^n \frac{\partial f_j}{\partial x_j} \right) \mathrm{d}x_1 \wedge \cdots \wedge \mathrm{d}x_n.\end{aligned}$$

Since the identity mapping

$$\mathrm{id} : M \setminus \partial M \to M \setminus \partial M$$

defines a coordinate system of $M \setminus \partial M$ compatible with its orientation, the last integral coincides with the Lebesgue integral

$$\int_{M \setminus \partial M} \left(\sum_{j=1}^n \frac{\partial f_j}{\partial x_j} \right) \mathrm{d}\boldsymbol{x}.$$

The fact that ∂M is a null set in $\mathbb{R}^n$ allows us to deduce

$$\int_{\partial M} \omega = \int_M \mathrm{Div}\, \boldsymbol{F}(\boldsymbol{x})\, \mathrm{d}\boldsymbol{x}.$$

□

Theorem 9.4.2 (Divergence theorem, second version). *Let M be a compact regular n-surface with boundary of class C^2 in $\mathbb{R}^n$ and let $\boldsymbol{F}$ be a vector field of class C^1 on an open neighborhood of M. Then*

$$\int_M \mathrm{Div}\, \boldsymbol{F}(\boldsymbol{x})\, \mathrm{d}\boldsymbol{x} = \int_{\partial M} \langle \boldsymbol{F}, \boldsymbol{N} \rangle \, \mathrm{d}V_{n-1}.$$

Proof. We will first prove the result while supposing that $\boldsymbol{\varphi} : \mathbb{A} \subset \mathbb{H}^n \to M$ defines a coordinate system of M compatible with the orientation and with the property that the support of $\boldsymbol{F} : M \to \mathbb{R}^n$ is contained in $\boldsymbol{\varphi}(\mathbb{A})$.

Take ω to be the differential form of degree $n-1$ of Theorem 9.4.1,

$$\omega := \sum_{j=1}^n (-1)^{j-1} f_j \, \mathrm{d}x_1 \wedge \cdots \wedge \widehat{\mathrm{d}x_j} \wedge \cdots \wedge \mathrm{d}x_n.$$

Then according to Theorem 9.4.1, it suffices to show that

$$\int_{\partial M} \omega = \int_{\partial M} \langle \boldsymbol{F}, \boldsymbol{N} \rangle \, \mathrm{d}V_{n-1}.$$

To do this, we consider the coordinate system $(\widehat{\mathbb{A}}, \widehat{\varphi})$ of ∂M defined by

$$\widehat{\mathbb{A}} := \{ s = (s_1, \ldots, s_{n-1}) \in \mathbb{R}^{n-1} \ : \ (0, s) \in \mathbb{A} \}$$

and

$$\widehat{\varphi} : \widehat{\mathbb{A}} \subset \mathbb{R}^{n-1} \to \mathbb{R}^n, \quad \widehat{\varphi}(s) := \varphi(0, s).$$

According to Theorem 8.4.1, $(\widehat{\mathbb{A}}, \widehat{\varphi})$ is a coordinate system of the boundary compatible with its orientation. Since the support of $\boldsymbol{F}$ in ∂M is contained in $\widehat{\varphi}(\widehat{\mathbb{A}})$, it follows from Definitions 7.3.2 and 7.2.3 that

$$\int_{\partial M} \langle \boldsymbol{F}, \boldsymbol{N} \rangle \, \mathrm{d}V_{n-1}$$

can be written as

$$\int_{\widehat{\mathbb{A}}} \sum_{j=1}^{n} (-1)^{j-1} f_j(\widehat{\varphi}(s)) \, \mathrm{d}x_1 \wedge \cdots \wedge \widehat{\mathrm{d}x_j} \wedge \cdots \wedge \mathrm{d}x_n \left(\frac{\partial \widehat{\varphi}}{\partial s_1}(s), \ldots, \frac{\partial \widehat{\varphi}}{\partial s_{n-1}}(s) \right) \mathrm{d}s,$$

or equivalently,

$$\int_{\widehat{\mathbb{A}}} \omega(\widehat{\varphi}(s)) \left(\frac{\partial \widehat{\varphi}}{\partial s_1}(s), \ldots, \frac{\partial \widehat{\varphi}}{\partial s_{n-1}}(s) \right) \mathrm{d}s = \int_{\partial M} \omega,$$

and this is what we wanted to prove.

Now for the general case. Since M is compact, it admits a finite atlas. Then the conclusion follows from Definition 7.3.2 and the previous case, proceeding as in the proof of Stokes's theorem (Theorem 9.3.1). □

In classical vector analysis, $n = 3$, and it is usual to denote by ∇ the operator

$$\nabla := \left(\frac{\partial}{\partial x}, \frac{\partial}{\partial y}, \frac{\partial}{\partial z} \right).$$

Thus, for a given vector field $\boldsymbol{F} = (f_1, f_2, f_3)$ of class C^1 on an open set of $\mathbb{R}^3$ we can write the formal identity

$$\nabla \cdot \boldsymbol{F} := \left(\frac{\partial}{\partial x}, \frac{\partial}{\partial y}, \frac{\partial}{\partial z} \right) \cdot (f_1, f_2, f_3) = \operatorname{Div} \boldsymbol{F}.$$

The divergence theorem can be expressed as follows (see Example 7.3.1 for different notations related to the surface integrals).

Theorem 9.4.3 (Gauss's theorem). *Let M be a compact regular 3-surface with boundary of class C^2 in $\mathbb{R}^3$ and let $\boldsymbol{F}$ be a vector field of class C^1 on a neighborhood of M. Then*

$$\int_{\partial M} \langle \boldsymbol{F}, \boldsymbol{N} \rangle \, \mathrm{d}S = \iiint_M (\nabla \cdot \boldsymbol{F})(x, y, z) \, \mathrm{d}(x, y, z).$$

Theorem 9.4.4 (Classical Stokes's theorem). *Let M be a compact and orientable regular 2-surface with boundary in $\mathbb{R}^3$ of class C^2 and let $\boldsymbol{F} = (f_1, f_2, f_3)$ be a vector field of class C^1 on an open neighborhood of M. Then*

$$\int_M \langle \nabla \times \boldsymbol{F},\ \boldsymbol{N}\rangle\ \mathrm{d}S = \int_{\partial M} \langle \boldsymbol{F}, \boldsymbol{T}\rangle\ \mathrm{d}s,$$

where for each $\boldsymbol{x} \in \partial M$, $\boldsymbol{T}(\boldsymbol{x})$ represents the unit tangent vector at the point $\boldsymbol{x}$ for any coordinate system $((a,b),\gamma)$, $\gamma : (a,b) \to \partial M$, compatible with the orientation such that $\boldsymbol{x} \in \gamma(a,b)$.

Proof. We consider the degree-1 differential form

$$\boldsymbol{\omega} = f_1\ \mathrm{d}x + f_2\ \mathrm{d}y + f_3\ \mathrm{d}z.$$

As we saw in Example 6.3.2,

$$\mathrm{d}\boldsymbol{\omega} = \left(\frac{\partial f_3}{\partial y} - \frac{\partial f_2}{\partial z}\right)\ \mathrm{d}y \wedge \mathrm{d}z + \left(\frac{\partial f_1}{\partial z} - \frac{\partial f_3}{\partial x}\right)\ \mathrm{d}z \wedge \mathrm{d}x + \left(\frac{\partial f_2}{\partial x} - \frac{\partial f_1}{\partial y}\right)\ \mathrm{d}x \wedge \mathrm{d}y.$$

That is, the exterior differential of $\boldsymbol{\omega}$ is the differential form of degree 2 associated to the curl of the vector field $\boldsymbol{F}$. According to Example 7.3.1 applied to the vector field $\boldsymbol{G} := \nabla \times \boldsymbol{F}$ and the differential form $\boldsymbol{\nu} := \mathrm{d}\boldsymbol{\omega}$ of degree 2, we have

$$\int_M \langle \nabla \times \boldsymbol{F},\ \boldsymbol{N}\rangle\ \mathrm{d}S = \int_M \mathrm{d}\boldsymbol{\omega}.$$

As we noted after Example 7.3.3, this last integral is usually represented as

$$\int_{\partial M} \boldsymbol{\omega} = \int_{\partial M} \langle \boldsymbol{F}, \boldsymbol{T}\rangle\ \mathrm{d}s.$$

The conclusion follows from the general form of Stokes's theorem, Theorem 9.3.1.

□

We point out that the classical theorems of Gauss and Stokes's just obtained above include the case in which the boundary of M has several connected components. The same applies to the version of Green's theorem obtained below.

As a consequence of Theorem 9.4.4 we can obtain an alternative proof of the identity provided in Example 6.5.2.

Corollary 9.4.1. *Let $\boldsymbol{F} = (f_1, f_2, f_3)$ be a vector field of class C^1 on an open set $U \subset \mathbb{R}^3$. Given a point $\boldsymbol{a} \in U$ and a unit vector $\boldsymbol{N} \in \mathbb{R}^3$, we denote by C_ε the boundary of the disk D_ε of radius ε, centered at $\boldsymbol{a}$ and contained in the plane that contains $\boldsymbol{a}$ and is orthogonal to $\boldsymbol{N}$. Suppose the disk is oriented according to the normal vector $\boldsymbol{N}$ and C_ε has the induced orientation. Then*

$$\langle (\nabla \times \boldsymbol{F})(\boldsymbol{a}), \boldsymbol{N}\rangle = \lim_{\varepsilon \to 0} \frac{1}{\pi \varepsilon^2} \int_{C_\varepsilon} \langle \boldsymbol{F}, \boldsymbol{T}\rangle\ \mathrm{d}s.$$

Proof. For every $\varepsilon > 0$ we have

$$\langle(\nabla \times F)(a), N\rangle = \frac{1}{\pi\varepsilon^2}\int_{D_\varepsilon} \langle(\nabla \times F)(a), N\rangle \; \mathrm{d}S,$$

since the integral on the right-hand side is the product of the constant

$$\langle(\nabla \times F)(a), N\rangle$$

and the area of the disk. Moreover, by the classical Stokes's theorem, Theorem 9.4.4,

$$\frac{1}{\pi\varepsilon^2}\int_{C_\varepsilon} \langle F, T\rangle \; \mathrm{d}s = \frac{1}{\pi\varepsilon^2}\int_{D_\varepsilon} \langle(\nabla \times F)(x), N\rangle \; \mathrm{d}S.$$

Consequently,

$$\left|\frac{1}{\pi\varepsilon^2}\int_{C_\varepsilon} \langle F, T\rangle - \langle(\nabla \times F)(a), N\rangle\right|$$

is less than or equal to

$$\frac{1}{\pi\varepsilon^2}\int_{D_\varepsilon} f \; \mathrm{d}S,$$

where

$$f(x) := |\langle(\nabla \times F)(x) - (\nabla \times F)(a), N\rangle|$$

(see Definition 7.3.3). Finally, from the Cauchy–Schwarz inequality we obtain

$$f(x) \leq \| \, (\nabla \times F)(x) - (\nabla \times F)(a) \, \|,$$

and hence

$$\frac{1}{\pi\varepsilon^2}\int_{D_\varepsilon} f \; \mathrm{d}S \leq \max\{\| \, (\nabla \times F)(x) - (\nabla \times F)(a) \, \| \, ; \|x - a\| \leq \varepsilon\},$$

which goes to zero as $\varepsilon \to 0$ because $\nabla \times F$ is a continuous vector field. The conclusion follows. □

We will give an explanation of the meaning of the curl as used in physics. Suppose that F represents the velocity field of a fluid. If γ is a path, then $\langle F(\gamma(t)), \gamma'(t)\rangle$ represents the component of the velocity of the fluid at the point $\gamma(t)$ in the direction tangent to the curve. In the case $\int_\gamma \langle F, T\rangle \, \mathrm{d}s > 0$, we can consider that the particles of fluid that are on γ tend to rotate in the sense that point the tangent vectors to the curve $\gamma'(t)$. If $\int_\gamma \langle F, T\rangle \, \mathrm{d}s < 0$, then the particles rotate in the opposite sense. We also refer to the integral $\int_\gamma \langle F, T\rangle \, \mathrm{d}s$ as the *circulation* of the vector field F along the curve γ.

Take a point $\boldsymbol{a}$ in the fluid and choose a unit vector $\boldsymbol{N}$ supported at that point. Let D_ε be the disk with radius ε centered at $\boldsymbol{a}$ and orthogonal to $\boldsymbol{N}$. In the boundary C_ε of the disk we consider the induced orientation assuming that the disk is oriented in such a way that $\boldsymbol{N}$ is the normal exterior vector. By Corollary 9.4.1,

$$\langle(\boldsymbol{\nabla}\times\boldsymbol{F})(\boldsymbol{a}),\boldsymbol{N}\rangle = \lim_{\varepsilon\to 0}\frac{1}{\pi\varepsilon^2}\int_{C_\varepsilon}\langle\boldsymbol{F},\boldsymbol{T}\rangle\,\mathrm{d}s.$$

We deduce that *the component of the curl of $\boldsymbol{F}$ in the direction of the unit vector $\boldsymbol{N}$ represents the circulation of $\boldsymbol{F}$ per unit area in the orthogonal plane to $\boldsymbol{N}$*. If $\langle(\boldsymbol{\nabla}\times\boldsymbol{F})(\boldsymbol{a}),\boldsymbol{N}\rangle \neq 0$, then the particles of the fluid *near $\boldsymbol{a}$* tend to rotate around $\boldsymbol{N}$ in the orthogonal plane to $\boldsymbol{N}$. The direction of rotation depends on the sign of $\langle(\boldsymbol{\nabla}\times\boldsymbol{F})(\boldsymbol{a}),\boldsymbol{N}\rangle$. Since the dot product $\langle(\boldsymbol{\nabla}\times\boldsymbol{F})(\boldsymbol{a}),\boldsymbol{N}\rangle$ attains its maximum value when

$$\boldsymbol{N} = \frac{(\boldsymbol{\nabla}\times\boldsymbol{F})(\boldsymbol{a})}{\|\,(\boldsymbol{\nabla}\times\boldsymbol{F})(\boldsymbol{a})\,\|},$$

we deduce that the rotation will be greatest in the plane orthogonal to the curl of the vector field at $\boldsymbol{a}$.

In the next result we assume, as usual, that M is oriented in such a way that the identity mapping on the interior of M defines a coordinate system compatible with the orientation of M. In that case, we can provide a completely rigorous proof of Green's theorem.

Theorem 9.4.5 (Green's theorem). *Let M be a compact and orientable regular 2-surface with boundary of class C^2 in $\mathbb{R}^2$ and let $\omega = P\,\mathrm{d}x + Q\,\mathrm{d}y$ be a differential form of degree 1 and class C^1 on an open neighborhood of M. Then*

$$\int_{\partial M}\omega = \iint_M\left(\frac{\partial Q}{\partial x}-\frac{\partial P}{\partial y}\right)\mathrm{d}(x,y).$$

The proof is a simple consequence of the fact that

$$\mathrm{d}\omega = \left(\frac{\partial Q}{\partial x}-\frac{\partial P}{\partial y}\right)\mathrm{d}x\wedge\mathrm{d}y.$$

We remind the reader that in the above situation, the induced orientation on ∂M is that for which on traversing ∂M, the set M remains to our left. See Sect. 8.5.

As required in practice, Stokes's theorem can be applied to more general sets than just compact regular surfaces with boundary and of class C^2 (for instance, a cube or a truncated cone). Instead of trying to develop a technically complicated general theory, we will content ourselves with showing how the theorem can be applied in such situations by means of concrete examples. On the other hand, these will show clearly the general method that one should follow.

We next give an example of a surface with boundary that is not compact but in which Stokes's theorem is valid.

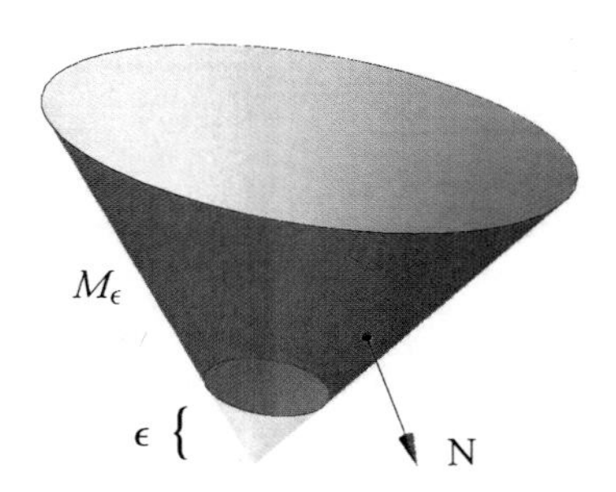

Fig. 9.3 Truncated cone of Example 9.4.1

Example 9.4.1. Consider the truncated cone (Fig. 9.3)

$$M := \{(x,y,z) \in \mathbb{R}^3 : x^2 + y^2 = z^2 \ ; \ 0 < z \leq 1\}.$$

For every $0 < \varepsilon < 1$, let

$$M_\varepsilon := \{(x,y,z) \in \mathbb{R}^3 : x^2 + y^2 = z^2 \ ; \ \varepsilon \leq z \leq 1\}.$$

Then M_ε is a regular surface with boundary. We suppose it is oriented so that the normal vector points toward the *exterior* of the bounded solid

$$\{(x,y,z) \in \mathbb{R}^3 : x^2 + y^2 \leq z^2 \ ; \ \varepsilon \leq z \leq 1\}.$$

We first obtain a parameterization of the surface that is compatible with its orientation. To do this, note that M_ε is the graph of a function. Instead of taking parameters $u := x, v := y$ as usual, we will take instead $u := y, v := x$ (a posteriori it will be clear that we need this reversal of the variables so that the coordinate system will be compatible with the orientation). More precisely, we consider as domain of parameterization

$$D_\varepsilon := \{(u,v) \in \mathbb{R}^2 : \varepsilon^2 \leq u^2 + v^2 \leq 1\}.$$

Now define

$$\boldsymbol{g}(u,v) := \left(v, u, \sqrt{u^2+v^2}\right).$$

Then

$$\left(\frac{\partial \boldsymbol{g}}{\partial u} \times \frac{\partial \boldsymbol{g}}{\partial v}\right)(u,v) = \begin{vmatrix} \boldsymbol{e}_1 & \boldsymbol{e}_2 & \boldsymbol{e}_3 \\ 0 & 1 & \frac{u}{\sqrt{u^2+v^2}} \\ 1 & 0 & \frac{v}{\sqrt{u^2+v^2}} \end{vmatrix}$$

$$= \left(\frac{v}{\sqrt{u^2+v^2}}, \frac{u}{\sqrt{u^2+v^2}}, -1\right).$$

Since the third coordinate is negative, it turns out that the normal vector

$$\boldsymbol{N} := \frac{\left(\frac{\partial g}{\partial u} \times \frac{\partial g}{\partial v}\right)}{\left\|\left(\frac{\partial g}{\partial u} \times \frac{\partial g}{\partial v}\right)\right\|}$$

points toward the exterior of the solid bounded by the surface, which means that the restriction of $\boldsymbol{g}$ to the interior of D_ε defines a coordinate system of $M_\varepsilon \setminus \partial M_\varepsilon$ that is compatible with its orientation. Moreover,

$$\partial M_\varepsilon = \boldsymbol{g}(\partial D_\varepsilon)$$

(see Criterion 8.6.1). Since the boundary of the oriented surface D_ε in $\mathbb{R}^2$ consists of two circles, the interior oriented clockwise and the exterior oriented counterclockwise, and moreover $\boldsymbol{g}$ conserves the orientation in the boundary by Criterion 8.6.1, it follows that

$$\partial M_\varepsilon = S_\varepsilon \cup S_1,$$

where

$$S_r := \{(x,y,r) \in \mathbb{R}^3 \ : x^2 + y^2 = r^2\}.$$

We can define two positively oriented coordinate systems of ∂M_ε (which cover the boundary with the exception of two points) by

$$\boldsymbol{\gamma} :]0,2\pi[\to \mathbb{R}^3 \ ; t \mapsto \boldsymbol{g}(\cos t, \sin t) = (\sin t,\ \cos t,\ 1)$$

and

$$\boldsymbol{\gamma}_\varepsilon :]0,2\pi[\to \mathbb{R}^3 \ ; t \mapsto \boldsymbol{g}(\varepsilon \sin t, \varepsilon \cos t) = (\varepsilon \cos t,\ \varepsilon \sin t,\ \varepsilon)\,.$$

In this way, the orientation in the boundary of M_ε is similar to that of Example 8.4.3.

Now let

$$\boldsymbol{F} := (F_1, F_2, F_3)$$

be a vector field of class C^1 on an open neighborhood of M. Since M_ε is compact, we can apply Stokes's theorem. That is,

$$\int_{M_\varepsilon} \boldsymbol{\nabla} \times \boldsymbol{F} \cdot \mathrm{d}\boldsymbol{S} = \int_{\boldsymbol{\gamma}} \boldsymbol{F} + \int_{\boldsymbol{\gamma}_\varepsilon} \boldsymbol{F}.$$

To evaluate the left-hand integral it suffices (see Proposition 9.1.1) to consider the coordinate system (consistent with the orientation) of $M_\varepsilon \setminus \partial M_\varepsilon$ defined by

$$\begin{aligned} \boldsymbol{\varphi}(r,t) &:= \boldsymbol{g}(r\cos t, r\sin t) \\ &= (r\sin t, r\cos t, r)\,, \end{aligned}$$

where $0 < t < 2\pi$ and $\varepsilon < r < 1$. That is,

$$\int_{M_\varepsilon} \nabla \times \boldsymbol{F} \cdot \mathrm{d}\boldsymbol{S} = \int_0^{2\pi} \int_\varepsilon^1 \left\langle \nabla \times \boldsymbol{F}(\boldsymbol{\varphi}(r,t)), \left(\frac{\partial \boldsymbol{\varphi}}{\partial r} \times \frac{\partial \boldsymbol{\varphi}}{\partial t} \right)(r,t) \right\rangle \, \mathrm{d}r \, \mathrm{d}t,$$

which converges, as $\varepsilon \to 0$, to

$$\int_0^{2\pi} \int_0^1 \left\langle \nabla \times \boldsymbol{F}(\boldsymbol{\varphi}(r,t)), \left(\frac{\partial \boldsymbol{\varphi}}{\partial r} \times \frac{\partial \boldsymbol{\varphi}}{\partial t} \right)(r,t) \right\rangle \, \mathrm{d}r \, \mathrm{d}t = \int_M \nabla \times \boldsymbol{F} \cdot \mathrm{d}\boldsymbol{S}$$

if that integral exists, for example in the case that F is C^1 on a neighborhood of the compact set $M \cup \{0\}$. This means that

$$\lim_{\varepsilon \to 0} \int_{M_\varepsilon} \nabla \times \boldsymbol{F} \cdot \mathrm{d}\boldsymbol{S} = \int_M \nabla \times \boldsymbol{F} \cdot \mathrm{d}\boldsymbol{S}.$$

On the other hand,

$$\left| \int_{\boldsymbol{\gamma}_\varepsilon} \boldsymbol{F} \right| = \left| \int_0^{2\pi} \langle \boldsymbol{F}(\boldsymbol{\gamma}_\varepsilon(t)), (-\varepsilon \sin t, \varepsilon \cos t, 0) \rangle \, \mathrm{d}t \right|$$

$$\leq \varepsilon \int_0^{2\pi} \|\boldsymbol{F}(\boldsymbol{\gamma}_\varepsilon(t))\| \, \mathrm{d}t,$$

and this implies

$$\lim_{\varepsilon \to 0} \int_{\boldsymbol{\gamma}_\varepsilon} \boldsymbol{F} = 0$$

in the case that F is continuous at $\{\mathbf{0}\}$. Finally, we have

$$\int_M \nabla \times \boldsymbol{F} \cdot \mathrm{d}\boldsymbol{S} = \int_{\boldsymbol{\gamma}} \boldsymbol{F}.$$

9.5 Stokes's Theorem on a Transformation of the k-Cube

The aim of this section is to present some additional examples of oriented but noncompact regular surfaces with boundary for which the general Stokes's theorem remains valid.

We first analyze the unit cube. Put $I := (0,1)^k \subset \mathbb{R}^k$, and for every $1 \leq j \leq k$,

$$I_{j,0} := \{\boldsymbol{t} \in \mathbb{R}^k \ : t_j = 0\ ; 0 < t_i < 1 \text{ if } i \neq j\}$$

and

$$I_{j,1} := \{\boldsymbol{t} \in \mathbb{R}^k \ : t_j = 1\ ; 0 < t_i < 1 \text{ if } i \neq j\}.$$

We recall that in Sect. 8.6 we introduced the cube without edges as the set

$$U := I \cup \bigcup_{j=1}^{k} (I_{j,0} \cup I_{j,1})$$

and

$$\partial U = \bigcup_{j=1}^{k} (I_{j,0} \cup I_{j,1}).$$

Clearly $I = U \setminus \partial U$ is an open set in $\mathbb{R}^k$, and ∂U is a C^∞ regular $(k-1)$-surface that is orientable. According to Sect. 8.6, that orientation is given by the unit normal vector $\boldsymbol{e}_j$ at the points of $I_{j,1}$ and by $-\boldsymbol{e}_j$ at the points of $I_{j,0}$. The problem is that U is not compact, since $\overline{U} = [0,1]^k \neq U$. We let

$$A := (0,1)^{k-1}.$$

Proposition 9.5.1. *Let ω be a differential form of degree $(k-1)$ and class C^1 on an open neighborhood of $\overline{U}$. Then*

$$\int_{\partial U} \omega = \int_U \mathrm{d}\omega.$$

Proof. Assume

$$\omega := \sum_{j=1}^{k} (-1)^{j-1} f_j(t_1,\ldots,t_k) \cdot \mathrm{d}t_1 \wedge \ldots \wedge \widehat{\mathrm{d}t_j} \wedge \ldots \wedge \mathrm{d}t_k$$

where f_j is a function of class C^1 on an open neighborhood of U. Then

$$\int_{\partial U} \omega = \sum_{j=1}^{k} \left(\int_{I_{j,0}} \omega + \int_{I_{j,1}} \omega \right).$$

We consider the vector field

$$\boldsymbol{F} := (f_1, f_2, \ldots, f_k).$$

A parameterization of $I_{j,a}$, where $a \in \{0,1\}$, is given by

$$\boldsymbol{\psi}_{j,a} : A \to \mathbb{R}^k \; ; \; \boldsymbol{\psi}_{j,a}(s_1, \ldots, s_{k-1}) = (s_1, \ldots, s_{j-1}, a, s_j, \ldots, s_{k-1}).$$

Since

$$\frac{\partial \boldsymbol{\psi}_{j,a}}{\partial s_1}(\boldsymbol{t}) \times \ldots \times \frac{\partial \boldsymbol{\psi}_{j,a}}{\partial s_{k-1}}(\boldsymbol{s}) = (-1)^{j-1} e_j,$$

some of the coordinate systems $(A, \boldsymbol{\psi}_{j,a})$ are not compatible with the orientation of the corresponding face $I_{j,a}$. We now concentrate in the case $a = 1$, so that the face $I_{j,1}$ is oriented according to the unit normal vector $\boldsymbol{e}_j$. If $(A, \boldsymbol{\psi}_{j,1})$ is compatible with the orientation of $I_{j,1}$ then

$$\int_{I_{j,1}} \omega = \int_A \langle \boldsymbol{F} \circ \boldsymbol{\psi}_{j,1}(\boldsymbol{s}), \boldsymbol{e}_j \rangle \, \mathrm{d}(s_1, \ldots, s_{k-1}).$$

Otherwise, we apply Lemma 5.2.3 to conclude that the coordinate system $(\boldsymbol{L}^{-1}(A), \boldsymbol{\psi}_{j,1} \circ \boldsymbol{L})$ is compatible with the orientation, where $\boldsymbol{L}(\boldsymbol{t}) = (t_1, \ldots, t_{k-2}, t_k, t_{k-1})$. Then, after a change of variables in the integral we also obtain

$$\int_{I_{j,1}} \omega = \int_{\boldsymbol{L}^{-1}(A)} \langle \underline{\mathrm{F}} \circ \boldsymbol{\psi}_{j,1} \circ \boldsymbol{L}(\boldsymbol{t}), \boldsymbol{e}_j \rangle \, \mathrm{d}(t_1, \ldots, t_{k-1}) = \int_A \langle \boldsymbol{F} \circ \boldsymbol{\psi}_{j,1}(\boldsymbol{s}), \boldsymbol{e}_j \rangle \, \mathrm{d}(s_1, \ldots, s_{k-1}).$$

A similar argument can be carried out in the case $a = 0$. Hence

$$\begin{aligned}
\int_{I_{j,0}} \omega + \int_{I_{j,1}} \omega &= \int_A \langle \boldsymbol{F} \circ \boldsymbol{\psi}_{j,1}(\boldsymbol{s}), \boldsymbol{e}_j \rangle \, \mathrm{d}(s_1, \ldots, s_{k-1}) \\
&\quad + \int_A \langle \boldsymbol{F} \circ \boldsymbol{\psi}_{j,0}(\boldsymbol{s}), -\boldsymbol{e}_j \rangle \, \mathrm{d}(s_1, \ldots, s_{k-1}) \\
&= \int_A \left(f_j \circ \boldsymbol{\psi}_{j,1} - f_j \circ \boldsymbol{\psi}_{j,0} \right)(\boldsymbol{s}) \, \mathrm{d}(s_1, \ldots, s_{k-1}).
\end{aligned}$$

On the other hand,

$$\mathrm{d}\omega = \left(\sum_{j=1}^{k} \frac{\partial f_j}{\partial t_j} \right) \cdot \mathrm{d}t_1 \wedge \ldots \wedge \mathrm{d}t_k$$

and hence

$$\int_U \mathrm{d}\omega = \sum_{j=1}^{k} \int_U \frac{\partial f_j}{\partial t_j}(\boldsymbol{t}) \, \mathrm{d}\boldsymbol{t}.$$

After evaluating the previous integrals using Fubini's Theorem and the Fundamental Theorem of Calculus in the j-th coordinate, we arrive to the conclusion. □

Corollary 9.5.1. *Let U be the unit cube in $\mathbb{R}^k$ without edges and $\boldsymbol{N}$ the vector field of unit normal vectors to ∂U defined by $\boldsymbol{N} = \boldsymbol{e}_j$ at points of $\boldsymbol{I}_{j,1}$ and $\boldsymbol{N} = -\boldsymbol{e}_j$ at points of $\boldsymbol{I}_{j,0}$. If $\boldsymbol{F}$ is a vector field of class C^1 on a neighborhood of the closed unit cube, then*

$$\int_U \mathrm{Div}\boldsymbol{F}(\boldsymbol{t})\mathrm{d}\boldsymbol{t} = \int_{\partial U} \boldsymbol{F}\cdot\boldsymbol{N}\,\mathrm{d}V_{k-1}.$$

Remark 9.5.1. The integral on the left-hand side, as a Lebesgue integral, coincides with

$$\int_{[0,1]^k} \mathrm{Div}\boldsymbol{F}(\boldsymbol{t})\mathrm{d}\boldsymbol{t} = \int_{(0,1)^k} \mathrm{Div}\boldsymbol{F}(\boldsymbol{t})\mathrm{d}\boldsymbol{t},$$

and although an abuse of notation, it is usual to write the right-hand-side integral as

$$\int_{\partial[0,1]^k} \boldsymbol{F}\cdot\boldsymbol{N}\,\mathrm{d}V_{k-1}.$$

Thus, Corollary 9.5.1 can be formulated for any vector field $\boldsymbol{F}$ of class C^1 on a neighborhood of $[0,1]^k$ as follows:

$$\int_{[0,1]^k} \mathrm{Div}\boldsymbol{F}(\boldsymbol{t})\mathrm{d}\boldsymbol{t} = \int_{\partial[0,1]^k} \boldsymbol{F}\cdot\boldsymbol{N}\,\mathrm{d}V_{k-1}.$$

Proposition 9.5.2. *Let M be an oriented regular k-surface with boundary. Suppose there is a mapping f of class C^2 on an open neighborhood W of the closed unit cube*

$$\boldsymbol{f}: W \to \mathbb{R}^n$$

such that $\boldsymbol{f}: U \to M$ is a bijection and $\boldsymbol{f}'$ has maximum rank at all points of U. If $\boldsymbol{\omega}$ is a differential form of degree $k-1$ and class C^1 on a neighborhood of $\overline{M}$, then

$$\int_{\partial M} \boldsymbol{\omega} = \int_M \mathrm{d}\boldsymbol{\omega}.$$

Proof. Again (and throughout this section) we denote by $I = (0,1)^k$ the open unit cube. According to Criterion 8.6.1, $\boldsymbol{f}(I) = M \setminus \partial M$ and $(I,\boldsymbol{f})$ is a coordinate system of $M \setminus \partial M$. First suppose that this coordinate system is compatible with the orientation of M. Define

$$\boldsymbol{\nu} := \boldsymbol{f}^*\boldsymbol{\omega},$$

which is a differential form of degree $k-1$ and class C^1 on a neighborhood of the closed unit cube. If $(A,\boldsymbol{\psi})$ is a coordinate system of ∂U compatible with its orientation, then $\boldsymbol{f} \circ \boldsymbol{\psi}$ defines a coordinate system of ∂M that preserves the orientation (see Criterion 8.6.1). Moreover, by Theorem 7.1.1,

$$\int_{(A,\boldsymbol{f}\circ\boldsymbol{\psi})} \boldsymbol{\omega} = \int_A (\boldsymbol{f}\circ\boldsymbol{\psi})^* \boldsymbol{\omega} = \int_A \boldsymbol{\psi}^* \boldsymbol{\nu} = \int_{(A,\boldsymbol{\psi})} \boldsymbol{\nu},$$

whence using a suitable atlas of ∂U we get

$$\int_{\partial M} \boldsymbol{\omega} = \int_{\partial U} \boldsymbol{\nu}.$$

From $\mathrm{d}\boldsymbol{\nu} = \boldsymbol{f}^*(\mathrm{d}\boldsymbol{\omega})$, we can also deduce analogously that

$$\int_M \mathrm{d}\boldsymbol{\omega} = \int_U \mathrm{d}\boldsymbol{\nu}.$$

Now it suffices to apply Proposition 9.5.1. To finish, suppose that the coordinate system $(I,\boldsymbol{f}_|)$ reverses the orientation of $M \setminus \partial M$. In that case, replace the function $\boldsymbol{f}$ by $\boldsymbol{g}(x_1,x_2,\ldots,x_k) := \boldsymbol{f}(x_2,x_1,\ldots,x_k)$ to obtain a new function $\boldsymbol{g}$ of class C^2 on a neighborhood of the unit cube with the property that $\boldsymbol{g} : U \to M$ is a bijection and $\boldsymbol{g}'$ has maximum rank. Moreover, $(I,\boldsymbol{g})$ is a coordinate system of $M \setminus \partial M$ compatible with its orientation. From the previous discussion it follows that also in this case

$$\int_M \mathrm{d}\boldsymbol{\omega} = \int_{\partial M} \boldsymbol{\omega}.$$

We observe that there are only two possibilities for the orientation of M, since it is connected. □

In order to check the validity of the divergence theorem in regions such as the full cylindrical pipe

$$K = \{(x,y,z) \in \mathbb{R}^3 : x^2 + y^2 \leq 1 \; ; \; 0 \leq z \leq 1\}$$

or the half-ball

$$K = \{(x,y,z) \in \mathbb{R}^3 : x^2 + y^2 + z^2 \leq 1 \; ; \; z \geq 0\},$$

which are not regular 3-surfaces with boundary in $\mathbb{R}^3$, observe that in both cases the set K can be obtained as the image of the closed unit cube by means of an adequate transformation (basically, cylindrical or spherical coordinates; see Examples 9.5.1 and 9.5.2 for the details).

From now on, we take $I = (0,1)^3$ (and hence $\overline{I} = [0,1]^3$) and we suppose that $K = \boldsymbol{f}(\overline{I})$, where

$$\boldsymbol{f} : W \to \mathbb{R}^3$$

is a mapping of class C^2 on an open neighborhood W of $\overline{I}$, which is injective and has strictly positive Jacobian on I.

Observe that K is not necessarily a regular 3-surface with boundary in $\mathbb{R}^3$. However, letting

$$I_\varepsilon := (\varepsilon, 1-\varepsilon)^3$$

and considering U_ε as the regular 3-surface with boundary in $\mathbb{R}^3$ obtained on eliminating the edges of $\overline{I}_\varepsilon$, then according to Proposition 9.5.2,

$$K_\varepsilon := \boldsymbol{f}(U_\varepsilon)$$

is a regular 3-surface with boundary in $\mathbb{R}^3$, for which Stokes's theorem is valid. We assume that $(I_\varepsilon, \boldsymbol{f})$ is a coordinate system compatible with the orientation of K_ε. That is, the orientation of K_ε is compatible with the canonical orientation of $\mathbb{R}^3$. Consequently, the boundary of K_ε is oriented according to the exterior normal vector.

In particular, if

$$\boldsymbol{\omega} := F_1 \cdot \mathrm{d}y \wedge \mathrm{d}z + F_2 \cdot \mathrm{d}z \wedge \mathrm{d}x + F_3 \cdot \mathrm{d}x \wedge \mathrm{d}y$$

is a differential form of degree 2 and class C^1 on an open neighborhood of K, then

$$\iiint_{K_\varepsilon} \mathrm{Div}\, \boldsymbol{F}(x,y,z)\, \mathrm{d}(x,y,z) = \int_{K_\varepsilon} \mathrm{d}\boldsymbol{\omega} = \int_{\partial K_\varepsilon} \boldsymbol{\omega} = \int_{\partial K_\varepsilon} \langle \boldsymbol{F}, \boldsymbol{N} \rangle \, \mathrm{d}S.$$

In this section *we aim to analyze what happens after taking limits in the previous identity as* $\varepsilon \to 0$.

Lemma 9.5.1. $\displaystyle\lim_{\varepsilon \to 0} \iiint_{K_\varepsilon} \mathrm{Div}\, \boldsymbol{F}(x,y,z)\, \mathrm{d}(x,y,z) = \iiint_K \mathrm{Div}\, \boldsymbol{F}(x,y,z)\, \mathrm{d}(x,y,z).$

Proof. Since F is continuous on the compact set K, there exists $C > 0$ a positive constant such that $|\mathrm{Div}\boldsymbol{F}(x,y,z)| \le C$ for every $(x,y,z) \in K$. Since

$$K \setminus K_\varepsilon = \boldsymbol{f}(\overline{I}) \setminus \boldsymbol{f}(U_\varepsilon) \subset \boldsymbol{f}(\overline{I} \setminus U_\varepsilon),$$

the change of variables theorem allows us to conclude that

$$\left| \iiint_{K_\varepsilon} \mathrm{Div}\, \boldsymbol{F}(x,y,z)\, \mathrm{d}(x,y,z) - \iiint_K \mathrm{Div}\, \boldsymbol{F}(x,y,z)\, \mathrm{d}(x,y,z) \right|$$

is less than or equal to

$$\iiint_{f(\overline{I}\setminus U_\varepsilon)} |\text{Div}\, \boldsymbol{F}(x,y,z)|\, \mathrm{d}(x,y,z) \le C \iiint_{\overline{I}\setminus U_\varepsilon} |J\boldsymbol{f}(x,y,z)|\, \mathrm{d}(x,y,z)$$

$$\le C\left(\max_{(x,y,z)\in\overline{I}} |J\boldsymbol{f}(x,y,z)|\right) \iiint_{\overline{I}\setminus U_\varepsilon} \mathrm{d}(x,y,z),$$

which goes to 0 as $\varepsilon \to 0$. □

Now we intend to analyze

$$\lim_{\varepsilon\to 0} \int_{\partial K_\varepsilon} \omega .$$

In some cases the topological boundary of K can be decomposed as the union (perhaps nondisjoint) of regular surfaces with boundary, and this limit can be expressed as a sum of flux integrals, which makes it possible to give an adequate formulation of the divergence theorem.

Keeping this objective in mind, we present the next three lemmas. Suppose that $\boldsymbol{f}$ satisfies the previous conditions. We note that the boundary of K_ε consists of six connected components, that is, the images under $\boldsymbol{f}$ of the faces (without edges) of I_ε.

The first lemma will allow us to analyze the faces of the cube whose image under f is a regular surface.

Lemma 9.5.2. *Let C be a face of the cube without edges, U, such that $\boldsymbol{f}(C)$ is a regular surface and assume that $\boldsymbol{f}(C)$ is oriented so that if $(V,\boldsymbol{\psi})$ is a coordinate system of C compatible with the orientation of the cube, then $(V,\boldsymbol{f}\circ\boldsymbol{\psi})$ is a positively oriented coordinate system of $\boldsymbol{f}(C)$. For $0<\varepsilon<\frac{1}{2}$, denote by J_ε the nearer of the two faces of I_ε parallel to C. Then*

$$\lim_{\varepsilon\to 0} \int_{\boldsymbol{f}(J_\varepsilon)} \omega = \int_{\boldsymbol{f}(C)} \omega .$$

Proof. Suppose that C is the face of the cube defined by $x_1 = 1$. Then J_ε is the face of I_ε defined by $x_1 = 1-\varepsilon$. Since the mapping

$$(u,v) \to (1,u,v)$$

defines a positively oriented coordinate system of C, we deduce that

$$(u,v) \to \boldsymbol{f}(1,u,v)$$

defines a positively oriented coordinate system of $\boldsymbol{f}(C) \subset \partial K$. Moreover, from Criterion 8.6.1 we also conclude that

$$(u,v) \to \boldsymbol{f}(1-\varepsilon,u,v)$$

defines a positively oriented coordinate system of $\boldsymbol{f}(J_\varepsilon)$. Consequently,

$$\int_{\boldsymbol{f}(J_\varepsilon)} \boldsymbol{\omega} = \int_\varepsilon^{1-\varepsilon} \int_\varepsilon^{1-\varepsilon} \left\langle \boldsymbol{F}(\boldsymbol{f}(1-\varepsilon,u,v)), \left(\frac{\partial \boldsymbol{f}}{\partial y} \times \frac{\partial \boldsymbol{f}}{\partial z}\right)(1-\varepsilon,u,v) \right\rangle du\, dv,$$

which converges as $\varepsilon \to 0$ to

$$\int_0^1 \int_0^1 \left\langle \boldsymbol{F}(\boldsymbol{f}(1,u,v)), \left(\frac{\partial \boldsymbol{f}}{\partial y} \times \frac{\partial \boldsymbol{f}}{\partial z}\right)(1,u,v) \right\rangle du\, dv = \int_{\boldsymbol{f}(C)} \boldsymbol{\omega}.$$

□

The second lemma gives information about what happens with the faces of the cube whose image by $\boldsymbol{f}$ is not a regular surface.

Lemma 9.5.3. *Suppose that* $\frac{\partial f}{\partial y} \times \frac{\partial f}{\partial z} = 0$ *on the face of the cube defined by* $x_1 = 0$. *Denote by* J_ε *the face of* I_ε *defined by* $x_1 = \varepsilon$. *Then*

$$\lim_{\varepsilon \to 0} \int_{\boldsymbol{f}(J_\varepsilon)} \boldsymbol{\omega} = 0.$$

Proof. There is a constant $C > 0$ such that

$$\| \boldsymbol{F}(x,y,z) \| \leq C$$

whenever $(x,y,z) \in K$. Moreover, if we let

$$M_\varepsilon := \sup\left\{ \left\| \left(\frac{\partial \boldsymbol{f}}{\partial y} \times \frac{\partial \boldsymbol{f}}{\partial z}\right)(x,y,z) \right\| : x = \varepsilon\, ; 0 \leq y,z \leq 1 \right\},$$

then it follows, from the uniform continuity of $\frac{\partial f}{\partial y} \times \frac{\partial f}{\partial z}$ in $\overline{I}$ and the hypothesis, that

$$\lim_{\varepsilon \to 0} M_\varepsilon = 0.$$

Finally, bearing in mind that

$$(u,v) \to \boldsymbol{f}(\varepsilon,u,v)$$

is a parameterization of $\boldsymbol{f}(J_\varepsilon) \subset \partial K_\varepsilon$, we obtain

$$\left| \int_{\boldsymbol{f}(J_\varepsilon)} \boldsymbol{\omega} \right| \leq \int_0^1 \int_0^1 \left| \left\langle \boldsymbol{F}(\boldsymbol{f}(\varepsilon,u,v)), \left(\frac{\partial \boldsymbol{f}}{\partial y} \times \frac{\partial \boldsymbol{f}}{\partial z}\right)(\varepsilon,u,v) \right\rangle \right| du\, dv$$

$$\leq C M_\varepsilon,$$

which goes to zero as $\varepsilon \to 0$. □

The third and final lemma will be useful when we have to work with a parameterization involving periodicity in some of the variables. This typically occurs when one changes to spherical or cylindrical coordinates.

Lemma 9.5.4. *Let $\boldsymbol{f}$ be periodic with period 1 with respect to a coordinate and let $J_{\varepsilon,0}, J_{\varepsilon,1}$ be the two parallel faces of I_ε in which that coordinate is constant. Then*

$$\lim_{\varepsilon \to 0} \left(\int_{\boldsymbol{f}(J_{\varepsilon,0})} \boldsymbol{\omega} + \int_{\boldsymbol{f}(J_{\varepsilon,1})} \boldsymbol{\omega} \right) = 0.$$

Proof. Suppose that

$$\boldsymbol{f}(x,0,z) = \boldsymbol{f}(x,1,z)$$

whenever $0 \le x, z \le 1$. Then $J_{\varepsilon,0}$ is the face of I_ε defined by $y = \varepsilon$ and $J_{\varepsilon,1}$ is the face of I_ε defined by $y = 1 - \varepsilon$.

The mappings

$$\boldsymbol{g}(u,v) := (u, \varepsilon, v) \; ; \; \boldsymbol{h}(u,v) := (v, 1-\varepsilon, u)$$

define positively oriented coordinate systems of $J_{\varepsilon,0}$ and $J_{\varepsilon,1}$ respectively, because

$$\frac{\partial \boldsymbol{g}}{\partial u} \times \frac{\partial \boldsymbol{g}}{\partial v} = -\boldsymbol{e}_2 \text{ y } \frac{\partial \boldsymbol{h}}{\partial u} \times \frac{\partial \boldsymbol{h}}{\partial v} = \boldsymbol{e}_2.$$

From Criterion 8.6.1, we deduce that the mappings

$$(u,v) \mapsto \boldsymbol{f}(u, \varepsilon, v) \; ; \; (u,v) \mapsto \boldsymbol{f}(v, 1-\varepsilon, u)$$

define positively oriented coordinate systems of $\boldsymbol{f}(J_{\varepsilon,0})$ and $\boldsymbol{f}(J_{\varepsilon,1})$ respectively. Consequently,

$$\int_{\boldsymbol{f}(J_{\varepsilon,0})} \boldsymbol{\omega} = \int_\varepsilon^{1-\varepsilon} \int_\varepsilon^{1-\varepsilon} \left\langle \boldsymbol{F}(\boldsymbol{f}(u,\varepsilon,v)), (\frac{\partial \boldsymbol{f}}{\partial x} \times \frac{\partial \boldsymbol{f}}{\partial z})(u,\varepsilon,v) \right\rangle du\, dv,$$

while

$$\int_{\boldsymbol{f}(J_{\varepsilon,1})} \boldsymbol{\omega} = \int_\varepsilon^{1-\varepsilon} \int_\varepsilon^{1-\varepsilon} \left\langle \boldsymbol{F}(\boldsymbol{f}(v,1-\varepsilon,u)), -(\frac{\partial \boldsymbol{f}}{\partial x} \times \frac{\partial \boldsymbol{f}}{\partial z})(v,1-\varepsilon,u) \right\rangle du\, dv$$

$$= \int_\varepsilon^{1-\varepsilon} \int_\varepsilon^{1-\varepsilon} \left\langle \boldsymbol{F}(\boldsymbol{f}(u,1-\varepsilon,v)), -(\frac{\partial \boldsymbol{f}}{\partial x} \times \frac{\partial \boldsymbol{f}}{\partial z})(u,1-\varepsilon,v) \right\rangle du\, dv.$$

Since

$$\boldsymbol{f}(u,0,v) = \boldsymbol{f}(u,1,v)$$

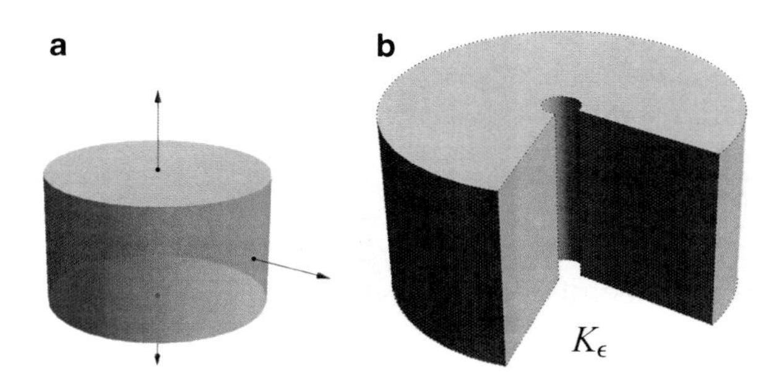

Fig. 9.4 (**a**) Full cylindrical pipe and (**b**) the set K_ε of Example 9.5.1

we easily deduce

$$\int_{f(J_{\varepsilon,0})} \omega + \int_{f(J_{\varepsilon,1})} \omega \to 0$$

as ε goes to 0. □

Example 9.5.1. Consider the full cylindrical pipe (Fig. 9.4)

$$K := \{(x,y,z) \in \mathbb{R}^3 \ : x^2 + y^2 \leq 1 \ ; \ 0 \leq z \leq 1\}.$$

Take

$$\boldsymbol{f} : \mathbb{R}^3 \to \mathbb{R}^3, \quad \boldsymbol{f}(r,s,t) := (r\cos(2\pi s), r\sin(2\pi s), t).$$

Then

$$K = \boldsymbol{f}(\bar{I}),$$

and the Jacobian of $\boldsymbol{f}$ is

$$\begin{vmatrix} \cos(2\pi s) & -2\pi r\sin(2\pi s) & 0 \\ \sin(2\pi s) & 2\pi r\cos(2\pi s) & 0 \\ 0 & 0 & 1 \end{vmatrix} = 2\pi r.$$

Hence $K_\varepsilon := \boldsymbol{f}(U_\varepsilon)$ is a regular 3-surface with boundary for which the general Stokes's theorem is valid.

If

$$\omega := F_1 \cdot \mathrm{d}y \wedge \mathrm{d}z + F_2 \cdot \mathrm{d}z \wedge \mathrm{d}x + F_3 \cdot \mathrm{d}x \wedge \mathrm{d}y$$

is a differential form of degree 2 and class C^1 on an open neighborhood of K, then we deduce from the comments in Sect. 8.6 and Lemma 9.5.1 that

$$\iiint_K \mathrm{Div}\, \boldsymbol{F}(x,y,z)\ \mathrm{d}(x,y,z) = \lim_{\varepsilon \to 0} \int_{\partial K_\varepsilon} \omega.$$

Observe that the topological boundary of K can be decomposed as

$$\partial^{\mathrm{top}} K = S_0 \cup S_1 \cup S_2,$$

where

$$S_j := \{(x,y,z) \in \mathbb{R}^3 \ : \ x^2+y^2 \leq 1 \ ; \ z=j\}$$

for $j=0,1$ are closed disck in horizontal planes and

$$S_2 := \{(x,y,z) \in \mathbb{R}^3 \ : \ x^2+y^2 = 1 \ ; \ 0 \leq z \leq 1\}$$

is a cylindrical surface with boundary. Each of three sets S_0, S_1, and S_2 is a regular 2-surface with boundary in $\mathbb{R}^3$. We provide S_0 with the orientation given by the normal vector $-\boldsymbol{e}_3$, S_1 with that given by the normal vector $\boldsymbol{e}_3$, and S_2 with that afforded by the vector field (of normal vectors at points of $S_2 \setminus \partial S_2$)

$$\boldsymbol{N}(x,y,z) = (x,y,0).$$

That is, the three surfaces are oriented according to the field of unit normal vectors pointing toward the outside of K. Although K is *not* a regular 3-surface with boundary, we commit an abuse of notation and write

$$\partial K := S_0 \cup S_1 \cup S_2$$

and

$$\int_{\partial K} \omega := \sum_{j=0}^{2} \int_{S_j} \omega.$$

As one easily discerns, our objective is to prove the identity

$$\lim_{\varepsilon \to 0} \int_{\partial K_\varepsilon} \omega = \int_{\partial K} \omega.$$

The boundary of K_ε consists of six connected components, namely, the images under $\boldsymbol{f}$ of the faces (without edges) of I_ε. We analyze each one of these components.

Image of the faces parallel to the plane $\{0\} \times \mathbb{R}^2$. First we fix $r=\varepsilon$. Since

$$\left\| \left(\frac{\partial \boldsymbol{f}}{\partial s} \times \frac{\partial \boldsymbol{f}}{\partial t} \right) (\varepsilon, s, t) \right\| = 2\pi\varepsilon,$$

from Lemma 9.5.3 we obtain, for $(I_\varepsilon)_{1,0} = \{(\varepsilon,s,t) : \ \varepsilon \leq s,t \leq 1-\varepsilon\}$, that

$$\lim_{\varepsilon \to 0} \int_{\boldsymbol{f}((I_\varepsilon)_{1,0})} \omega = 0.$$

Take now $r = 1-\varepsilon$. The parameterization

$$(s,t) \to \boldsymbol{f}(1-\varepsilon,s,t)$$

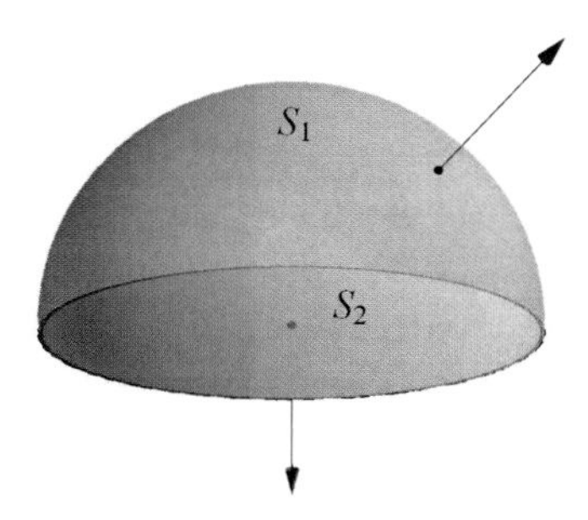

Fig. 9.5 Half-ball of Example 9.5.2

is positively oriented, and by Lemma 9.5.2,

$$\lim_{\varepsilon \to 0} \int_{f((I_\varepsilon)_{1,1})} \omega = \int_{S_2} \omega,$$

where

$$(I_\varepsilon)_{1,1} = \{(1-\varepsilon, s, t) : \varepsilon \leq s, t \leq 1-\varepsilon\}.$$

Image of the faces parallel to the plane $\mathbb{R} \times \{0\} \times \mathbb{R}$. In this case, take $s \in \{\varepsilon, 1-\varepsilon\}$. By Lemma 9.5.4,

$$\int_{f((I_\varepsilon)_{2,0})} \omega + \int_{f((I_\varepsilon)_{2,1})} \omega \to 0$$

as ε goes to 0, where

$$(I_\varepsilon)_{2,0} = \{(r, \varepsilon, t) : \varepsilon \leq r, t \leq 1-\varepsilon\}$$

and

$$(I_\varepsilon)_{2,1} = \{(r, 1-\varepsilon, t) : \varepsilon \leq r, t \leq 1-\varepsilon\}.$$

Image of the faces parallel to the plane $\mathbb{R} \times \mathbb{R} \times \{0\}$. Here, take $t = \varepsilon$ or $t = 1-\varepsilon$. By Lemma 9.5.2,

$$\lim_{\varepsilon \to 0} \int_{f((I_\varepsilon)_{3,1})} \omega = \int_{S_1} \omega$$

and

$$\lim_{\varepsilon \to 0} \int_{f((I_\varepsilon)_{3,0})} \omega = \int_{S_0} \omega,$$

where

$$(I_\varepsilon)_{3,0} = \{(r, s, \varepsilon) : \varepsilon \leq r, s \leq 1-\varepsilon\}$$

and

$$(I_\varepsilon)_{3,1} = \{(r, s, 1-\varepsilon) : \varepsilon \leq r, s \leq 1-\varepsilon\}.$$

Consequently,

$$\lim_{\varepsilon \to 0} \int_{\partial K_\varepsilon} \boldsymbol{\omega} = \int_{\partial K} \boldsymbol{\omega},$$

and we finally deduce that

$$\iiint_K \operatorname{Div} \boldsymbol{F}(x,y,z)\, \mathrm{d}(x,y,z) = \int_{\partial K} \boldsymbol{\omega}.$$

Example 9.5.2. Consider the half-ball (Fig. 9.5)

$$K := \{(x,y,z) \in \mathbb{R}^3 \ : \ x^2+y^2+z^2 \le 1 \ ; \ z \ge 0\}$$

and take

$$\boldsymbol{f}(r,s,t) := \left(r \sin\left(\frac{\pi}{2}\, s\right) \cos(2\pi t), r \sin\left(\frac{\pi}{2}\, s\right) \sin(2\pi t), r \cos\left(\frac{\pi}{2}\, s\right) \right).$$

Then $K = \boldsymbol{f}(I)$. The topological boundary of K can be decomposed as the union of two regular surfaces with boundary

$$\partial^{\text{top}}(K) = S_1 \cup S_2,$$

where

$$S_1 := \{(x,y,z) \in \mathbb{R}^3 \ : x^2+y^2+z^2 = 1 \ ; \ z \ge 0\}$$

and

$$S_2 := \{(x,y,z) \in \mathbb{R}^3 \ : x^2+y^2 \le 1 \ ; \ z = 0\}.$$

The surface S_1 is oriented according to the vector field

$$\boldsymbol{N}(x,y,z) = (x,y,z),$$

and S_2 according to the normal vector $-\boldsymbol{e}_3$.

Proceeding as in Example 9.5.1, we see that whenever

$$\boldsymbol{\omega} := F_1 \cdot \mathrm{d}y \wedge \mathrm{d}z + F_2 \cdot \mathrm{d}z \wedge \mathrm{d}x + F_3 \cdot \mathrm{d}x \wedge \mathrm{d}y$$

is a differential form of degree 2 and class C^1 on an open neighborhood of K, then

$$\iiint_K \operatorname{Div} \boldsymbol{F}(x,y,z)\, \mathrm{d}(x,y,z) = \int_{S_1} \boldsymbol{\omega} + \int_{S_2} \boldsymbol{\omega}.$$

In fact, it suffices to apply Lemma 9.5.3 to the faces of the cube $r=0$, $s=0$, Lemma 9.5.4 to the parallel faces $t=0$, $t=1$, and Lemma 9.5.2 to the faces $r=1$, $s=1$.

Example 9.5.3. Consider the truncated cone

$$K := \{(x,y,z) \in \mathbb{R}^3 \ : x^2+y^2 = z^2 \ ; \ 0 \le z \le 1\}$$

and the mapping $\boldsymbol{f} : \mathbb{R}^2 \to \mathbb{R}^2$ defined by

$$\boldsymbol{f}(t,r) := (r\cos(2\pi t), r\sin(2\pi t), r).$$

Then $K = \boldsymbol{f}(I)$ where $I := [0,1] \times [0,1]$ is the unit square. We already know that

$$M := K \setminus \{(0,0,0)\}$$

is a (noncompact) regular surface with boundary in $\mathbb{R}^3$. Moreover, suppose that M is oriented in such a way that the restriction of $\boldsymbol{f}$ to the interior of I defines a positively oriented coordinate system of M. Since

$$\left(\frac{\partial \boldsymbol{f}}{\partial t} \times \frac{\partial \boldsymbol{f}}{\partial r}\right)(t,r) = (2\pi r\cos(2\pi t), 2\pi r\sin(2\pi t), -2\pi r),$$

this means that $M \setminus \partial M$ is oriented according to the normal vector that points to the *exterior* of the solid bounded by M. Since $t \mapsto (t,1)$ is a parameterization of the boundary of the square (without vertices) that reverses the orientation, we deduce from Criterion 8.6.1 that

$$\boldsymbol{\gamma} : t \mapsto \boldsymbol{f}(t,1) = (\cos(2\pi t), \sin(2\pi t), 1), \quad 0 < t < 1,$$

is a parameterization of ∂M that reverses the orientation and covers all the boundary, except for one point. Now take

$$\omega = F_1 \cdot \mathrm{d}x + F_2 \cdot \mathrm{d}y + F_3 \cdot \mathrm{d}z$$

a differential form of degree 1 and class C^1 on a neighborhood of K. We want to show that

$$\int_{\partial M} \omega = \int_M \mathrm{d}\omega.$$

To do this, denote by I_ε the square $[\varepsilon, 1-\varepsilon]^2$ less its vertices and take

$$M_\varepsilon = \boldsymbol{f}(I_\varepsilon),$$

which is a regular surface with boundary in $\mathbb{R}^3$ for which the general Stokes's theorem is valid. In particular,

$$\int_{\partial M_\varepsilon} \omega = \int_{M_\varepsilon} \mathrm{d}\omega.$$

Also,

$$\int_{M_\varepsilon} \mathrm{d}\omega = \int_\varepsilon^{1-\varepsilon} \int_\varepsilon^{1-\varepsilon} \left\langle (\boldsymbol{\nabla} \times \boldsymbol{F})(\boldsymbol{f}(t,r)), \left(\frac{\partial \boldsymbol{f}}{\partial t} \times \frac{\partial \boldsymbol{f}}{\partial r}\right)(t,r) \right\rangle \mathrm{d}t\, \mathrm{d}r$$

converges as $\varepsilon \to 0$ to

$$\int_0^1 \int_0^1 \left\langle (\nabla \times \boldsymbol{F})(\boldsymbol{f}(t,r)), \left(\frac{\partial \boldsymbol{f}}{\partial t} \times \frac{\partial \boldsymbol{f}}{\partial r}\right)(t,r) \right\rangle \, \mathrm{d}t \, \mathrm{d}r = \int_M \mathrm{d}\boldsymbol{\omega}.$$

On the other hand, the boundary of M_ε consists of four connected components (images of the four faces, without vertices, of the square I_ε). Keeping in mind the facts

$$\boldsymbol{f}(0,r) = \boldsymbol{f}(1,r) \quad \text{and} \quad \frac{\partial \boldsymbol{f}}{\partial t}(t,0) = 0,$$

it follows, proceeding as in Lemmas 9.5.3 and 9.5.4 and considering

$$\boldsymbol{\gamma}_\varepsilon(t) := \boldsymbol{f}(t, 1-\varepsilon), \quad \varepsilon < t < 1-\varepsilon,$$

that

$$\begin{aligned}
\lim_{\varepsilon \to 0} \int_{\partial M_\varepsilon} \boldsymbol{\omega} &= -\lim_{\varepsilon \to 0} \int_{\boldsymbol{\gamma}_\varepsilon} \boldsymbol{\omega} \\
&= -\lim_{\varepsilon \to 0} \int_\varepsilon^{1-\varepsilon} \langle \boldsymbol{F}(\boldsymbol{\gamma}_\varepsilon(t)), (\boldsymbol{\gamma}_\varepsilon)'(t) \rangle \, \mathrm{d}t \\
&= -\int_0^1 \langle \boldsymbol{F}(\boldsymbol{\gamma}(t)), \boldsymbol{\gamma}'(t) \rangle \, \mathrm{d}t \\
&= \int_{\partial M} \boldsymbol{\omega}.
\end{aligned}$$

9.6 Appendix: Flux of a Gravitational Field

Let us consider an object of mass M located at the point

$$\boldsymbol{P} = (x_0, y_0, z_0).$$

For every $(x,y,z) \in \mathbb{R}^3$, let

$$\boldsymbol{r} = (x - x_0, y - y_0, z - z_0).$$

Then the force exercised by the point mass M on another point mass m located at $(x,y,z) \in \mathbb{R}^3 \setminus \boldsymbol{P}$ is

$$\boldsymbol{F}(x,y,z) = -\frac{GMm}{\|\boldsymbol{r}\|^3}\,\boldsymbol{r}.$$

We observe that the norm of $\boldsymbol{F}(x,y,z)$ is proportional to the inverse of the square of the distance between the two points. It is obvious that

$$\boldsymbol{F} : \mathbb{R}^3 \setminus \{\boldsymbol{P}\} \to \mathbb{R}^3$$

is a vector field of class C^∞. The aim of this section is to evaluate the flux of this vector field $\boldsymbol{F}$ across a compact regular surface of class C^2 that does not contain the point $\boldsymbol{P}$ and is the boundary of a regular 3-surface with boundary.

Proposition 9.6.1. *The divergence of the vector field* $\boldsymbol{F}$ *vanishes.*

Proof. We first evaluate

$$\frac{\partial F_1}{\partial x} = -GMm\left(\frac{1}{\|\boldsymbol{r}\|^3} - \frac{3(x - x_0)^2}{\|\boldsymbol{r}\|^5}\right),$$

$$\frac{\partial F_2}{\partial y} = -GMm\left(\frac{1}{\|\boldsymbol{r}\|^3} - \frac{3(y - y_0)^2}{\|\boldsymbol{r}\|^5}\right),$$

and

$$\frac{\partial F_3}{\partial z} = -GMm\left(\frac{1}{\|\boldsymbol{r}\|^3} - \frac{3(z - z_0)^2}{\|\boldsymbol{r}\|^5}\right).$$

From

$$(x - x_0)^2 + (y - y_0)^2 + (z - z_0)^2 = \|\boldsymbol{r}\|^2,$$

we obtain, on summing these partial derivatives,

$$\mathrm{Div}\,\boldsymbol{F} = 0.$$

□

Lemma 9.6.1. *Let B be the closed ball centered at a point $\boldsymbol{P}$ and with radius $R > 0$. The flux of the vector field $\boldsymbol{F}$ across ∂B is*

$$-4\pi GMm.$$

Proof. The unit normal vector to ∂B at a point $(x,y,z) \in \partial B$, pointing to the *exterior* of B, is given by

$$\boldsymbol{N} = \frac{\boldsymbol{r}}{\| \boldsymbol{r} \|}.$$

Hence, for

$$\boldsymbol{F} = -\frac{GMm}{\| \boldsymbol{r} \|^3}\,\boldsymbol{r},$$

it turns out that

$$\langle \boldsymbol{F}, \boldsymbol{N} \rangle = -GMm\frac{1}{\| \boldsymbol{r} \|^2} = -GMm\frac{1}{R^2}.$$

That is, $\langle \boldsymbol{F}, \boldsymbol{N} \rangle$ is constant on the sphere. Consequently, the flux of the vector field $\boldsymbol{F}$ across ∂B coincides with the product of such a constant with the area of ∂B. That is,

$$\text{Flux} = -4\pi GMm.$$

□

Theorem 9.6.1 (Gauss's law). *Let U be a regular 3-surface with boundary of class C^2 in $\mathbb{R}^3$ and assume that*

$$S := \partial U$$

is a compact regular surface that is oriented according to the normal vector pointing to the exterior of U.

(1) If $\boldsymbol{P} \notin U$, then the flux of $\boldsymbol{F}$ across S is zero.
(2) If $\boldsymbol{P} \in U \setminus S$, then the flux of $\boldsymbol{F}$ across S is $-4\pi GMm$.

Proof. We first suppose that $\boldsymbol{P} \notin U$. In this case, $\boldsymbol{F}$ is a vector field of class C^1 on a neighborhood of U, and by the divergence theorem, the flux of $\boldsymbol{F}$ across S coincides with

$$\iiint_U \mathrm{Div}\boldsymbol{F}(x,y,z)\ \mathrm{d}(x,y,z) = 0.$$

Next, suppose that $\boldsymbol{P} \in U \setminus S$. This time, it is not so easy to apply the divergence theorem, because $\boldsymbol{F}$ is not defined at the point $\boldsymbol{P} \in U$. To avoid this complication, take $R > 0$ small enough that the closed ball B centered at $\boldsymbol{P}$ and with radius R is contained in the interior of U (Fig. 9.6). Now consider

$$\Omega = \{(x,y,z) \in U \ : \| (x - x_0, y - y_0, z - z_0) \| \geq R\},$$

which is a regular 3-surface with boundary

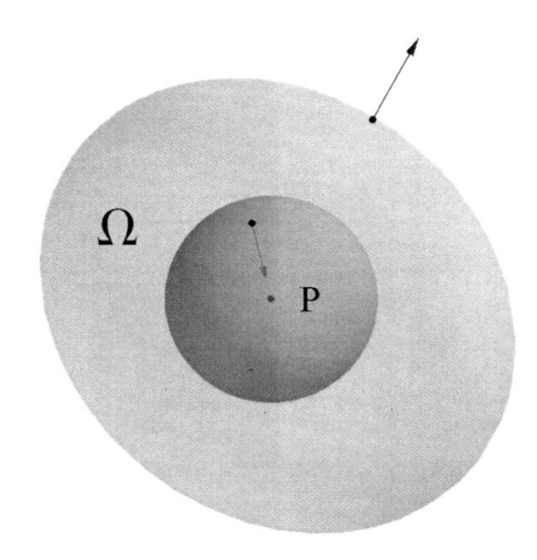

Fig. 9.6 Illustration of the proof of Theorem 9.6.1

$$\partial\Omega = S \cup \partial B.$$

As usual, the boundary of Ω is oriented according the normal vector exterior to Ω. This means that the orientation of $\partial\Omega$ at the points of S is given by the normal vector to S that points to the exterior of U, and the orientation of $\partial\Omega$ at points of ∂B is given by the normal vectors to the sphere that point to the *interior* of the ball.

So, if we consider in ∂B the orientation according to the normal vector *exterior* to the ball, then

$$\int_{\partial\Omega} \langle \boldsymbol{F}, \boldsymbol{N} \rangle \, \mathrm{d}S = \int_S \langle \boldsymbol{F}, \boldsymbol{N} \rangle \, \mathrm{d}S - \int_{\partial B} \langle \boldsymbol{F}, \boldsymbol{N} \rangle \, \mathrm{d}S.$$

On the other hand, since $\boldsymbol{F}$ is a vector field of class C^1 on Ω, we can apply the divergence theorem to conclude that

$$\int_{\partial\Omega} \langle \boldsymbol{F}, \boldsymbol{N} \rangle \, \mathrm{d}S = \iiint_{\Omega} \mathrm{Div}\boldsymbol{F}(x,y,z) \, \mathrm{d}(x,y,z) = 0.$$

Consequently, from the previous lemma,

$$\int_S \langle \boldsymbol{F}, \boldsymbol{N} \rangle \, \mathrm{d}S = \int_{\partial B} \langle \boldsymbol{F}, \boldsymbol{N} \rangle \, \mathrm{d}S = -4\pi GMm.$$

□

It should be noted that Gauss's law is usually formulated in terms of the electrical field generated by a point charge.

In what follows, we assume for simplicity that $m = 1$. Note that the flux of the gravitational field generated by a point mass M located at $\boldsymbol{P}$ in the interior of U depends only on M and not on the position $\boldsymbol{P}$. If we consider a series of point masses

$M_1, M_2, \ldots$ in the interior of U, the resulting vector field $\boldsymbol{F}$ is the sum of the vector fields generated by each one of the point masses, and the flux of $\boldsymbol{F}$ across S is given by

$$\sum_j -4\pi G M_j = -4\pi G M,$$

where

$$M = \sum_j M_j$$

is the total mass. Therefore, it is no surprise that *if $\boldsymbol{F}$ is the gravitational field generated by an object of total mass M located in the interior of U, then the flux of $\boldsymbol{F}$ across $S = \partial U$ equals $-4\pi G M$*. (see [11]).

9.7 Exercises

Exercise 9.7.1. For the regular surface

$$M := \{(x,y) \in \mathbb{R}^2 \; ; \; a^2 \leq x^2 + y^2 \leq b^2\}$$

with boundary, solve the following problems.

(1) Find a parameterization of the circles

$$x^2 + y^2 = a^2, \; x^2 + y^2 = b^2$$

that is compatible with the orientation of M.

(2) Using Green's formula, evaluate

$$\int_{\partial M} (x^3 - y) \, \mathrm{d}x + (\cos y + 2x) \, \mathrm{d}y.$$

Exercise 9.7.2. Let M be the part of the plane $x + y + z = 1$ whose projection onto the xy-plane is the closed disk centered at the origin and with radius 1. Suppose that M is oriented according to the normal vector $(1,1,1)$. Using Stokes's theorem, find

$$\int_{\partial M} x \, \mathrm{d}y + y \, \mathrm{d}z.$$

Exercise 9.7.3. Verify Stokes's theorem for the lower hemisphere of the unit sphere

$$z = -\sqrt{1 - x^2 - y^2}$$

and the vector field

$$\boldsymbol{F}(x, \; y, \; z) = (-y, \; x, \; z).$$

Exercise 9.7.4. Let M be the sphere with radius $a > 0$ centered at the origin and $\boldsymbol{F}$ a vector field of class C^2 on a neighborhood of the corresponding closed ball. Show that

$$\int_M \boldsymbol{\nabla} \times \boldsymbol{F} \cdot \mathrm{d}\boldsymbol{S} = 0.$$

Exercise 9.7.5. Let M be the triangle:

$$\{(x,y,z) \in \mathbb{R}^3 \; ; \; x + y + z = 1, \; x \geq 0, \; y \geq 0, \; z \geq 0\}$$

with the vertices $(1,0,0)$, $(0,1,0)$, and $(0,0,1)$ removed (Fig. 10.18). Verify the validity of Stokes's theorem for this surface.

Exercise 9.7.6. Find the work done in moving a particle counterclockwise around the (topological) boundary of the intersection of the plane $x+y+z=1$ with the first octant under the action of the force field

$$\boldsymbol{F}(x,y,z) = (x+xz^2,\ x,\ y).$$

Exercise 9.7.7. Verify the divergence theorem for the vector field

$$\boldsymbol{F}(x,y,z) = (x,\ 2y,\ z)$$

and the *closed surface* $M := M_1 \cup M_2$, where

$$M_1 := \{(x,y,z) \in \mathbb{R}^3\ ;\ x^2+y^2 = z^2,\ 0 \le z \le 2\}$$

and M_2 is the closed disk with radius 2 in the plane $z = 2$ and centered at $(0,0,2)$.

Exercise 9.7.8. Using Stokes's theorem, find the z-coordinate of the center of gravity of the hemisphere

$$M = \{(x,y,z) \in \mathbb{R}^3;\ x^2+y^2+z^2 = R^2, z \ge 0\}$$

assuming that the surface density is constant.

Exercise 9.7.9. Consider the vector field

$$\boldsymbol{F}(x,y,z) = (x,y,z)$$

and let M be the part of the paraboloid $z = 4-x^2-y^2$ that satisfies $z \ge 0$. Assuming that M is oriented so that the third coordinate of the normal vector is positive, find the relation between the flux of the vector field $\boldsymbol{F}$ across M and the flux of $\boldsymbol{F}$ across the disk D in the xy-plane centered at the origin, of radius 2 and oriented according to the normal vector $(0,0,-1)$.

Chapter 10
Solved Exercises

10.1 Solved Exercises of Chapter 1

1.3.3. Check that

$$\text{Div (Curl } \mathbf{F}) = 0$$

for the vector field

$$\mathbf{F}(x,y,z) = \left(x^2 z \ , \ x \ , \ 2yz\right).$$

Solution: The components of the vector field are

$$f_1(x,y,z) = x^2 z, \ f_2(x,y,z) = x, \ f_3(x,y,z) = 2yz.$$

Hence

$$\text{Curl } \mathbf{F} = \begin{vmatrix} \mathbf{e_1} & \mathbf{e_2} & \mathbf{e_3} \\ \frac{\partial}{\partial x} & \frac{\partial}{\partial y} & \frac{\partial}{\partial z} \\ x^2 z & x & 2yz \end{vmatrix} = (2z, x^2, 1).$$

That is, the components of Curl $\mathbf{F} = (g_1, g_2, g_3)$ are given by $g_1(x,y,z) = 2z$, $g_2(x,y,z) = x^2$, $g_3(x,y,z) = 1$. Consequently,

$$\begin{aligned} \text{Div(Curl } \mathbf{F}) &= \frac{\partial g_1}{\partial x} + \frac{\partial g_2}{\partial y} + \frac{\partial g_3}{\partial z} \\ &= 0. \end{aligned}$$

1.3.5. If $\mathbf{F}, \mathbf{G} : \mathbb{R}^3 \to \mathbb{R}^3$ are vector fields of class C^1, prove that

$$\text{Div}(\mathbf{F} \times \mathbf{G}) = \langle \text{Curl } \mathbf{F} \ , \ \mathbf{G} \rangle - \langle \mathbf{F} \ , \ \text{Curl } \mathbf{G} \rangle .$$

A. Galbis and M. Maestre, *Vector Analysis Versus Vector Calculus*, Universitext,
DOI 10.1007/978-1-4614-2200-6_10, © Springer Science+Business Media, LLC 2012

Solution: We put $\mathbf{F} = (f_1, f_2, f_3)$ and $\mathbf{G} = (g_1, g_2, g_3)$. According to Definition 1.2.11, we have

$$\langle \text{Curl}\, \mathbf{F}\,,\, \mathbf{G} \rangle = g_1 \left(\frac{\partial f_3}{\partial y} - \frac{\partial f_2}{\partial z} \right) + g_2 \left(\frac{\partial f_1}{\partial z} - \frac{\partial f_3}{\partial x} \right) + g_3 \left(\frac{\partial f_2}{\partial x} - \frac{\partial f_1}{\partial y} \right).$$

Analogously,

$$\langle \mathbf{F}\,,\, \text{Curl}\, \mathbf{G} \rangle = f_1 \left(\frac{\partial g_3}{\partial y} - \frac{\partial g_2}{\partial z} \right) + f_2 \left(\frac{\partial g_1}{\partial z} - \frac{\partial g_3}{\partial x} \right) + f_3 \left(\frac{\partial g_2}{\partial x} - \frac{\partial g_1}{\partial y} \right).$$

Moreover, from Definition 1.1.4 it follows that

$$\mathbf{F} \times \mathbf{G} = (h_1, h_2, h_3),$$

where

$$h_1 = f_2 g_3 - f_3 g_2\,,\; h_2 = f_3 g_1 - f_1 g_3\,,\; h_3 = f_1 g_2 - f_2 g_1,$$

and from Definition 1.2.10 we have

$$\text{Div}(\mathbf{F} \times \mathbf{G}) = \frac{\partial h_1}{\partial x} + \frac{\partial h_2}{\partial y} + \frac{\partial h_3}{\partial z}.$$

Finally, it is routine (although a tedious calculation) to check that

$$\frac{\partial h_1}{\partial x} + \frac{\partial h_2}{\partial y} + \frac{\partial h_3}{\partial z} = \langle \text{Curl}\, \mathbf{F}\,,\, \mathbf{G} \rangle - \langle \mathbf{F}\,,\, \text{Curl}\, \mathbf{G} \rangle.$$

10.2 Solved Exercises of Chapter 2

2.8.3 Integrate the vector field

$$\boldsymbol{F}(x,y) = (y^2, -2xy)$$

along the triangle with vertices $(0,0),(1,0),(0,1)$, oriented counterclockwise.

Solution: The triangle is the union of three segments. A parameterization of the segment from $(0,0)$ to $(1,0)$ is

$$\gamma_1 : [0,1] \to \mathbb{R}^2, \quad \gamma_1(t) = (t,0).$$

A parameterization of the segment from $(1,0)$ to $(0,1)$ is

$$\gamma_2 : [0,1] \to \mathbb{R}^2, \quad \gamma_2(t) = (1-t)(1,0) + t(0,1) = (1-t,t).$$

Instead of the segment from $(0,1)$ to $(0,0)$ we consider the opposite path (the segment from $(0,0)$ to $(0,1)$), which is easier. This is

$$\gamma_3 : [0,1] \to \mathbb{R}^2, \quad \gamma_3(t) = (0,t).$$

Then the path along which we have to integrate the vector field $\boldsymbol{F}$ is the union of three paths,

$$\gamma = \gamma_1 \cup \gamma_2 \cup (-\gamma_3),$$

and consequently,

$$\int_\gamma \boldsymbol{F} = \int_{\gamma_1} \boldsymbol{F} + \int_{\gamma_2} \boldsymbol{F} - \int_{\gamma_3} \boldsymbol{F}.$$

We now evaluate

$$\int_{\gamma_1} \boldsymbol{F} = \int_{\gamma_1} y^2 \, dx - 2xy \, dy$$

$$= \int_0^1 0 \, dt = 0$$

and also

$$\int_{\gamma_3} F = 0.$$

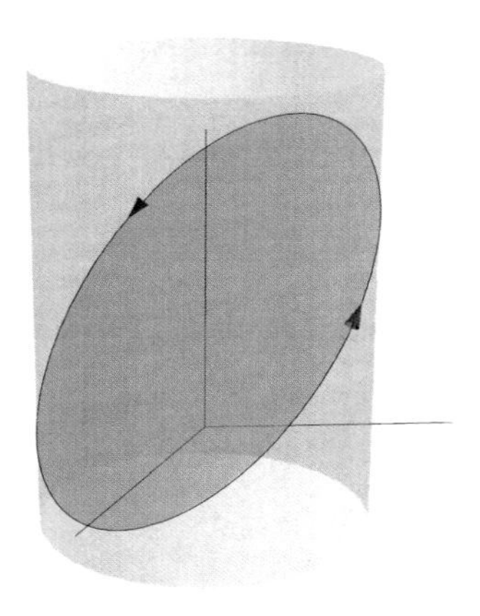

Fig. 10.1 Path of Exercise 2.8.4

Moreover,

$$\int_{\gamma_2} F = \int_{\gamma_2} y^2\,dx - 2xy\,dy$$

$$= \int_0^1 \left(t^2(-1) - 2(1-t)t\right) dt$$

$$= \int_0^1 (t^2 - 2t)\,dt = \frac{1}{3} - 1.$$

Finally,

$$\int_{\gamma} F = \frac{1}{3} - 1.$$

2.8.4 (1) Find a path γ whose trajectory is the intersection of the cylinder $x^2+y^2=1$ with the plane $x+y+z=1$ and with the additional properties that the initial (and final) point is $(0,-1,2)$ and the projection onto the plane XY is oriented counterclockwise (Fig. 10.1).

(2) Evaluate

$$\int_{\gamma} xy\,dx \;+\; yz\,dy \;-\; x\,dz.$$

Solution: (1) The intersection of the cylinder with the plane is an ellipse in $\mathbb{R}^3$ whose projection onto the XY-plane is the unit circle. Consequently, we first parameterize the unit circle in such a way that the initial and final point is $(0,-1)$ and it is oriented counterclockwise. That is, we put

$$x = \cos t \;;\; y = \sin t \quad \left(-\frac{\pi}{2} \le t \le \frac{3\pi}{2}\right).$$

From the equation of the plane we deduce

$$z = 1 - x - y = 1 - \cos t - \sin t.$$

The path we are looking for is

$$\gamma : \left[-\frac{\pi}{2}, \frac{3\pi}{2}\right] \to \mathbb{R}^3, \quad \gamma(t) = (\cos t\ ,\ \sin t\ ,\ 1 - \cos t - \sin t).$$

(2) We now define

$$\omega = xy\,\mathrm{d}x + yz\,\mathrm{d}y - x\,\mathrm{d}z.$$

We could evaluate

$$\int_\gamma \omega = \int_{-\frac{\pi}{2}}^{\frac{3\pi}{2}} \omega(\gamma(t))(\gamma'(t))\,\mathrm{d}t,$$

but a more efficient method consists in making a *formal* substitution in the integral

$$x = \cos t\ ;\ y = \sin t, \quad z = 1 - \cos t - \sin t$$

and also

$$\mathrm{d}x = (-\sin t)\,\mathrm{d}t, \quad \mathrm{d}y = \cos t\,\mathrm{d}t, \quad \mathrm{d}z = (\sin t - \cos t)\,\mathrm{d}t$$

to obtain

$$\int_\gamma \omega = \int_{-\frac{\pi}{2}}^{\frac{3\pi}{2}} \left(-2\sin^2 t\cos t - \sin t\cos^2 t + \cos^2 t\right)\mathrm{d}t.$$

Moreover,

$$\int_{-\frac{\pi}{2}}^{\frac{3\pi}{2}} \sin^2 t\ \cos t\ \mathrm{d}t = \frac{\sin^3 t}{3}\bigg|_{-\frac{\pi}{2}}^{\frac{3\pi}{2}} = 0,$$

$$\int_{-\frac{\pi}{2}}^{\frac{3\pi}{2}} -\sin t\ \cos^2 t\ \mathrm{d}t = \frac{\cos^3 t}{3}\bigg|_{-\frac{\pi}{2}}^{\frac{3\pi}{2}} = 0,$$

and

$$\int_{-\frac{\pi}{2}}^{\frac{3\pi}{2}} \cos^2 t\ \mathrm{d}t = \int_{-\frac{\pi}{2}}^{\frac{3\pi}{2}} \frac{1+\cos(2t)}{2}\,\mathrm{d}t = \frac{t}{2} + \frac{\sin(2t)}{4}\bigg|_{-\frac{\pi}{2}}^{\frac{3\pi}{2}} = \pi.$$

Finally,

$$\int_\gamma \omega = \pi.$$

2.8.6 Find a path whose trajectory is the intersection of the upper hemisphere of the sphere with radius $2a$ (Fig. 10.2)

$$x^2 + y^2 + z^2 = 4a^2$$

with the cylinder

$$x^2 + (y-a)^2 = a^2.$$

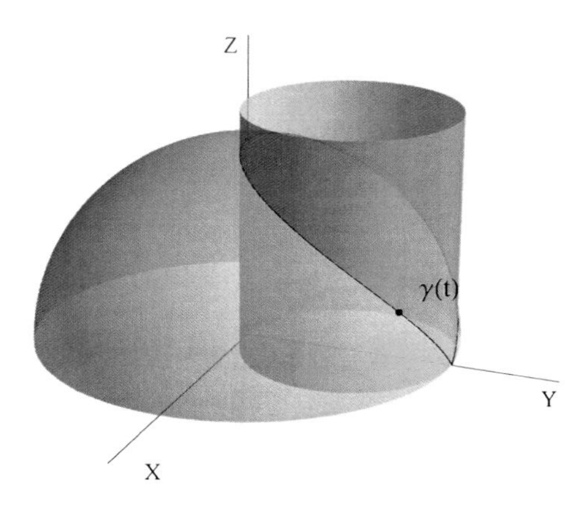

Fig. 10.2 Path of Exercise 2.8.6

Solution: If (x,y,z) is a point on that trajectory, then (x,y) is on the circle

$$x^2+(y-a)^2=a^2.$$

Hence,

$$x = a\cos t, \quad y-a=a\sin t$$

for some value of t between 0 and 2π. After substituting in the equation of the sphere, we obtain

$$a^2\cos^2 t+(a+a\sin t)^2+z^2=4a^2.$$

That is,

$$z^2=2a^2(1-\sin t),$$

and from the fact that $z\geq 0$ (since we are considering the upper hemisphere), we conclude that

$$z=a\sqrt{2(1-\sin t)}.$$

Consequently, the sought-after path is

$$\boldsymbol{\gamma}:[0,2\pi]\to\mathbb{R}^3$$

defined by

$$\boldsymbol{\gamma}(t)=\Big(a\cos t\ ,\ a+a\sin t\ ,\ a\sqrt{2(1-\sin t)}\Big).$$

2.8.10 (1) Show that the vector field

$$\boldsymbol{F}(x,y)=(\mathrm{e}^x(\sin(x+y)+\cos(x+y))+1\ ,\ \mathrm{e}^x\cos(x+y))$$

is conservative on $\mathbb{R}^2$ and find a potential.
(2) Evaluate the line integral

$$\int_{\boldsymbol{\gamma}}\boldsymbol{F},$$

where

$$\boldsymbol{\gamma} : [0,\pi] \to \mathbb{R}^2, \quad \boldsymbol{\gamma}(t) = \left(\sin(\pi e^{\sin(t)}), \cos^5(t)\right).$$

Solution: (1) Writing

$$P(x,y) = e^x(\sin(x+y) + \cos(x+y)) + 1$$

and

$$Q(x,y) = e^x \cos(x+y),$$

it follows that

$$\frac{\partial Q}{\partial x} = e^x(\cos(x+y) - \sin(x+y)) = \frac{\partial P}{\partial y}.$$

Hence $\boldsymbol{F}$ is conservative. If $f : \mathbb{R}^2 \to \mathbb{R}$ is a potential function of $\boldsymbol{F}$, then

$$\frac{\partial f}{\partial x} = e^x(\sin(x+y) + \cos(x+y)) + 1 \tag{10.1}$$

and

$$\frac{\partial f}{\partial y} = e^x \cos(x+y). \tag{10.2}$$

By integrating (10.2) with respect to y we obtain

$$f(x,y) = e^x \sin(x+y) + \varphi(x).$$

We now compute the derivative with respect to x and substitute in (10.1) to get

$$e^x(\sin(x+y) + \cos(x+y)) + \varphi'(x) = e^x(\sin(x+y) + \cos(x+y)) + 1,$$

whence it follows that $\varphi'(x) = 1$. Finally, we take $\varphi(x) = x$, and we conclude that

$$f(x,y) = e^x \sin(x+y) + x$$

is a potential function for $\boldsymbol{F}$.
(2) The computation of the line integral (2) using Definition 2.2.1 could be rather cumbersome due to the expression for the path $\boldsymbol{\gamma}$. Fortunately, since the vector field is conservative, we can apply Theorem 2.5.1 to get

$$\int_{\boldsymbol{\gamma}} \boldsymbol{F} = f(\boldsymbol{\gamma}(\pi)) - f(\boldsymbol{\gamma}(0)).$$

Since $\boldsymbol{\gamma}(0) = (0,1)$ and $\boldsymbol{\gamma}(\pi) = (0,-1)$, we have

$$\int_{\boldsymbol{\gamma}} \boldsymbol{F} = f(0,-1) - f(0,1) = -2\sin 1.$$

2.8.12 For the vector field

$$\boldsymbol{F} = (f_1, f_2) : \mathbb{R}^2 \setminus \{(0,0)\} \to \mathbb{R}^2$$

defined by

$$f_1(x,y) = \frac{-y}{x^2+y^2}, \quad f_2(x,y) = \frac{x}{x^2+y^2}:$$

(1) Show that

$$\frac{\partial f_2}{\partial x}(x,y) = \frac{\partial f_1}{\partial y}(x,y)$$

for all $(x,y) \in \mathbb{R}^2 \setminus \{(0,0)\}$.
(2) Let $\boldsymbol{\gamma}$ be the unit circle oriented counterclockwise. Show that

$$\int_{\boldsymbol{\gamma}} \boldsymbol{F} = 2\pi.$$

Is this fact a contradiction to Poincaré's lemma?
(3) Argue whether this statement is true: for every closed and piecewise C^1 path $\boldsymbol{\alpha} : [a,b] \to \mathbb{R}^2 \setminus \{(0,0)\}$ such that $\alpha_1(t) \geq 0$ for all $t \in [a,b]$, the following holds:

$$\int_{\boldsymbol{\alpha}} \boldsymbol{F} = 0.$$

(4) Evaluate $\int_{\boldsymbol{\gamma}} \boldsymbol{F}$, where

$$\boldsymbol{\gamma}(t) = \left(\cos(t), \sin^7(t)\right), \quad -\frac{\pi}{2} \leq t \leq \frac{\pi}{2}.$$

Solution: (1) An easy computation gives

$$\frac{\partial f_1}{\partial y}(x,y) = \frac{\partial f_2}{\partial x}(x,y) = \frac{y^2-x^2}{(x^2+y^2)^2}$$

for every $(x,y) \in \mathbb{R}^2 \setminus \{(0,0)\}$.
(2) A parameterization of the unit circle is

$$\boldsymbol{\gamma} : [0,2\pi] \to \mathbb{R}^2, \quad \boldsymbol{\gamma}(t) = (\cos t\ ,\ \sin t).$$

By substituting *formally*

$$x = \cos t\ , y = \sin t\ , \mathrm{d}x = -\sin t\ \mathrm{d}t\ , \mathrm{d}y = \cos t\ \mathrm{d}t,$$

we obtain

$$\int_{\gamma} \boldsymbol{F} = \int_{\gamma} f_1(x,y)\,\mathrm{d}x + f_2(x,y)\,\mathrm{d}y$$
$$= \int_0^{2\pi} \frac{\sin^2 t + \cos^2 t}{\sin^2 t + \cos^2 t}\,\mathrm{d}t = 2\pi. \tag{10.3}$$

The fact that the line integral is different from zero is not a contradiction to Poincaré's lemma, since the open set $\mathbb{R}^2 \setminus \{(0,0)\}$ is not starlike.
(3) The open set

$$U := \mathbb{R}^2 \setminus \Big( ]-\infty, 0] \times \{0\} \Big)$$

is starlike with respect to the point $(1,0)$. The set U is obtained by deleting the nonpositive part of the X axis. Now condition (1) and Poincaré's lemma permits the conclusion that $\boldsymbol{F}$ is a conservative vector field in U. That is, there exists a C^1 function

$$f : U \subset \mathbb{R}^2 \to \mathbb{R}$$

such that

$$\nabla f(x,y) = \boldsymbol{F}(x,y)$$

for all $(x,y) \in U$. If $\boldsymbol{\alpha} = (\alpha_1, \alpha_2) : [a,b] \to \mathbb{R}^2 \setminus \{(0,0)\}$ is a closed and piecewise C^1 path such that $\alpha_1(t) \geq 0$ for all $t \in [a,b]$, then it follows that $\boldsymbol{\alpha}([a,b]) \subset U$, and because $\boldsymbol{F}$ is conservative in U, we get

$$\int_{\boldsymbol{\alpha}} \boldsymbol{F} = 0.$$

(4) The trajectory of the path $\boldsymbol{\gamma}$ is contained in U. Since $\boldsymbol{F}$ is conservative in U, the value of the line integral does not change after replacing $\boldsymbol{\gamma}$ by any other path contained in U with initial point $\boldsymbol{\gamma}(-\frac{\pi}{2}) = (0,-1)$ and final point $\boldsymbol{\gamma}(\frac{\pi}{2}) = (0,1)$. We can consider, for instance, the semicircle

$$\boldsymbol{\beta} : \left[-\frac{\pi}{2}, \frac{\pi}{2}\right] \to \mathbb{R}^2, \quad \boldsymbol{\beta}(t) = (\cos t, \sin t),$$

which gives an easier expression for the line integral (Fig. 10.3). In fact, proceeding as in (10.3) we obtain

$$\int_{\gamma} \boldsymbol{F} = \int_{\boldsymbol{\beta}} \boldsymbol{F} = \int_{-\frac{\pi}{2}}^{\frac{\pi}{2}} \mathrm{d}t = \pi.$$

2.8.14 Let $\boldsymbol{\gamma}$ be the path whose trajectory is the union of the graph of $y = x^3$ from $(0,0)$ to $(1,1)$ and the segment from $(1,1)$ to $(0,0)$. Using Green's theorem, evaluate (Fig. 10.4)

$$\int_{\gamma} (x^2 + y^2)\,\mathrm{d}x + (2xy + x^2)\,\mathrm{d}y.$$

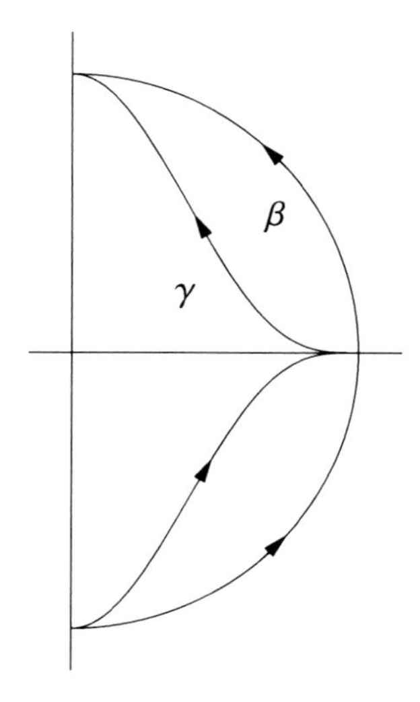

Fig. 10.3 The integral of a conservative field along $\boldsymbol{\gamma}$ coincides with the integral along $\boldsymbol{\beta}$

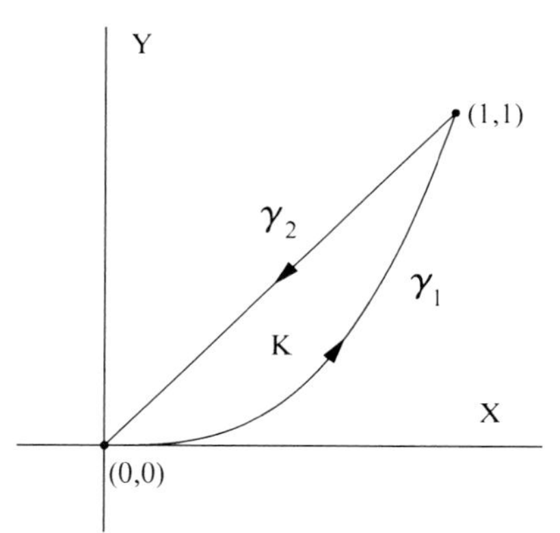

Fig. 10.4 Path of Exercise 2.8.14

Solution: $\boldsymbol{\gamma} = \boldsymbol{\gamma}_1 \cup \boldsymbol{\gamma}_2$ is the boundary, oriented counterclockwise, of the region of type I

$$K := \{(x,y) \in \mathbb{R}^2, \quad 0 \leq x \leq 1 \,,\, x^3 \leq y \leq x\}.$$

Writing

$$P(x,y) = x^2 + y^2, \quad Q(x,y) = 2xy + x^2,$$

Green's theorem gives

$$\begin{aligned}\int_{\boldsymbol{\gamma}} P(x,y)\ \mathrm{d}x + Q(x,y)\ \mathrm{d}y &= \iint_K \left(\frac{\partial Q}{\partial x} - \frac{\partial P}{\partial y}\right)\ \mathrm{d}(x,y)\\ &= \int_0^1 \left(\int_{x^3}^x 2x\ \mathrm{d}y\right)\ \mathrm{d}x\\ &= \int_0^1 2x(x - x^3)\ \mathrm{d}x = \frac{4}{15}.\end{aligned}$$

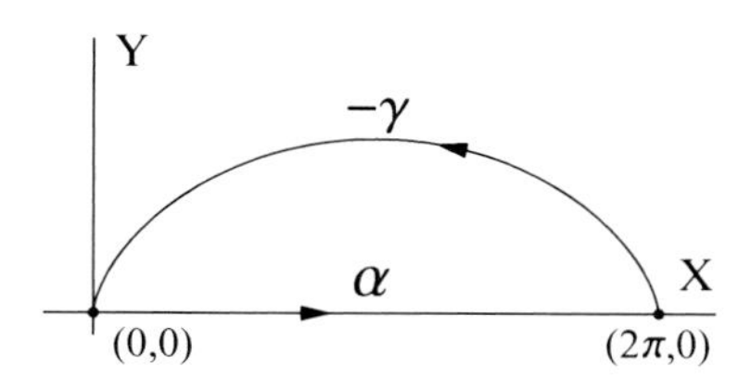

Fig. 10.5 Path of Exercise 2.8.17

2.8.16 Let K be a region of type I or type II and let $\boldsymbol{\gamma}$ be a path whose trajectory is the boundary of K oriented counterclockwise. Then

$$\text{area}\ (K) = \int_{\boldsymbol{\gamma}} \frac{1}{2}(-y\ \mathrm{d}x + x\ \mathrm{d}y) = \int_{\boldsymbol{\gamma}} -y\ \mathrm{d}x = \int_{\boldsymbol{\gamma}} x\ \mathrm{d}y.$$

Solution: The choice $P(x,y) = -\frac{y}{2}$ and $Q(x,y) = \frac{x}{2}$ produces the following line integral when Green's theorem is applied:

$$\int_{\boldsymbol{\gamma}} P\ \mathrm{d}x + Q\ \mathrm{d}y = \iint_K \left(\frac{\partial Q}{\partial x} - \frac{\partial P}{\partial y}\right)\ \mathrm{d}(x,y)$$

$$= \iint_K \mathrm{d}(x,y) = \text{area}\ (K).$$

2.8.17 Evaluate the area bounded by the cycloid $\boldsymbol{\gamma} : [0,2\pi] \to \mathbb{R}^2$,

$$\boldsymbol{\gamma}(t) = (at - a\sin(t), a - a\cos(t))$$

$(a > 0)$, and the X-axis (Fig. 10.5).

Solution: We observe that $\boldsymbol{\gamma}(0) = (0,0)$ and $\boldsymbol{\gamma}(2\pi) = (a2\pi,0)$. Moreover,

$$\gamma_1(t) := at - a\sin(t)$$

is strictly increasing on $[0,2\pi]$, while

$$\gamma_2(t) := a - a\cos(t) > 0,$$

whenever $0 < t < 2\pi$. Hence the trajectory of the path $\boldsymbol{\gamma}$ and the Xaxis limit a region K of type I with boundary, oriented counterclockwise,

$$\boldsymbol{\alpha} \cup (-\boldsymbol{\gamma}),$$

where $\boldsymbol{\alpha}$ is the segment from $(0,0)$ to $(a2\pi,0)$. We put

$$\boldsymbol{\omega} = \frac{1}{2}(-y\ \mathrm{d}x + x\ \mathrm{d}y).$$

From Exercise 2.8.16, the area of K is given by

$$A := \int_{\alpha} \omega - \int_{\gamma} \omega.$$

Because $y = 0$ on the segment from $(0,0)$ to $(a2\pi, 0)$, we get

$$\int_{\alpha} \omega = 0$$

and

$$A = -\frac{1}{2} \int_{\gamma} -y\,\mathrm{d}x + x\,\mathrm{d}y.$$

In order to evaluate this line integral we substitute

$$x = a(t - \sin t), \;\; \mathrm{d}x = a(1 - \cos t)\mathrm{d}t,$$

$$y = a(1 - \cos t),\; \mathrm{d}y = a \sin t \mathrm{d}t,$$

Then

$$A = a^2 \int_0^{2\pi} \left(1 - \cos t - t\frac{\sin t}{2}\right)\,\mathrm{d}t.$$

Since $\displaystyle\int_0^{2\pi} \cos t\;\mathrm{d}t = 0$ and integrating by parts yields

$$\int_0^{2\pi} -t \sin t\;\mathrm{d}t = t\cos t\Big|_0^{2\pi} - \int_0^{2\pi} \cos t\;\mathrm{d}t = 2\pi,$$

we conclude that

$$A = 3\pi a^2.$$

2.8.19 Evaluate the area limited by the circles

$$C_1 := \left\{(x,y) \in \mathbb{R}^2 \;:\; x^2 + y^2 = a^2\right\}$$

and

$$C_2 := \left\{(x,y) \in \mathbb{R}^2 \;:\; x^2 + y^2 = 2ax\right\} \;(a > 0).$$

Solution: The circle C_1 is centered at the origin and has radius a, while the circle C_2 is centered at the point $(a,0)$ and also has radius a. The two circles meet at the solutions of the following system of equations

$$\begin{cases} x^2 + y^2 = a^2, \\ x^2 + y^2 = 2ax. \end{cases} \tag{10.4}$$

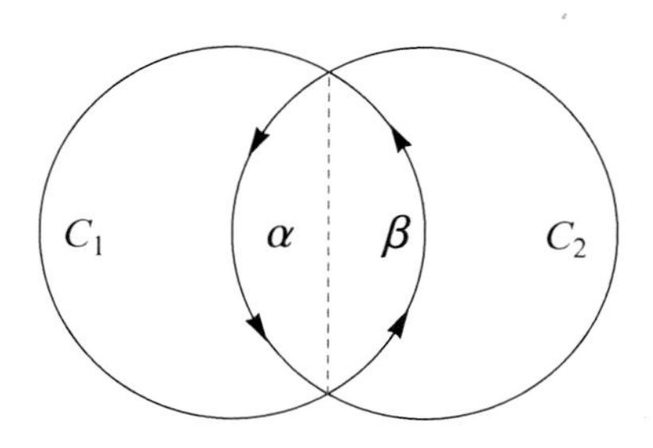

Fig. 10.6 Positively oriented boundary of the region limited by two circles

From $a^2 = 2ax$ we deduce $x = \frac{a}{2}$; hence $y = \pm\frac{\sqrt{3}}{2}a$. Consequently, the vertical line $x = \frac{a}{2}$ decomposes the region limited by the two circles into two pieces (Fig. 10.6). It is a region of type II whose boundary, oriented counterclockwise, is given by

$$\boldsymbol{\gamma} = \boldsymbol{\alpha} \cup \boldsymbol{\beta}.$$

The trajectory of $\boldsymbol{\alpha}$ is the semicircle located to the left of the line $x = \frac{a}{2}$, and an appropriate parameterization is

$$\boldsymbol{\alpha} : \left[\frac{2\pi}{3}, \frac{4\pi}{3}\right] \to \mathbb{R}^2, \quad \boldsymbol{\alpha}(t) = (a + a\cos t, a\sin t).$$

Also, the trajectory of $\boldsymbol{\beta}$ is the semicircle located to the right of the line $x = \frac{x}{2}$, and an appropriate parameterization is

$$\boldsymbol{\beta} : \left[-\frac{\pi}{3}, \frac{\pi}{3}\right] \to \mathbb{R}^2, \quad \boldsymbol{\beta}(t) = (a\cos t, a\sin t).$$

Since

$$\int_{\boldsymbol{\alpha}} -y\,\mathrm{d}x + x\,\mathrm{d}y = \int_{\frac{2\pi}{3}}^{\frac{4\pi}{3}} a^2(1 + \cos t)\,\mathrm{d}t$$

$$= a^2\left(\frac{2\pi}{3} - \sqrt{3}\right)$$

and

$$\int_{\boldsymbol{\beta}} -y\,\mathrm{d}x + x\,\mathrm{d}y = \int_{-\frac{\pi}{3}}^{\frac{\pi}{3}} a^2\,\mathrm{d}t = a^2\frac{2\pi}{3},$$

we conclude that the area is

$$A = \frac{1}{2}\int_{\boldsymbol{\gamma}} -y\,\mathrm{d}x + x\,\mathrm{d}y = a^2\left(\frac{2\pi}{3} - \frac{\sqrt{3}}{2}\right).$$

10.3 Solved Exercises of Chapter 3

3.5.2 Show that the cone

$$S := \{(x,y,z) \in \mathbb{R}^3 \ ; x^2+y^2 = z^2 \ , \ z \geq 0\}$$

is not a regular surface, but $S \setminus \{(0,0,0)\}$, the cone without the *vertex*, is.

Solution: Let us assume that S is a regular surface. According to Proposition 3.1.1, there exist an open neighborhood $W \subset \mathbb{R}^2$ of $(0,0)$, an open neighborhood U of $(0,0,0)$ in $\mathbb{R}^3$, and a function of class C^1,

$$g : W \subset \mathbb{R}^2 \to \mathbb{R},$$

such that

(a) $S \cap U = \{(x,y,g(x,y)), \quad (x,y) \in W\}$

or

(b) $S \cap U = \{(x,g(x,z),z)), \quad (x,z) \in W\}$

or

(c) $S \cap U = \{(g(y,z),y,z), \quad (y,z) \in W\}$.

Since the projections onto the xz- and yz-planes are not injective on $S \cap U$, the cases (b) and (c) cannot occur. Consequently,

$$S \cap U = \{(x,y,g(x,y)), \quad (x,y) \in W\}.$$

In particular,

$$g(x,y) = \sqrt{x^2+y^2}$$

for every $(x,y) \in W$; hence g does not admit partial derivatives at the point $(0,0)$, which is a contradiction, since g is a function of class C^1. Consequently, S is not a regular surface. However, $S \setminus \{(0,0,0)\}$ is a regular surface, since it is the graph of the function of class C^1

$$g : \mathbb{R}^2 \setminus \{(0,0)\} \to \mathbb{R}.$$

3.5.3 Determine whether the intersection of the cone (Fig. 10.7)

$$x^2+y^2 = z^2$$

with the plane

$$x+y+z = 1$$

is a regular curve.

Solution. The intersection of the cone with the plane can be described as

$$M := \boldsymbol{\Phi}^{-1}\{(0,0)\},$$

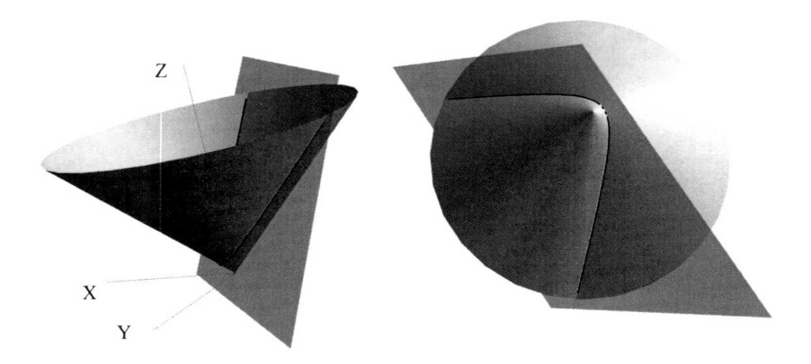

Fig. 10.7 Intersection of a cone and a plane

where $\boldsymbol{\Phi} : \mathbb{R}^3 \to \mathbb{R}^2$ is the function defined by

$$\boldsymbol{\Phi}(x,y,z) := (x^2 + y^2 - z^2 \,,\; x + y + z - 1).$$

Since $\boldsymbol{\Phi}$ is a function of class C^∞, it suffices (by Proposition 3.2.1) to check that the Jacobian matrix of $\boldsymbol{\Phi}$ has rank 2 at each point $(x,y,z) \in M$. That Jacobian matrix is

$$A := \begin{pmatrix} 2x & 2y & -2z \\ 1 & 1 & 1 \end{pmatrix},$$

and it has rank 2 in $U := \mathbb{R}^3 \setminus \{(t,t,-t) : t \in \mathbb{R}\}$. Since M is contained in U, we conclude that the rank of the matrix is 2 at each point $(x,y,z) \in M$, and consequently, M is a regular curve (regular surface of dimension 1) of class C^∞.

3.5.4 Let

$$\boldsymbol{\gamma} : [a,b] \to U \subset \mathbb{R} \times (0,+\infty)$$

be a path such that $\boldsymbol{\gamma}([a,b])$ is a level curve. Show that the set obtained by rotating $\boldsymbol{\gamma}([a,b])$ around the x-axis is a level surface (Fig. 10.8).

Solution: Let $f : U \subset \mathbb{R}^2 \to \mathbb{R}$ be a C^1 function on the open set U, and let $c \in \mathbb{R}$ be such that

$$\boldsymbol{\gamma}([a,b]) = \{(x,y) \in U, \quad f(x,y) = c\}.$$

If we select an arbitrary point $(x_0,y_0) \in \boldsymbol{\gamma}([a,b])$ and rotate $(x_0,y_0,0) \in \mathbb{R}^3$ through an angle s about the x-axis, we get the point with coordinates

$$x = x_0, \quad y = y_0 \cos s, \quad z = y_0 \sin s.$$

That is, the x-coordinate of the point does not change, and we make a rotation of angle s in the yz-plane of the point with coordinates $y = y_0$, $z = 0$. Since $y_0 > 0$, the points of the yz-plane of the form

$$(y_0 \cos s, y_0 \sin s)$$

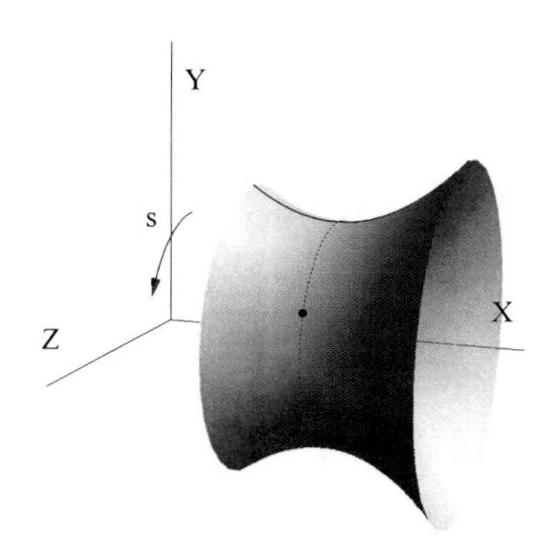

Fig. 10.8 Surface of revolution of Exercise 3.5.4

are precisely the points (y,z) satisfying

$$y_0 = \sqrt{y^2+z^2}.$$

We consider the continuous function

$$\boldsymbol{h}: \mathbb{R}^3 \to \mathbb{R}^2, \quad \boldsymbol{h}(x,y,z) = \left(x, \sqrt{y^2+z^2}\right)$$

and $V := \boldsymbol{h}^{-1}(U)$, which is an open subset of $\mathbb{R}^3$. Finally, we define

$$g := f \circ \boldsymbol{h} : V \subset \mathbb{R}^3 \to \mathbb{R}.$$

Then the set obtained after rotating $\boldsymbol{\gamma}([a,b])$ about the x-axis is

$$M = \{(x,y,z) \in V\ ; g(x,y,z) = c\}.$$

The obtained level surface is known as a *surface of revolution*, and as we have just proven, it is described by the equation

$$f\left(x, \sqrt{y^2+z^2}\right) = c.$$

3.5.8 Let $0 < r < R$.
(1) Show that the torus obtained by rotation of the circle

$$x^2 + (y-R)^2 = r^2$$

about the x-axis is a regular surface.
(2) Find a coordinate system (Fig. 10.9).

Solution: (1) We define

$$f: \mathbb{R}^2 \to \mathbb{R}, \quad f(x,y) = x^2 + (y-R)^2.$$

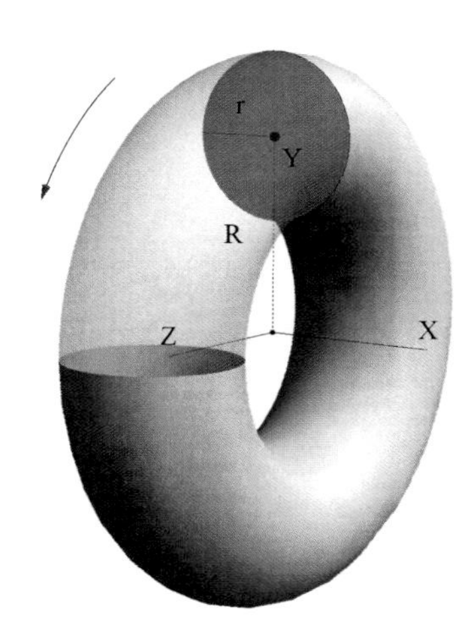

Fig. 10.9 Torus of Exercise 3.5.8

Then the torus is the level surface

$$M = \left\{(x,y,z) \in \mathbb{R}^3,\ f\left(x, \sqrt{y^2+z^2}\right) = r^2\right\}.$$

In order to conclude that M is a regular surface, it suffices to check that

$$g(x,y,z) := x^2 + \left(\sqrt{y^2+z^2} - R\right)^2 = x^2+y^2+z^2+R^2-2R\sqrt{y^2+z^2}$$

has gradient different from zero on an open set containing M. Put

$$U := \mathbb{R}^3 \setminus \left(\{(0,y,z) :\ y^2+z^2 = R^2\} \cup \{(0,0,0)\}\right),$$

which is an open set containing M. Since $\boldsymbol{\nabla} g(x,y,z)$ is the vector

$$\left(2x,\ 2y - \frac{2Ry}{\sqrt{y^2+z^2}},\ 2z - \frac{2Rz}{\sqrt{y^2+z^2}}\right),$$

it follows that $\boldsymbol{\nabla} g$ does not vanish on the open set U containing M.
(2) We first fix a parameterization of the circle. For instance

$$\boldsymbol{\gamma} = (\gamma_1, \gamma_2) : (0,2\pi) \to \mathbb{R}^2\ ;\ \boldsymbol{\gamma}(t) = (r\cos t, R + r\sin t).$$

If we rotate a point $(\gamma_1(t), \gamma_2(t), 0)$ an angle s about the x-axis, we obtain the point with coordinates

$$x = \gamma_1(t), \quad y = \gamma_2(t)\cos s, \quad z = \gamma_2(t)\sin s.$$

This suggests the definition

$$\varphi : \mathbb{R}^2 \to M \subset \mathbb{R}^3, \quad \varphi(s,t) = \big(\gamma_1(t), \gamma_2(t)\cos s, \gamma_2(t)\sin s\big).$$

The mapping φ is of class C^1 on $\mathbb{R}^2$. If

$$\varphi(s_1,t_1) = \varphi(s_2,t_2),$$

then $\gamma_1(t_1) = \gamma_1(t_2)$ and

$$\gamma_2(t_1)(\cos s_1, \sin s_1) = \gamma_2(t_2)(\cos s_2, \sin s_2).$$

Evaluating norms in the last identity, we deduce $\gamma_2(t_1) = \gamma_2(t_2)$. Hence $\boldsymbol{\gamma}(t_1) = \boldsymbol{\gamma}(t_2)$ and $(\cos s_1, \sin s_1) = (\cos s_2, \sin s_2)$. From the injectivity of $\boldsymbol{\gamma}$ and the properties of the functions sin and cos, it follows that $t_1 = t_2$ and $s_1 - s_2$ is a multiple of 2π. In particular, φ is injective on each open set

$$W_c := \ (c, c+2\pi) \ \times \ (0, 2\pi) \subseteq \mathbb{R}^2 \ \ (c \in \mathbb{R}).$$

We show that

$$(W_c, \ \varphi)$$

is a coordinate system. Since $\varphi : W_c \to \varphi(W_c)$ is a continuous bijection, it suffices (according to Corollary 3.3.1) to check that $\varphi'(s,t)$ has rank 2, or equivalently,

$$\left(\frac{\partial \varphi}{\partial s} \times \frac{\partial \varphi}{\partial t}\right)(s,t) \neq (0,0,0)$$

for every (s,t). Since,

$$\left(\frac{\partial \varphi}{\partial s} \times \frac{\partial \varphi}{\partial t}\right)(s,t) = \gamma_2(t)\big(-\gamma_2'(t), \gamma_1'(t)\cos s, \gamma_1'(t)\sin s\big)$$

has norm

$$\begin{aligned}\left\|\left(\tfrac{\partial \varphi}{\partial s} \times \tfrac{\partial \varphi}{\partial t}\right)(s,t)\right\|^2 &= \gamma_2(t)^2\big(\gamma_2'(t)^2 + \gamma_1'(t)^2\big) \\ &= r^2\big(R + r\sin t\big)^2 > 0,\end{aligned}$$

we are done.

3.5.9 Find the tangent plane to the paraboloid (Fig. 10.10) $z = x^2 + y^2$ at the point $(\frac{1}{2}, \frac{1}{2}, \frac{1}{2})$.

Solution: The paraboloid can be described as

$$M = \Phi^{-1}\{(0)\}$$

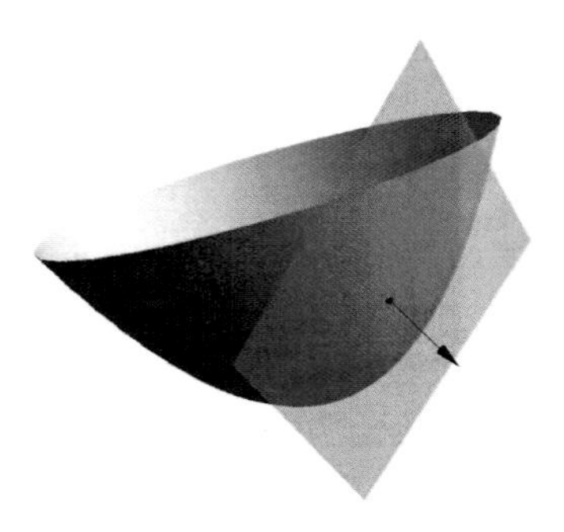

Fig. 10.10 Tangent plane to a paraboloid

for $\Phi(x,y,z) = z - x^2 - y^2$. By Proposition 3.4.1, the vector

$$\nabla\Phi\left(\frac{1}{2},\frac{1}{2},\frac{1}{2}\right) = (-1,-1,1)$$

is orthogonal to the tangent plane we are asked to find. Hence, the equation of the tangent plane is

$$\left\langle (x,y,z) - \left(\frac{1}{2},\frac{1}{2},\frac{1}{2}\right) , (-1,-1,1) \right\rangle = 0,$$

that is,

$$x + y - z = \frac{1}{2}.$$

An alternative approach consists in evaluating a basis for the plane determined by the tangent vectors to the paraboloid. To do this, we consider the following parameterization of the surface:

$$\varphi : \mathbb{R}^2 \to M, \quad \varphi(s,t) = (s\ ,\ t\ ,\ s^2 + t^2).$$

Since $\varphi(\frac{1}{2},\frac{1}{2}) = (\frac{1}{2},\frac{1}{2},\frac{1}{2})$, it turns out that a basis of the tangent plane is given by the vectors

$$\frac{\partial\varphi}{\partial s}\left(\frac{1}{2},\frac{1}{2}\right) = (1,0,1)$$

and

$$\frac{\partial\varphi}{\partial t}\left(\frac{1}{2},\frac{1}{2}\right) = (0,1,1).$$

Consequently, the equation of the tangent plane to M is given by

$$\begin{vmatrix} x-\frac{1}{2} & y-\frac{1}{2} & z-\frac{1}{2} \\ 1 & 0 & 1 \\ 0 & 1 & 1 \end{vmatrix} = 0.$$

Evaluating the previous determinant, we again get

$$x + y - z = \frac{1}{2}.$$

10.4 Solved Exercises of Chapter 4

4.4.2 Let M be a regular surface of class C^1 in $\mathbb{R}^3$ and suppose that (W,φ) is a coordinate system of M,

$$\varphi : W \subset \mathbb{R}^2 \to M \subset \mathbb{R}^3,$$

with the property that $M \setminus \varphi(W)$ can be decomposed as the union of a finite family of subsets of M each of which is the image of a null set through some parameterization of M. Then

$$\text{area}\ M = \text{area}\ (W,\varphi).$$

Solution: By hypothesis there is a finite family $\{(A_j,\varphi_j) : \ j = 1,\dots,m\}$ of coordinate systems of M, and for each $j = 1,\dots,m$, there is a null set $N_j \subset A_j$ such that

$$M \setminus \varphi(W) = \bigcup_{1\le j\le m} \varphi_j(N_j).$$

Put $K_1 = N_1$, and for $2 \le j \le m$,

$$K_j := \varphi_j^{-1}\Big(\varphi_j(N_j) \setminus \bigcup_{i<j} \varphi_i(N_i)\Big) \subset N_j.$$

It could happen that some K_j is empty, but after selecting the nonempty sets, it turns out that

$$\{\varphi(W), \varphi_j(K_j)\}$$

is a partition of M. According to Definition 4.2.3,

$$\text{area}\ M = \text{area}\ (W,\varphi) + \sum_j \text{area}\ (K_j,\varphi_j).$$

To conclude, it suffices to observe that

$$\text{area}\ (K_j,\varphi_j) = \iint_{K_J} \sqrt{\det(\varphi_j'^T(s,t) \circ \varphi_j'(s,t))}\mathrm{d}(s,t) = 0,$$

since K_j is a null set.

4.4.4 Find the area of the part of the plane

$$\frac{x}{a} + \frac{y}{b} + \frac{z}{c} = 1$$

(a,b,c are positive constants) that lies in the first octant (Fig. 10.11).

Solution: We have to evaluate the area of

$$M := \Big\{(x,y,z) \in \mathbb{R}^3, \quad x > 0,\ y > 0,\ z > 0,\ \frac{x}{a} + \frac{y}{b} + \frac{z}{c} = 1\Big\}.$$

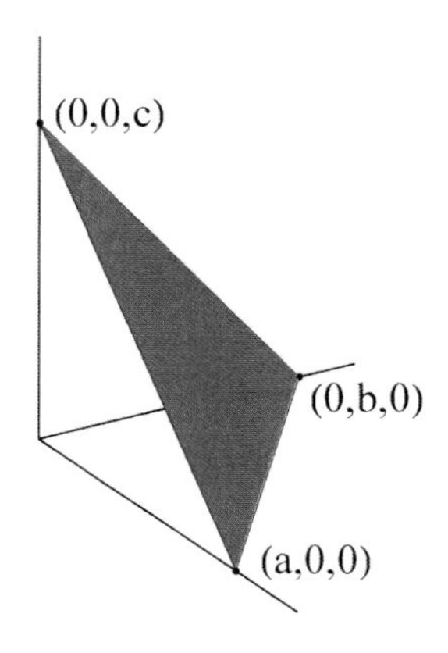

Fig. 10.11 Triangle of Exercise 4.4.4

To this end, we first parameterize the surface, taking as parameters

$$s := \frac{x}{a}, \quad t := \frac{y}{b}.$$

That is,

$$x = as, \quad y = bt, \quad z = c(1 - s - t).$$

The domain of the parameters is obtained by observing that the three coordinates x, y, z are strictly positive, and consequently,

$$s > 0,\ t > 0,\ 1 - s - t > 0.$$

So we consider the coordinate system (U, φ), where

$$U = \{(s,t) \in \mathbb{R}^2, \quad s > 0,\ t > 0,\ s + t < 1\}$$

and

$$\varphi : U \subset \mathbb{R}^2 \to M, \quad \varphi(s,t) = \big(as,\ bt,\ c(1 - s - t)\big).$$

Since $\varphi(U) = M$, it turns out that

$$\text{area}\, M = \iint_U \left\| \frac{\partial \varphi}{\partial s} \times \frac{\partial \varphi}{\partial t}(s,t) \right\| \mathrm{d}(s,t).$$

Finally, evaluate

$$\frac{\partial \varphi}{\partial s} \times \frac{\partial \varphi}{\partial t}(s,t) = \begin{vmatrix} \boldsymbol{e}_1 & \boldsymbol{e}_2 & \boldsymbol{e}_3 \\ a & 0 & -c \\ 0 & b & -c \end{vmatrix}$$

$$= (cb\ , ac,\ ab).$$

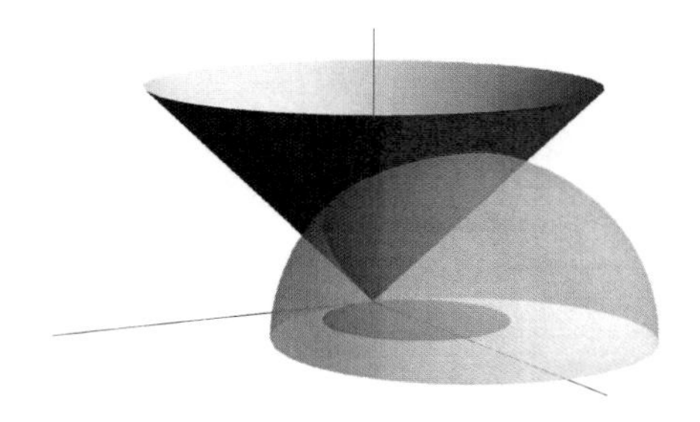

Fig. 10.12 Part of the cone inside the sphere

If we define

$$\lambda := (c^2b^2 + a^2c^2 + a^2b^2)^{1/2},$$

we obtain

$$\text{area } M = \lambda \int_0^1 \left(\int_0^{1-s} \mathrm{d}t \right) \mathrm{d}s = \frac{\lambda}{2} .$$

4.4.5 Evaluate the area of the part of the cone $x^2 + y^2 = z^2$ that lies above the plane $z = 0$ and inside the sphere $x^2 + y^2 + z^2 = 4ax$ with $a > 0$.

Solution: The part of the cone above the plane $z = 0$ and inside the sphere (Fig. 10.12)

$$(x - 2a)^2 + y^2 + z^2 = 4a^2$$

is described by

$$M := \{(x, y, z) \in \mathbb{R}^3, \quad x^2 + y^2 = z^2 \, , \, z > 0 \, , \, x^2 + y^2 + z^2 < 4ax\}.$$

The equation of the cone suggests a parameterization of the surface, while the two inequalities will determine the domain of the parameters. We note that the intersection of the cone with the horizontal plane $z = \rho$ is a circle with radius ρ whose projection onto the xy-plane admits the following parameterization:

$$x = \rho \cos\theta, \quad y = \rho \sin\theta.$$

Now $\rho > 0$, and from the inequality $x^2 + y^2 + z^2 < 4ax$ we deduce

$$\rho < 2a\cos\theta.$$

In particular, $\cos\theta > 0$, which means that we have to consider

$$-\frac{\pi}{2} < \theta < \frac{\pi}{2}.$$

Briefly, $M = \varphi(W)$, where

$$W := \left\{(\rho, \theta), \quad -\frac{\pi}{2} < \theta < \frac{\pi}{2} \, , \, 0 < \rho < 2a \, \cos\theta\right\}$$

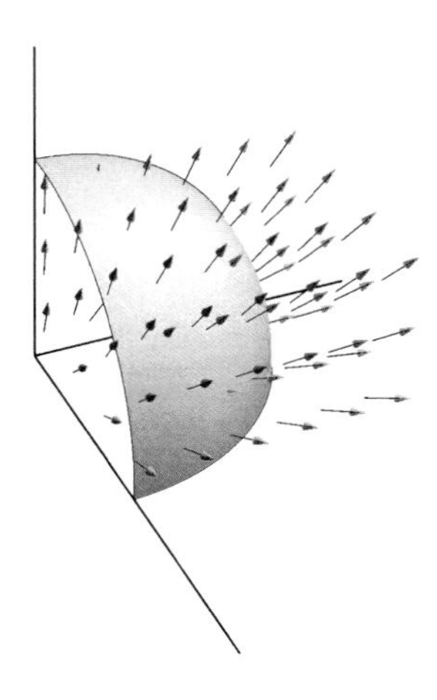

Fig. 10.13 Vector field of Exercise 4.4.7

and

$$\varphi(\rho,\theta) = \big(\rho\cos\theta,\ \rho\sin\theta,\ \rho\big).$$

Now evaluate

$$\frac{\partial\varphi}{\partial\rho} \times \frac{\partial\varphi}{\partial\theta}(\rho,\theta) = \rho\big(-\cos\theta, -\sin\theta, 1\big).$$

Since

$$\left\|\frac{\partial\varphi}{\partial\rho} \times \frac{\partial\varphi}{\partial\theta}(\rho,\theta)\right\| = \rho\sqrt{2} > 0,$$

we conclude that (W,φ) is a coordinate system of M and

$$\begin{aligned}\text{area } M &= \int_{-\frac{\pi}{2}}^{\frac{\pi}{2}} \Big(\int_0^{2a\cos\theta} \rho\sqrt{2}\,\mathrm{d}\rho\Big)\,\mathrm{d}\theta \\ &= 2a^2\sqrt{2}\int_{-\frac{\pi}{2}}^{\frac{\pi}{2}} \cos^2\theta\ \mathrm{d}\theta \\ &= \pi a^2\sqrt{2}.\end{aligned}$$

4.4.7 Determine the flux of the vector field

$$\boldsymbol{F}(x,y,z) = (x,y,2z)$$

across the part of the sphere

$$x^2+y^2+z^2 = 1$$

that lies in the first octant (Fig. 10.13).

Solution: Begin by writing the portion of the sphere in the first octant as

$$M := \{(x,y,z) \in \mathbb{R}^3 \ ; \ x^2 + y^2 + z^2 = 1 \ , \ x > 0 \ , \ y > 0 \ , \ z > 0\}.$$

We consider the coordinate system $(U, \boldsymbol{\varphi})$, where

$$\boldsymbol{\varphi} : U = \left(0, \frac{\pi}{2}\right) \times \left(0, \frac{\pi}{2}\right) \to M$$

is defined by

$$\boldsymbol{\varphi}(s,t) = (\sin s \cos t, \ \sin s \sin t, \ \cos s)$$

and U is chosen so that $\boldsymbol{\varphi}(U) = M$. First we evaluate

$$\frac{\partial \boldsymbol{\varphi}}{\partial s} \times \frac{\partial \boldsymbol{\varphi}}{\partial t}(s,t) = \begin{vmatrix} e_1 & e_2 & e_3 \\ \cos s \cos t & \cos s \sin t & -\sin s \\ -\sin s \sin t & \sin s \cos t & 0 \end{vmatrix}$$

$$= \sin s \ \boldsymbol{\varphi}(s,t).$$

Using the fact that $\boldsymbol{\varphi}(s,t)$ is a point of the unit sphere, we see that

$$\left\| \frac{\partial \boldsymbol{\varphi}}{\partial s} \times \frac{\partial \boldsymbol{\varphi}}{\partial t}(s,t) \right\| = \sin s,$$

and for $(x,y,z) = \boldsymbol{\varphi}(s,t)$,

$$\boldsymbol{N}(x,y,z) := \frac{\frac{\partial \boldsymbol{\varphi}}{\partial s} \times \frac{\partial \boldsymbol{\varphi}}{\partial t}(s,t)}{\| \frac{\partial \boldsymbol{\varphi}}{\partial s} \times \frac{\partial \boldsymbol{\varphi}}{\partial t}(s,t) \|} = (x,y,z)$$

is a normal vector to M directed outward from the sphere. From

$$\langle \boldsymbol{F}(x,y,z) \ , \ \boldsymbol{N}(x,y,z) \rangle = x^2 + y^2 + 2z^2 = 1 + z^2$$

and $z = \cos s$ we obtain the following expression for the flux of the vector field in the direction of the unit vector $\boldsymbol{N}$:

$$\text{Flux} = \iint_U (1 + \cos^2 s) \left\| \frac{\partial \boldsymbol{\varphi}}{\partial s} \times \frac{\partial \boldsymbol{\varphi}}{\partial t}(s,t) \right\| \mathrm{d}(s,t)$$

$$= \frac{\pi}{2} \int_0^{\frac{\pi}{2}} (1 + \cos^2 s) \sin s \ \mathrm{d}s = \frac{2\pi}{3}.$$

An alternative solution is to consider M as the graph of a function and to take as parameters the coordinates x, y, that is,

$$x = u, \quad y = v, \quad z = \sqrt{1 - u^2 - v^2}.$$

More precisely, we consider the coordinate system (D, ψ), where

$$D := \{(u,v) \in \mathbb{R}^2, \quad u > 0\ ,\ v > 0\ ,\ u^2 + v^2 < 1\}$$

and

$$\psi : D \to M, \quad \psi(u,v) = \left(u, v, \sqrt{1 - u^2 - v^2}\right).$$

We obtain

$$\text{Flux} = \iint_D \begin{vmatrix} f_1 & f_2 & f_3 \\ \frac{\partial x}{\partial u} & \frac{\partial y}{\partial u} & \frac{\partial z}{\partial u} \\ \frac{\partial x}{\partial v} & \frac{\partial y}{\partial v} & \frac{\partial z}{\partial v} \end{vmatrix} \, \mathrm{d}(u,v)$$

$$= \iint_D \begin{vmatrix} u & v & 2\sqrt{1 - u^2 - v^2} \\ 1 & 0 & \frac{-u}{\sqrt{1-u^2-v^2}} \\ 0 & 1 & \frac{-v}{\sqrt{1-u^2-v^2}} \end{vmatrix} \, \mathrm{d}(u,v)$$

$$= \iint_D \frac{2 - u^2 - v^2}{\sqrt{1 - u^2 - v^2}} \, \mathrm{d}(u,v).$$

The last integral is evaluated after a change to polar coordinates to get

$$\text{Flux} = \frac{\pi}{2} \int_0^1 \frac{2 - \rho^2}{\sqrt{1 - \rho^2}} \, \rho \, \mathrm{d}\rho \quad (\rho = \sin t)$$

$$= \frac{\pi}{2} \int_0^{\frac{\pi}{2}} (1 + \cos^2 t) \sin t \, \mathrm{d}t = \frac{2\pi}{3}.$$

We note that the flux just obtained has the same sign as the flux obtained in the previous solution. This is due to the fact that the two coordinate systems define the same field $\boldsymbol{N}$ of unit normal vectors to the sphere.

10.5 Solved Exercises of Chapter 5

5.3.3 Determine whether the following coordinate systems of the cylindrical surface

$$S := \{(x,y,z) \in \mathbb{R}^3, \quad x^2+y^2=1\ ,\ 0<z<1\}$$

define the same orientation in the tangent space to S at each common point

$$(x_0,y_0,z_0) = \boldsymbol{\varphi}(s_0,t_0) = \boldsymbol{\psi}(u_0,v_0):$$

(1) $\boldsymbol{\varphi} : (0,2\pi)\times(0,1) \to S, \quad \boldsymbol{\varphi}(s,t) = (\cos s, \sin s, t)$.
(2) $\boldsymbol{\psi} : (0,1)\times(0,1) \to S, \quad \boldsymbol{\psi}(u,v) = \left(u, \sqrt{1-u^2}, v\right)$.

Solution: We have to determine whether the two vectors

$$\frac{\partial \boldsymbol{\varphi}}{\partial s} \times \frac{\partial \boldsymbol{\varphi}}{\partial t}(s_0,t_0)$$

and

$$\frac{\partial \boldsymbol{\psi}}{\partial u} \times \frac{\partial \boldsymbol{\psi}}{\partial v}(u_0,v_0)$$

point in the same direction. But

$$\frac{\partial \boldsymbol{\varphi}}{\partial s} \times \frac{\partial \boldsymbol{\varphi}}{\partial t}(s_0,t_0) = \begin{vmatrix} \boldsymbol{e}_1 & \boldsymbol{e}_2 & \boldsymbol{e}_3 \\ -\sin s_0 & \cos s_0 & 0 \\ 0 & 0 & 1 \end{vmatrix}$$

$$= (\cos s_0, \sin s_0, 0) = (x_0, y_0, 0)$$

and

$$\frac{\partial \boldsymbol{\psi}}{\partial u} \times \frac{\partial \boldsymbol{\psi}}{\partial v}(u_0,v_0) = \begin{vmatrix} \boldsymbol{e}_1 & \boldsymbol{e}_2 & \boldsymbol{e}_3 \\ 1 & \frac{-u_0}{\sqrt{1-u_0^2}} & 0 \\ 0 & 0 & 1 \end{vmatrix}$$

$$= \left(\frac{-u_0}{\sqrt{1-u_0^2}}, -1, 0\right) = \frac{-1}{\sqrt{1-u_0^2}}(x_0,y_0,0).$$

Hence the two coordinate systems define different orientations.

5.3.5 Prove that any level surface M in $\mathbb{R}^3$ is orientable.

Solution: By hypothesis there is $\Phi : U \subset \mathbb{R}^3 \to \mathbb{R}$ of class C^1 on the open set U such that $\nabla\Phi(x) \neq 0$ for every $x \in \Phi^{-1}(0) = M$. We know that $\{\nabla\Phi(x)\}$ is a basis of N_xM for every $x \in M$. Hence

$$F(x) := \frac{\nabla \Phi(x)}{||\nabla \Phi(x)||}$$

is a continuous vector field of unit normal vectors.

10.6 Solved Exercises of Chapter 6

6.7.2 Let $\boldsymbol{F} = (F_1, F_2, F_3)$ and $\boldsymbol{G} = (G_1, G_2, G_3)$ be vector fields on an open set $U \subset \mathbb{R}^3$ and let $\boldsymbol{\omega}$, $\boldsymbol{\varphi}$ be the associated differential forms of degree 1. Find the relation between the vector field $\boldsymbol{F} \times \boldsymbol{G}$ and the vector field associated to the differential form of degree 2, $\boldsymbol{\omega} \wedge \boldsymbol{\varphi}$.

Solution: We have

$$\boldsymbol{\omega} = F_1 \, \mathrm{d}x + F_2 \, \mathrm{d}y + F_3 \, \mathrm{d}z, \qquad \boldsymbol{\varphi} = G_1 \, \mathrm{d}x + G_2 \, \mathrm{d}y + G_3 \, \mathrm{d}z,$$

and hence

$$\boldsymbol{\omega} \wedge \boldsymbol{\varphi} = \big(F_2 G_3 - F_3 G_2\big) \, \mathrm{d}y \wedge \mathrm{d}z + \big(F_3 G_1 - F_1 G_3\big) \, \mathrm{d}z \wedge \mathrm{d}x$$
$$+ \big(F_1 G_2 - F_2 G_1\big) \, \mathrm{d}x \wedge \mathrm{d}y,$$

which is the 2-form associated to the vector field $\boldsymbol{F} \times \boldsymbol{G}$.

6.7.5 Consider the following differential forms in $\mathbb{R}^3$:

(1) $\boldsymbol{\omega} = \mathrm{d}x - z\mathrm{d}y$;
(2) $\boldsymbol{\nu} = (x^2 + y^2 + z^2) \, \mathrm{d}x \wedge \mathrm{d}z + (xyz) \, \mathrm{d}y \wedge \mathrm{d}z$.

Evaluate

$$\mathrm{d}\boldsymbol{\omega} \ , \ \boldsymbol{\omega} \wedge \mathrm{d}\boldsymbol{\omega} \ , \ \mathrm{d}\boldsymbol{\nu} \ , \ \boldsymbol{\omega} \wedge \boldsymbol{\nu} .$$

Solution: (a) $\mathrm{d}\boldsymbol{\omega} = -\mathrm{d}z \wedge \mathrm{d}y = \mathrm{d}y \wedge \mathrm{d}z$.

(b) $\boldsymbol{\omega} \wedge \mathrm{d}\boldsymbol{\omega} = \mathrm{d}x \wedge \mathrm{d}y \wedge \mathrm{d}z$.

(c) $\mathrm{d}\boldsymbol{\nu} = (yz - 2y) \, \mathrm{d}x \wedge \mathrm{d}y \wedge \mathrm{d}z$.

(d) $\boldsymbol{\omega} \wedge \boldsymbol{\nu} = \big(xyz + z(x^2 + y^2 + z^2)\big) \, \mathrm{d}x \wedge \mathrm{d}y \wedge \mathrm{d}z$.

6.7.6 For the differential form

$$\boldsymbol{\omega} = \mathrm{d}x + (x^2 + y^2) \, \mathrm{d}y - \sin x \, \mathrm{d}z,$$

find the associated vector field $\boldsymbol{F}$ and show that

$$\boldsymbol{\nabla} \times \boldsymbol{F}$$

is the vector field associated to the differential form $\mathrm{d}\boldsymbol{\omega}$.

Solution: The vector field $\boldsymbol{F} : \mathbb{R}^3 \to \mathbb{R}^3$ is

$$\boldsymbol{F}(x, y, z) = (1, x^2 + y^2, -\sin x).$$

Then

$$\nabla \times \boldsymbol{F} = \begin{vmatrix} \boldsymbol{e}_1 & \boldsymbol{e}_2 & \boldsymbol{e}_3 \\ \frac{\partial}{\partial x} & \frac{\partial}{\partial y} & \frac{\partial}{\partial z} \\ 1 & x^2+y^2 & -\sin x \end{vmatrix} = (0, \cos x, 2x).$$

On the other hand,

$$\begin{aligned} \mathrm{d}\boldsymbol{\omega} &= 2x\ \mathrm{d}x \wedge \mathrm{d}y - \cos x\ \mathrm{d}x \wedge \mathrm{d}z \\ &= 0 \cdot \mathrm{d}y \wedge \mathrm{d}z + \cos x\ \mathrm{d}z \wedge \mathrm{d}x + 2x\ \mathrm{d}x \wedge \mathrm{d}y. \end{aligned}$$

6.7.8 Let

$$\boldsymbol{\omega} = xz\ \mathrm{d}y - y\ \mathrm{d}x, \quad \boldsymbol{\nu} = x^3\ \mathrm{d}z + \mathrm{d}x$$

and

$$\boldsymbol{\varphi}(s,t) = (\cos s, \sin s, t)$$

be given. Evaluate each of

$$\boldsymbol{\varphi}^*(\boldsymbol{\omega}), \quad \boldsymbol{\varphi}^*(\mathrm{d}\boldsymbol{\omega}), \quad \boldsymbol{\varphi}^*(\boldsymbol{\omega} \wedge \boldsymbol{\nu}), \quad \boldsymbol{\varphi}^*(\boldsymbol{\omega} \wedge \mathrm{d}\boldsymbol{\nu}).$$

Solution: Denote by (s,t) the coordinates of an arbitrary point of $\mathbb{R}^2$ and by (x,y,z) the coordinates of a point of $\mathbb{R}^3$, so that a basis of $\Lambda^1(\mathbb{R}^2)$ is $\{\mathrm{d}s,\ \mathrm{d}t\}$, while a basis of $\Lambda^1(\mathbb{R}^3)$ is $\{\mathrm{d}x,\ \mathrm{d}y,\ \mathrm{d}z\}$. If

$$\boldsymbol{\varphi} = (\varphi_1, \varphi_2, \varphi_3),$$

then

$$\begin{cases} \boldsymbol{\varphi}^*(\mathrm{d}x)(s,t) = \mathrm{d}\varphi_1(s,t) = -\sin s\ \mathrm{d}s, \\ \boldsymbol{\varphi}^*(\mathrm{d}y)(s,t) = \mathrm{d}\varphi_2(s,t) = \cos s\ \mathrm{d}s, \\ \boldsymbol{\varphi}^*(\mathrm{d}z)(s,t) = \mathrm{d}t. \end{cases}$$

Now put $f(x,y,z) = xz$ and $g(x,y,z) = y$. Then

$$\begin{aligned} \boldsymbol{\varphi}^*(\boldsymbol{\omega})(s,t) &= f\big(\boldsymbol{\varphi}(s,t)\big)\boldsymbol{\varphi}^*(\mathrm{d}y) - g\big(\boldsymbol{\varphi}(s,t)\big)\boldsymbol{\varphi}^*(\mathrm{d}x) \\ &= (t\cos^2 s + \sin^2 s)\ \mathrm{d}s \end{aligned}$$

and hence

$$\boldsymbol{\varphi}^*(\mathrm{d}\boldsymbol{\omega})(s,t) = \mathrm{d}(\boldsymbol{\varphi}^*\boldsymbol{\omega}) = -\cos^2 s\ \mathrm{d}s \wedge \mathrm{d}t.$$

Analogously,

$$\boldsymbol{\varphi}^*(\boldsymbol{\nu})(s,t) = -\sin s\ \mathrm{d}s + \cos^3 s\ \mathrm{d}t,$$

and we deduce that

$$\boldsymbol{\varphi}^*(\boldsymbol{\omega} \wedge \boldsymbol{\nu})(s,t) = \big(\boldsymbol{\varphi}^*(\boldsymbol{\omega}) \wedge \boldsymbol{\varphi}^*(\boldsymbol{\nu})\big)(s,t) = (t\cos^2 s + \sin^2 s)\cos^3 s\ \mathrm{d}s \wedge \mathrm{d}t.$$

Finally, we observe that $\boldsymbol{\omega} \wedge \mathrm{d}\boldsymbol{v} \in \Lambda^3(\mathbb{R}^3)$, and consequently,

$$\boldsymbol{\varphi}^*(\boldsymbol{\omega} \wedge \mathrm{d}\boldsymbol{v}) \in \Lambda^3(\mathbb{R}^2) = \{0\}.$$

10.7 Solved Exercises of Chapter 7

7.4.1 Find $\int_M \omega$, where

$$\omega = x^2 \, \mathrm{d}y \wedge \mathrm{d}z - y \, \mathrm{d}x \wedge \mathrm{d}z$$

and M is the portion of the plane $x+y+z=2$ whose projection onto the xy-plane is the triangle with vertices $(0,0),(0,1),(1,1)$, oriented according to the normal vector $(1,1,1)$ (Fig. 10.14).

Solution: We consider the coordinate system $(W,\boldsymbol{\varphi})$ defined by

$$\boldsymbol{\varphi}(u,v) = \big(u,\ v,\ 2-u-v\big),$$

where

$$W := \{(u,v) \in \mathbb{R}^2, \quad 0<u<1,\ u<v<1\}$$

is the projection of M onto the xy-plane. Since the cross product $\frac{\partial \boldsymbol{\varphi}}{\partial u} \times \frac{\partial \boldsymbol{\varphi}}{\partial v}$ evaluates as

$$\frac{\partial \boldsymbol{\varphi}}{\partial u} \times \frac{\partial \boldsymbol{\varphi}}{\partial v} = \begin{vmatrix} \boldsymbol{e}_1 & \boldsymbol{e}_2 & \boldsymbol{e}_3 \\ 1 & 0 & -1 \\ 0 & 1 & -1 \end{vmatrix} = (1,1,1),$$

it turns out that this coordinate system is compatible with the orientation of M. Also $\boldsymbol{\varphi}(W) = M$. Since $\boldsymbol{\omega}$ is the differential form of degree 2 associated to the vector field

$$\boldsymbol{F}(x,y,z) = (x^2, y, 0),$$

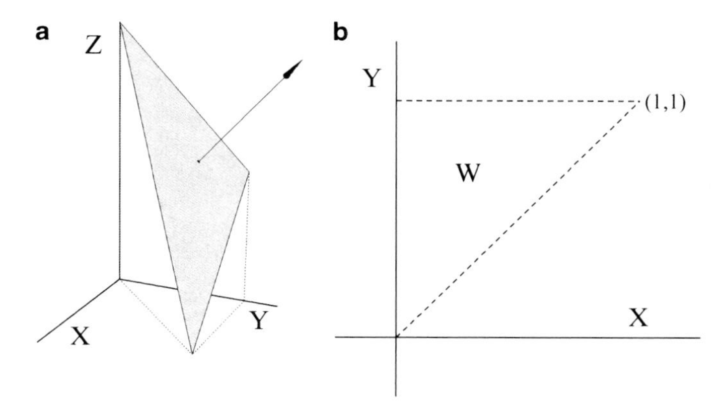

Fig. 10.14 (**a**) Triangle of Exercise 7.4.1; (**b**) projection onto the xy-plane

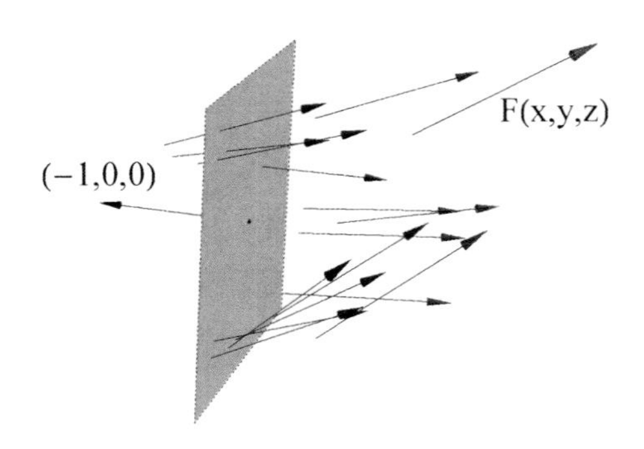

Fig. 10.15 Vector field of Exercise 7.4.5

we get

$$\int_M \omega = \iint_W \left\langle \boldsymbol{F}(\boldsymbol{\varphi}(u,v)),\ \frac{\partial \boldsymbol{\varphi}}{\partial u} \times \frac{\partial \boldsymbol{\varphi}}{\partial v} \right\rangle \, \mathrm{d}(u,v)$$

$$= \iint_W (u^2+v)\, \mathrm{d}(u,v)$$

$$= \int_0^1 \left(\int_u^1 (u^2+v)\, \mathrm{d}v \right) \mathrm{d}u = \frac{5}{12}.$$

7.4.5 Find the flux of the vector field $\boldsymbol{F}(x,y,z) = (1,\ xy,\ z^2)$ through the square in the yz-plane defined by $0 < y < 2$, $-1 < z < 1$ and oriented according to the normal vector $(-1,0,0)$ (Fig. 10.15).

Solution: The surface

$$M = \{(0,y,z)\ :\ 0 < y < 2, -1 < z < 1\}$$

is oriented according to the vector field $\boldsymbol{N}(x,y,z) = (-1,0,0)$. Since

$$\langle \boldsymbol{F}(x,y,z), \boldsymbol{N}(x,y,z) \rangle = -1,$$

the flux is

$$\text{flux} = -\text{area}(M) = -4.$$

Alternatively, we can consider the coordinate system $(A,\boldsymbol{\varphi})$ of M defined by $A = \{(s,t) \in \mathbb{R}^2\ :\ 0 < s < 2, -1 < t < 1\}$ and $\boldsymbol{\varphi}(s,t) = (0,s,t)$. From

$$\frac{\partial \boldsymbol{\varphi}}{\partial s} \times \frac{\partial \boldsymbol{\varphi}}{\partial t} = \begin{vmatrix} \boldsymbol{e}_1 & \boldsymbol{e}_2 & \boldsymbol{e}_3 \\ 0 & 1 & 0 \\ 0 & 0 & 1 \end{vmatrix}$$

$$= (1,0,0)$$

$$= -\boldsymbol{N}\Big(\boldsymbol{\varphi}(s,t)\Big),$$

we obtain that (A, φ) is incompatible with the chosen orientation of M. Consequently, we consider $B = \{(t,s) \in \mathbb{R}^2 \ : \ 0 < s < 2, -1 < t < 1\}$ and

$$\boldsymbol{\psi}(t,s) = \boldsymbol{\varphi}(s,t) = (0,s,t).$$

From Lemma 5.2.2, $(B, \boldsymbol{\psi})$ is a coordinate system of M compatible with the orientation. Hence, keeping in mind that $\boldsymbol{F}(\boldsymbol{\psi}(t,s)) = (1,0,t^2)$, $\frac{\partial \boldsymbol{\psi}}{\partial t} = (0,0,1)$, and $\frac{\partial \boldsymbol{\psi}}{\partial s} = (0,1,0)$, we conclude that the flux is

$$\begin{aligned} \text{flux} &= \iint_B \begin{vmatrix} 1 & 0 & t^2 \\ 0 & 0 & 1 \\ 0 & 1 & 0 \end{vmatrix} \, \mathrm{d}(t,s) \\ &= -\iint_B \mathrm{d}(t,s) = -4. \end{aligned}$$

The negative sign in this calculation of the flux is explained by the fact that the vector field $\boldsymbol{F}$ *flows in the opposite direction* to that given by the vector $(-1,0,0)$.

10.8 Solved Exercises of Chapter 8

8.7.1 Show that the portion of the cylinder $\{(x,y,z) \in \mathbb{R}^3 : x^2+y^2=a^2, 0 \leq z < 1\}$ is a regular surface with boundary (Fig. 10.16).

Solution: We will apply Theorem 8.3.2. To do this we take

$$S := \{(x,y,z) \in \mathbb{R}^3, \quad x^2+y^2=a^2,\ z<1\},$$

which is a regular surface of class C^∞, and define $\varphi : \mathbb{R}^3 \to \mathbb{R}$ by

$$\varphi(x,y,z) = -z.$$

If $(x,y,z) \in S \cap \varphi^{-1}(\{0\})$, then

$$\nabla\varphi(x,y,z) = (0,0,-1)$$

is not orthogonal to the surface S at (x,y,z), since the orthogonal vectors to the cylinder are horizontal. Consequently,

$$M := \{(x,y,z) \in S \ : \ \varphi(x,y,z) \leq 0\}$$

is a regular surface with boundary and

$$\partial M = \{(x,y,z) \in \mathbb{R}^3, \quad x^2+y^2=a^2,\ z=0\}.$$

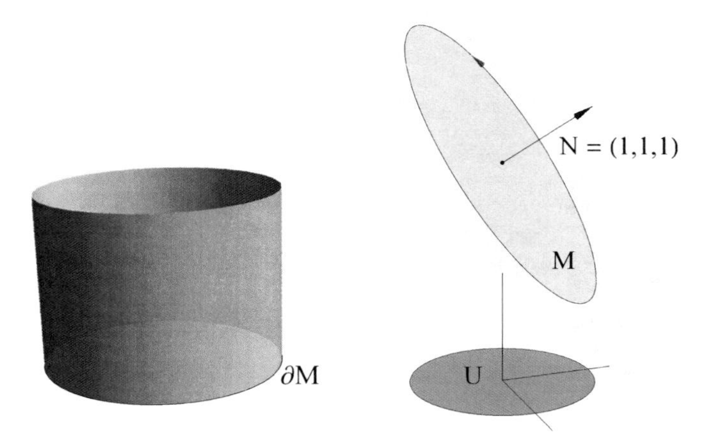

Fig. 10.16 (**a**) Cylinder of Exercise 8.7.1; (**b**) disk of Exercise 8.7.2

Alternatively, we could apply Criterion 8.3.1 instead of Theorem 8.3.2. Indeed, it suffices to consider

$$U := \{(x,y,z) \in \mathbb{R}^3 : z < 1\},$$

which is an open set in $\mathbb{R}^3$, and the mappings $\boldsymbol{\Phi}, f : U \subset \mathbb{R}^3 \to \mathbb{R}$ defined by

$$\boldsymbol{\Phi}(x,y,z) = x^2 + y^2 - a^2, \quad f(x,y,z) = -z.$$

Then $\nabla\boldsymbol{\Phi}(x,y,z) \neq (0,0,0)$ for every $(x,y,z) \in \boldsymbol{\Phi}^{-1}(\{0\})$, and

$$(\boldsymbol{\Phi}, f)'(x,y,z) = \begin{bmatrix} 2x & 2y & 0 \\ 0 & 0 & -1 \end{bmatrix}$$

has rank 2 at each point of

$$M := \{(x,y,z) \in \mathbb{R}^3 \ : \ \boldsymbol{\Phi}(x,y,z) = 0, f(x,y,z) \leq 0\}.$$

By Criterion 8.3.1, M is a regular surface with boundary and

$$\partial M = \{(x,y,z) \in \mathbb{R}^3, \quad \boldsymbol{\Phi}(x,y,z) = 0, \ f(x,y,z) = 0\}.$$

8.7.2 Let M be the portion of the plane $x + y + z = 1$ whose projection onto the xy-plane is the closed disk centered at the origin and with radius 1 (Fig. 10.16).

(1) Show that M is a regular surface with boundary.
(2) If M is oriented according to the normal vector $(1,1,1)$ and we consider the induced orientation in ∂M, find

$$\int_{\partial M} x \, \mathrm{d}y + y \, \mathrm{d}z.$$

Solution: (1) Let S be the plane $x + y + z = 1$ and $\varphi(x,y,z) = x^2 + y^2 - 1$. Then

$$M = \{(x,y,z) \in S, \quad \varphi(x,y,z) \leq 0\}.$$

Since

$$\nabla\varphi(x,y,z) = (2x, \ 2y, \ 0)$$

is not a multiple of the vector $(1,1,1)$, which is orthogonal to the plane S, it follows from Theorem 8.3.2 that M is a regular surface with boundary and

$$\partial M = \{(x,y,z) \in \mathbb{R}^3; \quad x + y + z = 1, \ x^2 + y^2 = 1\}.$$

(2) Take

$$S := \{(x,y) \in \mathbb{R}^2 \ ; \ x^2 + y^2 \leq 1\},$$

which is a regular surface with boundary in $\mathbb{R}^2$ whose boundary is the unit circle. The mapping

$$\boldsymbol{f} : S \to M, \quad \boldsymbol{f}(x,y) = (x,\ y,\ 1-x-y),$$

is a C^1 bijection, and since

$$\frac{\partial \boldsymbol{f}}{\partial x} \times \frac{\partial \boldsymbol{f}}{\partial y} = \begin{vmatrix} \boldsymbol{e}_1 & \boldsymbol{e}_2 & \boldsymbol{e}_3 \\ 1 & 0 & -1 \\ 0 & 1 & -1 \end{vmatrix} = (1,1,1),$$

we deduce that the differential $\mathrm{d}\boldsymbol{f}(x,y)$ has maximum rank at all points and the restriction of $\boldsymbol{f}$ to the open unit disk defines a coordinate system of $M \setminus \partial M$ that is compatible with its orientation. By Criterion 8.6.1, if we define

$$\boldsymbol{\gamma} : [0,2\pi] \to M \subset \mathbb{R}^3$$

as the image under $\boldsymbol{f}$ of the positively oriented circle

$$\boldsymbol{\gamma}(t) := \boldsymbol{f}(\cos t,\ \sin t) = \big(\cos t,\ \sin t,\ 1-\cos t - \sin t\big),$$

then

$$\big((0,2\pi),\ \boldsymbol{\gamma}\big)$$

is a coordinate system of the boundary ∂M that is compatible with its orientation. Finally, since

$$\partial M = \boldsymbol{\gamma}\big([0,2\pi]\big),$$

we conclude, for $\boldsymbol{\omega} = x\,\mathrm{d}y + y\,\mathrm{d}z$, that

$$\int_{\partial M} \boldsymbol{\omega} = \int_{\boldsymbol{\gamma}} \boldsymbol{\omega} = \int_0^{2\pi} \langle (0,\cos t,\sin t), (-\sin t,\cos t,\sin t - \cos t)\rangle \ \mathrm{d}t$$

$$= \int_0^{2\pi} \big(1 - \sin t \cos t\big)\ \mathrm{d}t = 2\pi.$$

10.9 Solved Exercises of Chapter 9

9.7.1 For the regular surface

$$M := \{(x,y) \in \mathbb{R}^2 \; ; \; a^2 \leq x^2 + y^2 \leq b^2\}$$

with boundary (Fig. 10.17):

(1) Find a parameterization of the circles

$$x^2 + y^2 = a^2, \; x^2 + y^2 = b^2$$

that is compatible with the orientation of M.

(2) Using Green's formula, evaluate

$$\int_{\partial M} (x^3 - y)\,\mathrm{d}x + (\cos y + 2x)\,\mathrm{d}y.$$

Solution: (1) According to Criterion 8.5.1, we parameterize the circles so that as we travel along them, we keep the set M to our left. That is, we consider

$$\gamma_1 : [0, 2\pi] \to \mathbb{R}^2; \; \gamma_1(t) = (b\cos t, b\sin t)$$

and

$$\gamma_2 : [0, 2\pi] \to \mathbb{R}^2; \; \gamma_2(t) = (a\cos t, -a\sin t).$$

Then

$$\partial M = \gamma_1([0, 2\pi]) \cup \gamma_2([0, 2\pi])$$

and

$$\big((0, 2\pi), \gamma_1\big), \big((0, 2\pi), \gamma_2\big)$$

are two coordinate systems of ∂M compatible with the orientation induced by that of M.

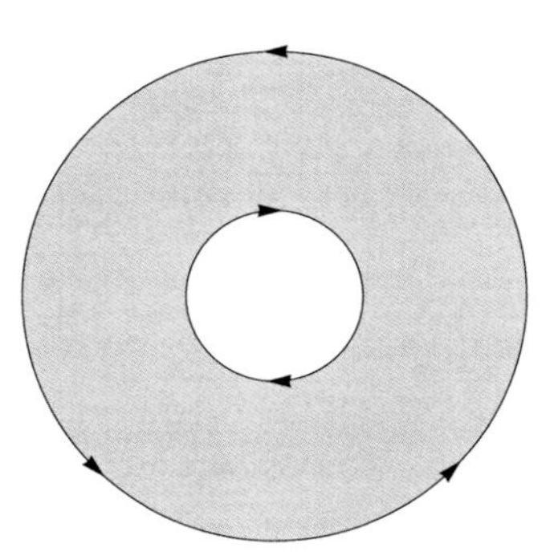

Fig. 10.17 Regular surface with boundary of Exercise 9.7.1

(2) We apply Green's formula (Theorem 9.4.5) with

$$P(x,y) = x^3 - y,\ Q(x,y) = \cos y + 2x.$$

Accordingly,

$$\int_{\partial M} (x^3 - y)\,\mathrm{d}x + (\cos y + 2x)\,\mathrm{d}y$$

coincides with

$$\iint_M \left(\frac{\partial Q}{\partial x} - \frac{\partial P}{\partial y}\right) \mathrm{d}(x,y) = \iint_M 3\,\mathrm{d}(x,y)$$

$$= 3\,\mathrm{area}(M)$$

$$= 3\pi(b^2 - a^2).$$

9.7.3 Verify Stokes's theorem for the lower hemisphere of the unit sphere

$$z = -\sqrt{1 - x^2 - y^2}$$

and the vector field

$$\boldsymbol{F}(x,\ y,\ z) = (-y,\ x,\ z).$$

Solution: Suppose that the sphere is oriented according to the exterior normal vector. We define

$$M := \{(x,y,z) \in \mathbb{R}^3\ ;\ x^2 + y^2 + z^2 = 1,\ z \le 0\}.$$

Consider the coordinate system $(D, \boldsymbol{\psi})$, where

$$D := \{(u,v) \in \mathbb{R}^2, \quad u^2 + v^2 < 1\}$$

and

$$\boldsymbol{\psi} : D \to M, \quad \boldsymbol{\psi}(u,v) = \left(u,\ v,\ -\sqrt{1 - u^2 - v^2}\right).$$

Now evaluate the cross product

$$\left(\frac{\partial \boldsymbol{\psi}}{\partial u} \times \frac{\partial \boldsymbol{\psi}}{\partial v}\right)(u,v) = \begin{vmatrix} \boldsymbol{e}_1 & \boldsymbol{e}_2 & \boldsymbol{e}_3 \\ 1 & 0 & \frac{u}{\sqrt{1-u^2-v^2}} \\ 0 & 1 & \frac{v}{\sqrt{1-u^2-v^2}} \end{vmatrix}$$

$$= \left(\frac{-u}{\sqrt{1-u^2-v^2}},\ \frac{-v}{\sqrt{1-u^2-v^2}},\ 1\right).$$

The third coordinate is positive, which means that the normal vector obtained points toward the interior of the sphere. Or equivalently, the coordinate system $(D,\boldsymbol{\psi})$ *reverses* the orientation of M. From $\boldsymbol{\psi}(D) = M \setminus \partial M$ we conclude that

$$\int_M \nabla \times \boldsymbol{F} \cdot \mathrm{d}\boldsymbol{S} = -\int_{(D,\boldsymbol{\psi})} \nabla \times \boldsymbol{F} \cdot \mathrm{d}\boldsymbol{S}.$$

Since

$$(\nabla \times \boldsymbol{F})(x,y,z) = \begin{vmatrix} \boldsymbol{e}_1 & \boldsymbol{e}_2 & \boldsymbol{e}_3 \\ \frac{\partial}{\partial_x} & \frac{\partial}{\partial_y} & \frac{\partial}{\partial_z} \\ -y & x & z \end{vmatrix} = (0,\ 0,\ 2),$$

we obtain

$$\begin{aligned}\int_M \nabla \times \boldsymbol{F} \cdot \mathrm{d}\boldsymbol{S} &= -\iint_D \left\langle (\nabla \times \boldsymbol{F})(\boldsymbol{\psi}(u,v)),\ \left(\frac{\partial \boldsymbol{\psi}}{\partial u} \times \frac{\partial \boldsymbol{\psi}}{\partial v}\right)(u,v) \right\rangle \,\mathrm{d}(u,v) \\ &= -\iint_D 2 \,\mathrm{d}(u,v) = -2\pi.\end{aligned}$$

We now check that also

$$\int_{\partial M} \boldsymbol{F} \cdot \mathrm{d}\boldsymbol{s} = \int_{\partial M} -y \,\mathrm{d}x + x \,\mathrm{d}y = -2\pi.$$

In fact, the boundary of M is the unit circle in the xy-plane,

$$\partial M = \{(x,y,z) \in \mathbb{R}^3 \ ;\ x^2 + y^2 = 1,\ z = 0\}.$$

This circle is oriented so that if we travel along it with our head toward the outside of the sphere, then the lower hemisphere is to our left. Or in other words, ∂M is oriented *clockwise*. We can see this more formally as follows.

Take the orientable regular 2-surface

$$S := \left\{(s,t) \ ;\ \frac{\pi}{2} \le s < \pi,\ 0 < t < 2\pi\right\}$$

in $\mathbb{R}^2$ and

$$\boldsymbol{\varphi} : S \to M \ ;\ \boldsymbol{\varphi}(s,t) = \big(\sin s \cos t,\ \sin s \sin t,\ \cos s\big).$$

We observe that the oriented boundary of S is the vertical segment from $(\frac{\pi}{2}, 2\pi)$ to $(\frac{\pi}{2}, 0)$, which can be parameterized as $t \mapsto (\frac{\pi}{2}, 2\pi - t)$. Since the restriction of $\boldsymbol{\varphi}$ to $S \setminus \partial S$ defines a coordinate system of $M \setminus \partial M$ that is compatible with its orientation, we apply Criterion 8.6.1 to conclude that

$$\boldsymbol{\gamma}(t) = \boldsymbol{\varphi}\left(\frac{\pi}{2}, 2\pi - t\right) = (\cos t,\ -\sin t,\ 0),\ \ 0 < t < 2\pi,$$

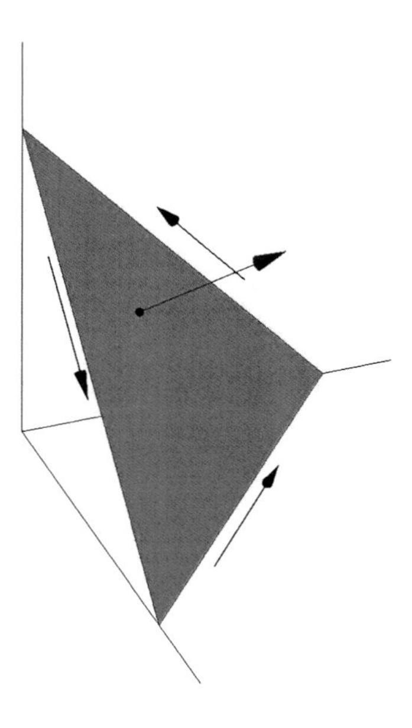

Fig. 10.18 Triangle of Exercise 9.7.5

defines a coordinate system of ∂M that is compatible with its orientation. Finally, from $\partial M = \boldsymbol{\gamma}([0,2\pi])$ we obtain

$$\int_{\partial M} -y\,\mathrm{d}x + x\,\mathrm{d}y = \int_{\boldsymbol{\gamma}} -y\,\mathrm{d}x + x\,\mathrm{d}y = -2\pi.$$

9.7.5 Let M be the triangle

$$\{(x,y,z) \in \mathbb{R}^3\ ;\ x+y+z=1,\ x \geq 0,\ y \geq 0\ z \geq 0,\}$$

with the vertices $(1,0,0)$, $(0,1,0)$, and $(0,0,1)$ removed (Fig. 10.18). Verify the validity of Stokes's theorem for this surface.

Solution: We suppose that the orientation of the plane $x+y+z=1$ is defined by the vector $(1,1,1)$. Consider the subset D of $\mathbb{R}^2$ consisting of the triangle

$$\{(x,y) \in \mathbb{R}^2, \quad x+y \leq 1,\ x \geq 0,\ y \geq 0\}$$

with the vertices $(0,0),(0,1),(1,0)$ removed. Also take

$$\boldsymbol{f} : \mathbb{R}^2 \to \mathbb{R}^3$$

to be the mapping defined by

$$\boldsymbol{f}(x,y) = (x,\ y,\ 1-x-y).$$

This is a C^∞ function, and the rank of f' is 2, which is maximal on $\mathbb{R}^2$. Thus D is a regular 2-surface with boundary in $\mathbb{R}^2$ and $M = \boldsymbol{f}(D)$ is a regular 2-surface with boundary in $\mathbb{R}^3$. As we proved in Exercise 8.7.2,

$$\frac{\partial f}{\partial x} \times \frac{\partial f}{\partial y} = \begin{vmatrix} \boldsymbol{e}_1 & \boldsymbol{e}_2 & \boldsymbol{e}_3 \\ 1 & 0 & -1 \\ 0 & 1 & -1 \end{vmatrix} = (1,1,1),$$

which means that $(\boldsymbol{f}, D \setminus \partial D)$ is a coordinate system of $M \setminus \partial M$ compatible with its orientation. According to Criterion 8.6.1, if $\boldsymbol{\gamma}$ is a positively oriented parameterization of ∂D, then $\boldsymbol{f} \circ \boldsymbol{\gamma}$ is a positively oriented parameterization of ∂M. Let $\boldsymbol{\omega}$ be a differential form of degree 2 and class C^1 on a neighborhood of M and define

$$\boldsymbol{\nu} := \boldsymbol{f}^* \boldsymbol{\omega}.$$

According to Green's formula from Chap. 2,

$$\int_{\partial D} \boldsymbol{\nu} = \int_D \mathrm{d}\,\boldsymbol{\nu}.$$

Since by Theorem 7.1.1 and Proposition 6.4.2,

$$\int_{\partial M} \boldsymbol{\omega} = \int_{\partial D} \boldsymbol{\nu}, \quad \int_M \mathrm{d}\,\boldsymbol{\omega} = \int_D \mathrm{d}\,\boldsymbol{\nu},$$

we can conclude that

$$\int_{\partial M} \boldsymbol{\omega} = \int_M \mathrm{d}\,\boldsymbol{\omega}.$$

9.7.7 Verify the divergence theorem for the vector field

$$\boldsymbol{F}(x,y,z) = (x,\ 2y,\ z)$$

and the *closed surface* $M := M_1 \cup M_2$, where

$$M_1 := \{(x,y,z) \in \mathbb{R}^3, \quad x^2 + y^2 = z^2,\ 0 \le z \le 2\}$$

and M_2 is the closed disk of radius 2 in the plane $z = 2$ centered at $(0,0,2)$.

Solution: $\tilde{M}_1 := M_1 \setminus \{(0,0,0)\}$ and M_2 are regular surfaces with boundary, and we suppose they are oriented according the vector field $\boldsymbol{N}$ of unit normal vectors pointing to the exterior of the region bounded by M. According to Example 9.4.1, a coordinate system of $\tilde{M}_1 \setminus \partial \tilde{M}_1$ compatible with the orientation is $(A, \boldsymbol{\varphi})$, where

$$A = \{(u,v) \in \mathbb{R}^2 \ :\ 0 < u^2 + v^2 < 4\}$$

and

$$\boldsymbol{\varphi} : A \subset \mathbb{R}^2 \to \mathbb{R}^3, \quad \boldsymbol{\varphi}(u,v) = \left(v, u, \sqrt{u^2+v^2}\right).$$

On the other hand, a coordinate system of $M_2 \setminus \partial M_2$ is $(B, \boldsymbol{\psi})$, where

$$B := \{(u,v) \in \mathbb{R}^2 \ : \ u^2+v^2 < 4\}$$

and

$$\boldsymbol{\psi} : B \subset \mathbb{R}^2 \to \mathbb{R}^3, \quad \boldsymbol{\psi}(u,v) = (u,v,2).$$

Since

$$\left(\frac{\partial \boldsymbol{\psi}}{\partial u} \times \frac{\partial \boldsymbol{\psi}}{\partial v}\right)(u,v) = (0,0,1) = \boldsymbol{N}(\boldsymbol{\psi}(u,v)),$$

it turns out that the coordinate system $(B, \boldsymbol{\psi})$ of M_2 is compatible with the orientation. Note that

$$\boldsymbol{\varphi}(A) = \tilde{M}_1 \setminus \partial \tilde{M}_1 \text{ and } \boldsymbol{\psi}(B) = M_2 \setminus \partial M_2.$$

Now evaluate

$$\int_M \langle \boldsymbol{F}, \boldsymbol{N} \rangle \, \mathrm{d}S := \int_{\tilde{M}_1} \langle \boldsymbol{F}, \boldsymbol{N} \rangle \, \mathrm{d}S + \int_{M_2} \langle \boldsymbol{F}, \boldsymbol{N} \rangle \, \mathrm{d}S.$$

Since $\boldsymbol{F}(\boldsymbol{\varphi}(u,v)) = (v, 2u, \sqrt{u^2+v^2})$, we have

$$\int_{\tilde{M}_1} \langle \boldsymbol{F}, \boldsymbol{N} \rangle \, \mathrm{d}S = \iint_A \begin{vmatrix} v & 2u & \sqrt{u^2+v^2} \\ 0 & 1 & \frac{u}{\sqrt{u^2+v^2}} \\ 1 & 0 & \frac{v}{\sqrt{u^2+v^2}} \end{vmatrix} \mathrm{d}(u,v)$$

$$= \iint_A \frac{u^2}{\sqrt{u^2+v^2}} \, \mathrm{d}(u,v)$$

$$= \left(\int_0^2 \rho^2 \mathrm{d}\rho\right)\left(\int_0^{2\pi} \cos^2 \theta \mathrm{d}\theta\right) = \frac{8\pi}{3}.$$

Keeping in mind that

$$\langle \boldsymbol{F}(\boldsymbol{\psi}(u,v)), \boldsymbol{N}(\boldsymbol{\psi}(u,v)) \rangle = 2,$$

we deduce

$$\int_{M_2} \langle \boldsymbol{F}, \boldsymbol{N} \rangle \, \mathrm{d}S = 2 \, \mathrm{area}(M_2) = 8\pi.$$

Consequently,

$$\int_M \langle \boldsymbol{F}, \boldsymbol{N} \rangle \, \mathrm{d}S = \frac{32}{3}\pi.$$

Finally, evaluate

$$\operatorname{Div} \boldsymbol{F}(x,y,z) = 4,$$

and denoting by

$$\Omega := \{(x,y,z) \in \mathbb{R}^3, \quad x^2+y^2 \leq z^2,\ 0 \leq z \leq 2\}$$

the region bounded by M, we obtain (using cylindrical coordinates $x = \rho\cos\theta$; $y = \rho\sin\theta$; $z = z$)

$$\iiint_\Omega \operatorname{Div} \boldsymbol{F}(x,y,z)\ \mathrm{d}(x,y,z) = 4\int_0^2 \left(\int_0^z \left(\int_0^{2\pi} \rho\,\mathrm{d}\theta\right)\mathrm{d}\rho\right)\mathrm{d}z$$

$$= 4\pi\int_0^2 z^2\,\mathrm{d}z = \frac{32}{3}\pi.$$

We have verified that

$$\int_M \langle \boldsymbol{F}, \boldsymbol{N}\rangle\,\mathrm{d}S = \iiint_\Omega \operatorname{Div} \boldsymbol{F}(x,y,z)\ \mathrm{d}(x,y,z).$$

References

1. T.M. Apostol; *Mathematical Analysis*, 2nd edition, Addison-Wesley Publishing Co., Reading, Mass.–London–Don Mills, Ont., 1974.
2. T.M. Apostol; *Calculus Vol. 2: Multi-variable Calculus and Linear Algebra with Applications,* second edition, Blaisdell Publishing Co., Ginn and Co., Mass.–Toronto, Ont.–London, 1969.
3. R.G. Bartle; *The Elements of Real Analysis,* second edition, John Wiley and Sons, New York–London–Sidney, 1976.
4. J.C. Burkill, H. Burkill; *A Second Course in Mathematical Analysis,* Cambridge University Press, Cambridge–New York, 1980.
5. M.P. do Carmo; *Differential Geometry of Curves and Surfaces,* Prentice-Hall, Inc., Englewood Cliffs, N.J., 1976.
6. M.P. do Carmo; *Differential Forms and Applications,* Universitext, Springer-Verlag, Berlin, 1994.
7. H. Cartan; *Differential Forms,* Houghton Mifflin Co., Boston, Mass 1970.
8. S. Dineen; *Functions of Two Variables,* Chapman and Hall, London 1995.
9. C.H. Edwards; *Advanced Calculus of Several Variables,* Academic Press (a subsidiary of Harcourt Brace Jovanovich), New York–London, 1973.
10. W. Fleming; *Functions of Several Variables,* Undergraduate Texts in Mathematics, Springer-Verlag, New York–Heidelberg, 1977.
11. A.P. French; *Newtonian Mechanics,* Norton, 1971.
12. E. Gaughan; *Introduction to Analysis,* fourth edition. Brooks/Cole Publishing Co., Pacific Grove, CA, 1993.
13. R.E. Larson, R.P. Hostetler, B.H. Edwards; *Calculus with Analytic Geometry,* D.C. Heath and Company, Lexington, 1994.
14. J.E. Marsden, A.J. Tromba; *Vector Calculus,* third edition, Freeman and Company, New York, 2003.
15. W. Rudin; *Principles of Mathematical Analysis,* third edition, McGraw-Hill Book Co, New York–Auckland–Düsseldorf, 1976.
16. H. Samelson, *Orientability of Hypersurfaces in* $\mathbb{R}^n$, P.A.M.S. 22 (1969), 301–302.
17. M. Spivak; *Calculus on Manifolds. A Modern Approach to Classical Theorems of Advanced Calculus.* W. A. Benjamin, Inc., New York–Amsterdam, 1965.
18. K.R. Stromberg; *Introduction to Classical Real Analysis.* Wadsworth International Group, Belmont CA, 1981.

A. Galbis and M. Maestre, *Vector Analysis Versus Vector Calculus*, Universitext,
DOI 10.1007/978-1-4614-2200-6,

List of Symbols

DOI 10.1007/978-1-4614-2200-6, © Springer Science+Business Media, LLC 2012

Index

A. Galbis and M. Maestre, *Vector Analysis Versus Vector Calculus*, Universitext,
DOI 10.1007/978-1-4614-2200-6, © Springer Science+Business Media, LLC 2012

Printed by Publishers' Graphics LLC
BT20130108.19.20.43